DISCARD

Creating Networks in Chemistry
The Founding and Early History of Chemical Societies in Europe

 Chemical Heritage Foundation

European Association for Chemical and Molecular Sciences

molecuien • mensen • meningen KNCV

SOCIEDADE
PORTUGUESA
DE QUÍMICA

Science History Tours
Naperville, IL, USA
Director: Yvonne Twomey

Science History Tours is a non-profit organization
promoting education in the history of science.

GESELLSCHAFT
DEUTSCHER CHEMIKER

Creating Networks in Chemistry
The Founding and Early History of Chemical Societies in Europe

Edited by

Anita Kildebæk Nielsen
Center for Electron Nanoscopy, Technical University of Denmark, Denmark

Soňa Štrbáňová
Institute of Contemporary History, Academy of Sciences of the Czech Republic, Prague, Czech Republic

RSCPublishing

Special Publication No. 313

ISBN: 978-0-85404-279-1

A catalogue record for this book is available from the British Library

Published by The Royal Society of Chemistry,
Thomas Graham House, Science Park, Milton Road,
Cambridge CB4 0WF, UK

Registered Charity Number 207890

For further information see our website at www.rsc.org

Preface

Background

Each book has its history. At the cradle of this one stood two associations of historians of science, two meetings and many enthusiastic people.

The first association to be mentioned is the *Working Party on the History of Chemistry* of the *European Association for Chemical and Molecular Sciences*,[1] which has been joining together history groups of the European national chemical societies and many individuals – professionals and lay historians of chemistry – not only from Europe but also from overseas. Since its constitution in 1976, it has organized numerous meetings and coordinated many other activities catalysing research into the European and world history of chemistry.

The establishment of the other, younger institution – the *European Society for the History of Science*[2] – in 2003 has become a momentous point of departure in bringing together the European historians of science. It was at its first conference "Science in Europe – Europe in Science" in Maastricht in November 2004 where the idea of this book started to bud and where approximately 100 scholars met face-to-face for the first time, including the two editors of this book. They both reported about chemical societies at the meeting; while Anita Kildebæk Nielsen focused on creation of disciplinary associations as part of their social demarcation, Soňa Štrbáňová was more oriented towards the disciplinary and intellectual context of their establishment and functioning in the international networks. In our conversations during the lunch breaks we realized how many puzzling problems have been connected with early development of the European chemical societies and scientific societies in general. Eventually we agreed that in order to find solutions and explanations we should organize a workshop on this topic which would attract other colleagues into

---

[1] Former *Working Party on the History of Chemistry* of *The Federation of the European Chemical Societies*. For details see http://www.euchems.org/Divisions/History/index.asp

[2] See http:// www.eshs.org

Creating Networks in Chemistry: The Founding and Early History of Chemical Societies in Europe
Edited by Anita Kildebæk Nielsen and Soňa Štrbáňová
© The Royal Society of Chemistry 2008

such studies. As a good opportunity for calling the workshop, the upcoming Fifth International Conference on the History of Chemistry in Portugal in 2005, organized by the *Working Party of History of Chemistry*, was the obvious suggestion. While still in Maastricht, we explained our idea to Professor Ernst Homburg, chairman of the *Working Party*, who readily accepted our project and has personally supported it ever since. And this is how everything started.

The Estoril Workshop and the Common Framework of the Workshop

The Fifth International Conference on the History of Chemistry took place in Estoril and Lisbon on 6th–10th September 2005 with the title "Chemistry, Technology and Society". The satellite workshop, which we decided to call *"European Chemical Societies – Comparative Analyses of Demarcations"*, materialized on 6th September in Estoril before the official start of the conference, enjoying the full support of the efficient Portuguese organizers.

The aim of the workshop was to show the establishment and early development of the European chemical societies in the period up to the First World War. In the directions to the potential speakers, we laid down a common set of questions to be applied for the description and analysis of the individual societies. These guidelines, as we hoped, helped to focus on certain problems and in the long run enabled trans-national comparisons of the circumstances in which the societies were constituted and functioned (See box at the end of the Preface).

An initial source of inspiration was the concept of "boundary-work" as formulated by the sociologist Thomas Gieryn.[3] Through his concept the establishments of a scientific society, in our case a chemical society, can be seen as a part of professionalization of chemistry as a discipline, a process that also included other aspects such as implementing advanced instruction and specialized research, rigorous control of competence through standardized examinations and establishment of predictable careers. One very important aspect was the formation of a professional autonomy that separated the professional chemists from amateurs and from practitioners of other scientific disciplines. The chemical societies provided one of the important institutional bases for this forming of disciplinary unity. The institutional and cultural surroundings helped to form a specialized self-consciousness and a unique social identity.

A focal question we were going to ask in the workshop was what we can learn about the process of professionalization of chemistry by studying the emergence of chemical societies in various European countries and their further development in the period before the First World War. In particular, it has been our ambition to explore in a comparative way the process of demarcation that inevitably takes place when a social institution of a scientific discipline is formed. Expansion, monopolization and protection of authority are three

---

[3] T. F. Gieryn (1983), "Boundary-Work and the Demarcation of Science from Non-Science : Strains and Interests in Professional Ideologies of Scientists", *American Sociological Review* **48**, 781–795.

important features of the process of professionalization of a scientific discipline. In the demarcation of the chemical societies, the European chemists were forced to make explicit their tacit assumptions of what chemistry was and what it was not. With focus on these so-called boundary-works one can reassess the professionalization of European chemistry in the nineteenth century as it evolved through the ideologies formulated in the chemical societies. The terminology has been applied to different degrees in the different chapters, and other important angles of analysis were discovered as the project evolved, making the perspectives of the book more wide-ranged.

The initial, over-ambitious idea was to include the chemical societies of most, if not all, European countries. However, due to other obligations, lack of source material or short notice of the workshop, we had to settle on a selection of countries. We were lucky enough to arouse the interest of a number of historians of science who were willing to elaborate the early history of one or more societies of chemists in a particular country. Those who came to Estoril certainly constituted an impressive group of internationally recognized scholars, as the list of contents bears witness.

We understood from the very beginning of this project that the materials of the workshop should be disseminated in some manner. We felt the great responsibility of such a job as valuable historical material on the different chemical societies in Europe transcending the initial "demarcation topic" was gathered for the workshop. It contained data on the circumstances in which the societies were constituted and functioned, their relations with universities and chemical industry, international contacts, *etc.* During the workshop the participants and observers also discussed the understanding of chemistry as a discipline in the latter half of the nineteenth century and the early twentieth century and the roles scientific societies played in the political and cultural environment of the various countries. This way the original idea of a comparative analysis of demarcations grew into a much more extensive elaboration of the foundation and early history of the chemical societies in Europe treating the issue of demarcation as one of the important problems.

When we deliberated about the possible product of the workshop we came to the decision that such a wealth of material might enable us to prepare not simple proceedings, but rather a book on European chemical societies with uniformly structured chapters and an extensive final part where an analysis and comparison across the geographical idea would be made. We were happy to announce at the workshop that the publishing house of The Royal Society of Chemistry was willing to accept this task.

Forerunners

There exist numerous analyses of different aspects of the history of chemistry in the European countries and some of them have considered at least peripherally the history of the chemical societies.[4]

---

[4] The reader will find the references in the individual chapters.

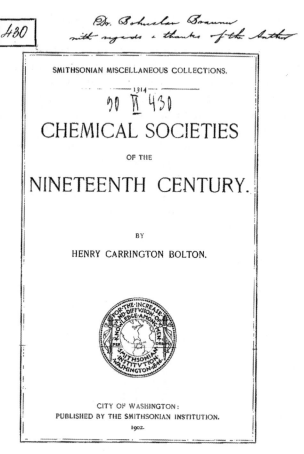

Figure 1 Title page of Bolton's brochure on chemical societies. The copy is dedicated to the Czech chemist Bohuslav Brauner, Bolton's friend, who helped Bolton to gather some data.

In 1902, the American chemist and historian of chemistry H. C. Bolton published a thin brochure as an attempt to produce an overview on all chemical societies in the world that existed in 1900.[5] His motivation was the "beginning of the new century [which] affords an opportune period for chronicling the progress of chemistry as shown by the organizations formed to foster its study and to stimulate its adherents".[6] The statistical data in the booklet (seat and date of founding, name of president and membership in 1900, serial publications and remarks) Bolton collected chiefly by correspondence with the officers of the societies. In the booklet, however, several countries and societies had been left out, and the contents were strictly descriptive (Figure 1).

---

[5] H. C. Bolton, *Chemical Societies of the Nineteenth Century*, Smithsonian Institution, Washington, 1902.

[6] *Ibid.*, p. 1.

Another important piece of literature that should be mentioned is a collected work by David Knight and Helge Kragh. As part of a large European Science Foundation project on the history of chemistry, the two historians published an anthology with chapters on the social history of chemistry in Europe from the French revolution to the First World War, entitled *The Making of the Chemist*.[7] The book comprises chapters on many of the same countries covered in this volume, and some of them also deal in part with the creation of national chemical societies. As the focus in *The Making of the Chemist* is more general we recommend seeking further information on the national scenes in that publication. Not all countries represented in our volume are, however, included in that treatise and it is our belief that by applying different perspectives, the two publications, that is *The Making of the Chemist* and the one we are presenting here, will supplement each other. In any case, we suppose that our book, which deals in such a complex way with the emergence and functioning of scientific societies, has no predecessors.

Contents of the Book

We have chosen the year 1914 as a natural historical boundary that changed to a large extent not only the concept of science but also Europe as such in a very complex way. However, this "final" date was not considered dogma in some instances (*e.g.* Portugal or Poland) where the delayed establishment of a national chemical society or other reasons justified the extended time limit of the narrative in certain chapters.

As already mentioned, we have attempted to make as consistent as possible the structure of the individual chapters because only such an approach enables trans-national analysis and comparisons. The guidelines we elaborated for the authors are shown in the Box and we must underline that all authors did their best to comply with the task. Even so, however, the attentive reader may find substantial differences in how the authors coped with their chapters. In fact, this is what we would expect dealing with strong, independent individualities and local histories whose circumstances differed so much and required specific approaches. We should also keep in mind that there have been great variations in sources available for the different countries and societies. Some societies have kept extensive archives and some have not or the archives have been lost in the course of time. Some societies have already published books on their history and some have not. For some countries there already existed materials elaborated in English, German or French, but in some cases the chapter in our book is the very first presentation of the history of a national chemical society in a language intelligible to most readers. The bottom line, though, is the commitment to the guidelines; therefore all chapters have been written in numerous editions in order to fit the framework and facilitate the international comparison. In consequence we are able to present fourteen chapters

---

[7] D. Knight and H. Kragh (eds.), *The Making of the Chemist*, Cambridge University Press, Cambridge, 1998.

portraying societies where chemists of all geographical corners of Europe could assemble and pursue their interests before the First World War.

There exist, of course, many ways in which we could have ordered the chapters in the book. At an early stage we considered ranking the chapters according to the year of founding – from the oldest societies to the youngest. But, as the chapters will reveal, it is difficult to state these dates in some cases, as the societies sometimes existed as parts of other associations or because the societies themselves did not agree upon the date of their establishment. A geographical ordering also seemed too subjective and sensitive to criticism. The reader will see that we are dealing in some cases with geographically defined societies, but in some instances with societies differentiated by language in a complicated socio-political environment. For example, the Austrian, Czech, Hungarian and some Polish societies all existed within the Austrian or Austro-Hungarian monarchy, but the social, political and cultural background was different in these geographically or linguistically defined cases. The same applies to Poland, which existed in our time span only culturally, not as a state. Therefore we decided to choose a neutral order by presenting the chapters in alphabetical order in accordance with the English version of the country's name.

The final comparative chapter of the book recapitulates and analyzes some of the main points in the "national" chapters; this chapter also raises questions that may well become a starting point for further studies of chemical societies or parallel studies devoted to other scientific associations.

Readerships

It has been our endeavour to produce a book benefiting various audiences. We hope that it will be read and used not only by the members of the chemical societies, and historians of chemistry, but also by historians of other scientific fields, general historians and teachers. But what can it offer to all those groups of readers?

The modern chemist might be attracted by a comprehensive insight into the past social life of his or her predecessors and their societies and by learning about chemists' problems and interests in various countries in the past. The historians of chemistry and/or other scientific disciplines, as well as general historians, could profit from the presentation of new facts on various scientific and professional associations closely related to the formation of modern chemistry and its institutions in various European countries. The historian of science will hopefully realize some of the many advantages of studying institutional history – not as a hagiography or memoirs of the great founding fathers, but as a means of obtaining unique information on the social and institutional development of science and its political, cultural and economic context. General historians after reading this book may realize more than before that history of science is an integral part of history and culture in each country and may use the material of this book to enrich their articles and monographs. Finally, teachers at the secondary and university levels may appreciate gaining new material for their lessons; they may use it in various ways to help their students to understand the complex nature of science, especially with regard to the emergence of scientific institutions. An extra bonus is the enormous amount of sources on which the narratives are based – references to books, journals, articles, reports and archives that may serve all historians and teachers.

Last but not Least

The final analytical chapter at the end of the book will hopefully demonstrate how a new insight into the national stories can be gained by taking up a comparative trans-national perspective. As already hinted at, we would be most happy if it could spur the initiative of similar comparative studies into the history of mathematics, physics, molecular biology or arts.

We are aware of many gaps in the information presented, and perhaps some errors exist. The reader might find inconsistencies in the individual chapters as well as in the book as a whole. Unfortunately there are also missing chapters on some countries with powerful chemical societies founded before 1914 (*i.e.* Italy and Spain). Nonetheless, we believe that the book will essentially enrich the general knowledge of the institutional history of chemistry and its professionalization as a whole, not just in the countries treated. What we consider crucial and new is the complex comparative perspective the book offers and we would be very pleased if it would instigate historians of other scientific disciplines to write similar parallel institutional histories for their disciplines.

Anita Kildebæk Nielsen and Soňa Štrbáňová
Hørsholm and Prague

Guidelines for the Individual Chapters Given to the Authors

Establishment

- ♦ When and how was the chemical society established?
- ♦ What arguments, if any, were provided to justify its establishment?
- ♦ Who were the initiators and what were their educational backgrounds and positions on the labour market?
- ♦ How was the society funded?

Members – numbers

- ♦ How many members did the society have at the point of establishment?
- ♦ How many members did it have in 1900 and 1914?
- ♦ What was the relation between the number of members and the number of persons with a degree in chemistry at the time of the establishment of the society and in 1914?

Members – who

- ♦ Who could become a member? Were the rules of admission altered within the time span under consideration?
- ♦ Were specific groups of chemical practitioners excluded from membership?
- ♦ Which were the dominant groups of members qualitatively and quantitatively?
- ♦ What was the educational background of the members and what were their positions in the labour market?
- ♦ What was the educational background of the executives and what were their positions in the labour market?

Everyday Routines

- ♦ What were the regular, irregular or annual events of the society?
- ♦ Who was invited to give papers and what topics were considered proper? Which topics were the most frequent?
- ♦ What were the society's relations with the surrounding society and what was its involvement in educational, political, cultural or social issues?
- ♦ What were the relations to other disciplinary and professional societies or learned academies at the national and international level?
- ♦ What was the balance between the idealized image of the society (given by the founders, statutes, *etc.*) and its work in reality?

Publications

- ♦ What publications were issued by the society (periodicals, treatises, textbooks, *etc.*) and what were their main contents?
- ♦ Who was the intended audience of the publications?

Demarcation

- ♦ Did the chemical society act with policies of expansion, monopolization and/or protection in relation to the question of professional autonomy? Did the strategy vary in different cases?
- ♦ What could the different groups of chemists gain by joining the society?

Contents

Creating Networks in Chemistry: The Founding and Early History of Chemical Societies in Europe
Edited by Anita Kildebæk Nielsen and Soňa Štrbáňová
© The Royal Society of Chemistry 2008

Chapter 9   THE NETHERLANDS: Keeping the Ranks Closed: The
Dutch Chemical Society, 1903–1914
Ernst Homburg

Chapter 10 NORWAY: A Group of Chemists in the Polytechnic Society in
Christiania. The Norwegian Chemical Society, 1893–1916
Bjørn Pedersen

**Chapter 13 RUSSIA: The Formation of the Russian Chemical Society
and Its History until 1914**
*Nathan M. Brooks, Masanori Kaji and
Elena Zaitseva*

**Chapter 14 SWEDEN: The Chemical Society in Sweden: Eclecticism in
Chemistry, 1883–1914**
Anders Lundgren

Acknowledgements

Such a book as this can be written only with the help and coordinated effort of many people and institutions. We were lucky enough to find those who have assisted with their knowledge, time, ideas, funds or other means of support. The succession in which they are mentioned here by no means reflects the magnitude of their credit.

The *Working Party on History of Chemistry* of the *European Association for Chemical and Molecular Sciences* (WP EuChemS) is a community of historians of chemistry who have been uniquely supportive in developing new ideas. Most of the authors of this book have been active in this group for many years and had met at the numerous conferences organized by the WP long before the idea of this book was born. Our thanks are due to this stimulating friendly circle led by its present chairman Professor Ernst Homburg (Maastricht University). We have been particularly fortunate that the workshop where the idea for this book came about was part of the history of chemistry conference organized by the WP in 2005 in Lisbon and Estoril (Portugal). Professor Maria Elvira Callapez and her wonderful team from several Portuguese institutions, the *Academia das Ciências de Lisboa*, Universidade Lusófona de Humanidades e Tecnologias, Laboratoria Chimico da Escola Politécnica de Lisboa, Museu da Farmácia and others, created a unique working atmosphere in which we were encouraged to start this project.

Our thanks are due to the institutions where we have been working for their versatile support and understanding in all phases of preparation of this book. This concerns the Prague Institute of Contemporary History of the Academy of Sciences of the Czech Republic and its director Dr Oldřich Tůma, as well as the Institute's Department of History of Science and its head Dr Antonín Kostlán; the History of Science Department at the University of Aarhus, the Saxo-Institute at the University of Copenhagen and associate professor Ole Hyldtoft, and BioCentrum and Center for Electron Nanoscopy at the Technical University of Denmark.

We owe tremendous gratitude to the sponsoring institutions which have generously contributed to the manufacturing costs of this book to make it accessible to wide circles of readers. These institutions are (in alphabetic order):

- *Česká společnost chemická* (Czech Chemical Society)
- Chemical Heritage Foundation

- *Gesellschaft Deutscher Chemiker* (German Chemical Society)
- *Gesellschaft Österreichischer Chemiker* (Austrian Chemical Society)
- *Kemisk Forening* (The Danish Chemical Society)
- *Koninklijke Nederlandse Chemische Vereniging* (Royal Dutch Chemical Society)
- *Magyar Kémikusok Egyesülete* (Hungarian Chemical Society)
- Science History Tours (Ms Yvonne Twomey)
- *Sociedade Portuguesa de Química* (Portuguese Chemical Society)

We must also show appreciation of the Prague Masaryk Institute – Archives of the Academy of Sciences of the Czech Republic – for making available without charge the cover picture. Special thanks go to Hana Barvíková, M.Sc., and to Vlasta Mádlová, M.Sc., who made a special effort in finding an appropriate photo in the vast collections of the Archives.

We thank for ideas, background materials and encouragement the numerous colleagues and institutions in various countries whose input has been acknowledged in the individual chapters.

Finally, we must emphasize that this book would not have been made possible without the extreme cooperative endeavour and goodwill of all co-authors who had to comply with the complicated topic and the guidelines. Thanks, of course, to Janet Freshwater, Annie Jacob, Katrina Harding and their colleagues at The Royal Society of Chemistry for their hard editing work and the patience with which they handled all our problems and delays.

Anita Kildebæk Nielsen and Soňa Štrbáňová

CHAPTER 1

AUSTRIA: Austrian Chemical Societies in the Last Decades of the Habsburg Monarchy, 1869–1914

W. GERHARD POHL

1.1 Chemistry Development and Chemical Education in Austria[1]

Chemical industries in the Habsburg monarchy in most cases were developed in connection with mining, brewing and metallurgical engineering. Many parts of Central Europe, which are today the Czech Republic, Slovakia, Hungary, parts of Poland, Romania or Ukraine, were in the nineteenth century parts of the Austrian Empire. After 1848, and particularly in the Dual Monarchy after 1867, Hungary succeeded to develop independently.

Bohemia and Moravia (the Czech Lands) belonged economically and intellectually to the most developed part of the Empire, in which Czech- and German-speaking people lived. At the beginning of the nineteenth century the language of education was German in all institutions of higher learning but after 1869 a parallel polytechnic and in 1882 a parallel university were established in Prague, in which the language of education and research was Czech. As the German University in Prague and the German technical universities in

---

[1] The basis for this chapter is the *Österreichische Chemiker-Zeitung* (*ÖCHZ*, Austrian Chemists Journal), which was founded in 1887 and became the official journal of the *Verein Österreichischer Chemiker in Wien* (VÖCH, Austrian Chemists Association in Vienna) in 1898. The complete archives of the VÖCH were lost when the association was eliminated in 1938 and during removals after the Second World War. See also references: Tumpel (1995), Markl (1997), Rosner (2004).

Creating Networks in Chemistry: The Founding and Early History of Chemical Societies in Europe
Edited by Anita Kildebæk Nielsen and Soňa Štrbáňová
© The Royal Society of Chemistry 2008

Prague and Brno were closely integrated with all other Austrian institutions of higher learning (in Vienna, Graz, Innsbruck, *etc.*) we consider the development of chemistry in these institutions, in which the language of instruction was German, as being part of the history of Austrian chemistry.

A considerable part of the rapidly developing chemical industry was owned or run by "Germans", often with their headquarters in Vienna, and therefore we also consider the development of this industry as part of the history of Austrian chemistry. Most factories were located in Bohemia and Lower Austria and produced a variety of acids, bases and salts, inorganic and organic dyes. In the first half of the nineteenth century the development of industry was supported by the state and polytechnic institutes were founded in Prague (1805), Graz (1811), Vienna (1815) and Brno (Brünn, 1849). Chemical education relevant for the industrial employment of graduates was established in these new institutes. Parallel with the increasing industrial production of chemicals mainly for the textile industry and ignition products like matches the number of students of chemistry doubled between the 1830s and the 1840s. Little theoretical research relevant for the development of chemistry was carried out before 1848 in Austria, neither in the universities nor in technical institutes.[2] Therefore, many chemists tried to finish their studies with a doctoral thesis abroad, mainly in Germany.

In the second half of the nineteenth century chemical education was modernized under the influence of Justus von Liebig.[3] Josef Redtenbacher in Prague and Vienna and Anton Schrötter in Graz and Vienna introduced in Austria Liebig's modern method of chemical education based on the laboratory work of all students. New chairs of chemistry were installed at universities and polytechnic institutes after the revolution of 1848. The organization of studies was reformed between 1849 and 1870. A third line of chemical education at a secondary level was established by the foundation of "Gewerbeschulen" (technical secondary schools for chemistry).

Research at Austrian universities was mainly concerned with organic chemistry, particularly with substances isolated from plants. Austrian chemists published their results mainly in German journals before 1848. After the foundation of the Austrian Academy of Sciences in 1847, the *Sitzungsberichte der kaiserlichen Wiener Akademie der Wissenschaften* (Proceedings of the Imperial Academy of Sciences in Vienna) were started in 1848 and Austrian chemists could then publish in that journal. Starting in 1880 chemical papers of the *Sitzungsberichte* were reprinted in the *Monatshefte für Chemie und verwandte Wissenschaften* (Monthly issues for chemistry and related sciences, today called Chemical Monthly. The titles "*Monatshefte*/Chemical Monthly" will be used throughout this paper).[4] This journal was well accepted in Austria and also abroad. Another journal, the *Berichte der Österreichischen Gesellschaft zur Förderung der chemischen Industrie* (Proceedings of the Austrian Society for the Advancement of Chemical Industry), was published in Prague from 1879 until 1898.

---

[2] Liebig (1838).
[3] Kernbauer (1997).
[4] Pohl and Soukup (2005).

Chemical research advanced in the last years of the nineteenth century, the industry grew and the number of chemistry students and graduates increased considerably. To support these developments and to make continued education possible, chemical associations seemed desirable.

1.2 Establishment of Chemical Societies

1.2.1 Predecessors of the Chemical Society

In 1869 the *Chemisch-Physikalische Gesellschaft* (CPG, Chemical-Physical Society) was founded by the chemist Heinrich Hlasiwetz and the physicists Josef Loschmidt, Josef Petzval and Josef Stefan. Aims of the association were to further the development of chemistry and physics and to disseminate chemical and physical knowledge. This was to be achieved by:

- lectures on research, demonstration of experiments and discussion of theoretical methods;
- support of research and edition of a journal both depending on the financial possibilities of the association;
- contacts with other scientific associations at home and abroad.

Written reports of the early period of the CPG were lost, but the association still exists today.

Presidents were always elected for one year. Among them were top scientists of Austria such as the above-mentioned founders, the physicists Ernst Mach and Ludwig Boltzmann, the chemists Adolf Lieben, Josef Maria Eder and Rudolf Wegscheider, the physiologist Siegmund Exner and others.

The main activity of the CPG was organizing lectures. At the beginning scientists from Vienna were lecturing, but, later on, speakers from abroad were also invited, such as Anton Lampa from Prague, Jacobus Henricus van't Hoff from Berlin, Marian von Smoluchowski from Cracow and George de Hevesy from Budapest. These names show the scientific level of the CPG lectures.

The success and high standard of the CPG lectures probably evoked the foundation of the *Wiener Verein zur Förderung des Physikalischen und Chemischen Unterrichts* (Viennese Association for the Advancement of Physical and Chemical Teaching) in 1895. This association was concerned with supporting chemistry and physics teachers at secondary schools. The first president of this association was the physicist Viktor von Lang. He was also president in 1900 when the association had 317 members.

1.3 The Association of Austrian Chemists

1.3.1 Founding the Association

Both associations mentioned before did not meet the necessities of chemists working in the industry, therefore a third association was needed. Since 1892, regular monthly meetings of Viennese chemists had been organized. Their participants decided to establish an association which should have mainly dealt with the professional situation of chemists. In February 1897 the First Austrian Day of Chemists (*Erster Österreichischer Chemikertag*) was set up. At this meeting a provisional governing group for a chemists' association was established, which in the following month received the permission of the authorities to start the association.

In June 1897 the *Verein Österreichischer Chemiker in Wien* (VÖCH, Association of Austrian Chemists in Vienna) was founded. The constituent assembly took place in the festive hall of the *Österreichischer Ingenieur- und Architekten-verein* (ÖIAV, Association of Austrian Engineers and Architects). The Technical College had granted the use of a lecture hall for future meetings of the VÖCH and a number of industrial enterprises had announced their financial support. So the starting conditions seemed to be favourable. In the beginning most of the members were Viennese chemists. Therefore the supplement *in Wien* (in Vienna) was added to the name. As the main centres of the chemical industry were in the provinces (Bohemia, Lower Austria, *etc.*) most chemists (about 60% of the members in the year 1900) were living outside Vienna. For this reason the supplement "in Vienna" was omitted from the name in 1901.

According to the concepts of the founders the VÖCH was to serve not only as a society in which information related to the development of chemistry as a science being discussed as it is in all scientific societies, but also as an organization representing college and university graduate chemists in their relationships with the state and their employers.

1.3.2 Management and Members

The first managing committee, consisting of eight persons, reflected the wish of the newly founded association to stay in contact with the authorities and to represent the chemists working in the industry: The president, Emerich Meißl, was director in the Imperial Ministry of Agriculture, the first vice-president, Professor Ernst Ludwig, was a member for the Upper Chamber of the Parliament and the first secretary, Karl Hazura, worked as chemist in the National Bank. The second vice-president, Josef Klaudy, worked in the *Technologisches Gewerbemuseum* (TGM, a higher technical school). Other members of the managing committee were directors or owners of industrial enterprises. It is remarkable that although the First Austrian Day of Chemists, where the foundation of the VÖCH originated, had mainly been attended by these technical chemists, the managing committee consisted of six persons with a Ph.D. and only two were graduates from the Technical College. It is unclear if the higher prestige of university graduates was the reason for giving them preference in the presidency. Additional members of the presidency besides the eight men of the managing committee were Hans Heger, editor of the *Pharmazeutische Post* (Pharmaceutical Post) and *Zeitschrift für Nahrungsmittel-Untersuchung und Hygiene* (Journal for Food Analysis and Hygiene), who initially printed two pages for VÖCH in his journal free of charge, and two

Table 1.1   Percentage of persons working in different branches of the Austrian chemical industry in 1902. Calculated from Tumpel (1995), 32.

Branch	Percentage
Chemical products (acids, bases, salts, solvents, *etc.*)	17
Coal, coke, coal gas	17
Pharmacies	11
Soaps, candles	10
Ignition products	9
Mineral oil, mineral pitch, bitumen	9
Fertilizers	5
Dyes	4
Fats, edible oils	4
Other	14

managers of important companies (Apollo Soap Company and Siemens & Halske). Meißl remained as president until his death in 1905 with an interruption of one year. He was followed by Rudolf Wegscheider, one of the most prominent Austrian chemists at that time, a professor of physical chemistry who remained as president until 1929.

The association was supposed to represent the interests of persons with academic training in the field of chemistry. The total number of these persons in the Habsburg monarchy is not exactly known, but it may have been 1000 to 2000 when the VÖCH was founded.

About 2% of all employees of the Habsburg monarchy (about 3.3 million in 1902) were employed by the chemical industry (not including the food industry).[5] The chemical companies amounted to about 1% of all industrial enterprises of the monarchy. Most of them were small (84% of the chemical firms employed between one and five persons). The numbers of chemical factories with more than 20 employees were 752 in 1901 and 967 in 1911.[6] Table 1.1 shows those branches of the chemical industry that employed the highest numbers of people.

The association was mainly financed from membership fees (the same for ordinary and extraordinary members) and to a lesser extent by contributions from the industry. Members living in Vienna paid twice as much as members from the province. Chemical companies paid three times the fee of the Viennese members.

The membership fees for members in Vienna were 6 Guldens (fl) in 1897 and increased to 7 Guldens in 1899; for members outside Vienna they were 3 Guldens in 1897, which increased to 5 Guldens in 1899. Later the Austrian currency changed from Gulden to Kronen (abbreviation K); 1 fl = 2 K. Membership fees were 15 K for members from Vienna and 12 K for all other members in 1908. Members from Germany had to pay an additional 2 K and

---

[5] *K.k. Statistische Zentralkommission* (ed.), Austrian statistics 75–1, issue 1902.
[6] Tumpel (1995), 31.

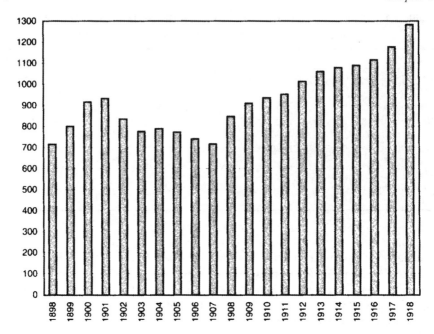

Figure 1.1 Number of members of the Association of Austrian Chemists between 1898 and 1918.

members from other foreign countries an additional 4 K for postage. Because the numbers of members decreased between 1902 and 1907, the membership fees for newly graduated chemists were reduced in 1908 to 8 K and the possibility was given to pay by four instalments of 2 K. This may have contributed to a considerable increase in the number of members during the following years.

There existed a special member category, known as the *"Gründer"* (founders). These were single persons or companies from the chemical branch contributing a higher membership fee (about 200 fl before 1900).

The total amount of membership fees was between 8500 and 14 400 Kronen (Austrian crowns) between 1901 and 1916. The sums are not well correlated with the number of members given in Figure 1.1. The amount decreased from 13 554 Kronen in 1910 to 13 381 Kronen in 1912, although the number of members increased from 937 in 1910 to 1013 in 1912. An additional income amounting to between 5 and 27% of the membership fees is documented in the ÖCHZ between 1901 and 1918 (without indicating the source).[7] Between one third and two thirds of the membership fees were necessary to cover the costs of the ÖCHZ. Founders contributed about 5 to 20% of the total income of VÖCH. The contribution amounted to about 15% in 1912 and reached a peak of 21.3% in 1914. It decreased to about 7% in 1918.

---

[7] Tumpel (1995), 47 (calculated from the VÖCH budget).

Table 1.2   Professional fields of the members of VÖCH. From Tumpel (1992), 45.

Self-employed[a]	224
Employees[b]	439
Chemistry teachers[c]	73
University teachers	52
Public servants	136
Without profession	10
Total	934

[a] People independent of an employer (own business, experts, writers, *etc.*)
[b] People employed in a factory, by the state, by some publisher, *etc.*
[c] Teachers in secondary schools.

Only chemists with sufficient academic training (at least six semesters' successful studies at the university or technical college) could become full members of VÖCH. Graduates from lower technical schools and students could become extraordinary members. Initially most members were graduates of the technical colleges, who had a somewhat lower rank compared to the graduates from the university. The rules of admission of full members, who had an academic degree, and "extraordinary" members, who had been trained in technical high schools, were not altered before 1900. No specific groups of chemical practitioners were explicitly prohibited from membership. Honorary membership has been awarded to outstanding chemists since 1903. Only two honorary memberships were awarded before the First World War: one in 1903 to the foremost chemist and inventor Carl Auer von Welsbach, and one in 1907 to Wilhelm Neuber, a Viennese factory owner, for his merits during the development of VÖCH.

Until 1916, the number of founders was between 14 and 17; in 1917 their number increased to 24 and in 1918 to 31. Until 1912, every year the numbers of ordinary and extraordinary members were compared in the minutes of the general assembly.

The professional fields of the VÖCH members are shown for the year 1901 in Table 1.2.

It follows from Table 1.2 that about 70% of the members were working in a private business. The extraordinary members amounted to about 10–18%. This means that mainly chemists with a higher degree were represented in the VÖCH, but the management of the association always also tried to support chemists with lower training (*e.g.* in secondary professional schools) with their specific requests.

In 1897 the association had about 200 members, in 1898 about 700 and in 1900 about 900. The number decreased to about 700 in 1907; since then we saw a sharp rise to about 1100 in 1914 and 1300 in 1918. In these years the numbers of extraordinary members (with no academic studies but only trained in technical high schools) were between 10 and 20% of the total number. In Figure 1.1, the numbers of members are given for the years 1898 to 1918.[8]

---

[8] Tumpel (1995), 43.

It is worth mentioning that information including the number of members in the year 1900 was sent to Bohuslav Brauner to Prague by the VÖCH secretary Karl Hazura and then most probably was forwarded to Henry Carrington Bolton from Columbia College NY.[9] Bolton used it in 1902 for his paper *Chemical Societies of the Nineteenth Century* with an overview of the world's chemical societies prepared for the 25th Anniversary Issue of the *Proceedings of the American Chemical Society*.[10] Bolton dedicated a copy of his publication to Bohuslav Brauner (Figure 1.2).[11] In a table showing the membership of the world's chemical societies, Bolton listed seven Austrian societies (four of them founded in Prague and described in Chapter 3) with a total of 3072 members (about 1500 of them in the Czech Lands). The Austrian societies ranked fourth after Germany, Great Britain and France (compare Table 1.3).

1.3.3   Everyday Routines

Among the problems to be addressed by the association were:

- Improving curricula of the *Technische Hochschule* (Technical College);
- Introducing the title "Ingenieur" for graduates (this was only granted by an imperial order in 1917);
- Improving patent laws;
- Improving the conditions for chemists in the industry;
- Furthering the interests of the chemical industry.

Several meetings of the VÖCH took place each year, starting in 1898. Minutes of these sessions were published in the journal *Österreichische Chemiker-Zeitung* (*ÖCHZ*). The report on the year 1897/1898[12] lists 13 plenary sessions. The meetings took place in lecture halls of the Technical College from 1897 until 1918. (Around 1870, the status of the technical college was changed. Originally established as *Polytechnisches Institut* in which a general technical education was offered, the college was converted according to the recommendations of the Czech chemist Carl Kořistka into the *Technische Hochschule* – Technical College – with new admission requirements, more specialization and stricter examinations.[13]) Starting in 1907, funds were collected for building a house for the VÖCH. Because of the First World War, however, the building plans were postponed. After the war the funds were lost due to the inflation and plans for the building were abandoned.

Standing orders authorized ordinary and extraordinary members to participate in discussions of the general assembly and to make proposals. These

---

[9] These pictures were received from Dr. Soňa Štrbáňová (Figure 1.2).
[10] Bolton (1902).
[11] Many thanks to Dr. Soňa Štrbáňová for supplying this information. The title page of Bolton's paper is shown in the preface.
[12] Meißl (1898).
[13] Kořistka (1863).

Figure 1.2   Postcard sent by K. Hazura, the first secretary of the Association of Austrian Chemists, to the Czech chemist Bohuslav Brauner from Vienna on May 30, 1900. The card contains current data on the Association. Brauner helped the American chemist Bolton to collect information on European chemical societies for Bolton's brochure *Chemical Societies of the Nineteenth Century* (Washington, 1902). The card is kept in the Museum of Czech Literature in Prague, Collection B. Brauner, Correspondence. The picture was given to the author by S. Štrbáňová.

Table 1.3   Membership numbers of chemical societies of the world in 1900
(from Bolton, 1902).

Country	No. of societies	No. of members
Austria[a]	7	3072
France	10	4065
Germany	10	7559
Great Britain	9	7550
USA	5	2379

[a] Including the Czech Lands.

proposals had to be treated by a committee or a commission (two thirds of their members had to be ordinary members) and after that put to the vote of the general assembly.

In October 1897 a general assembly established seven commissions to treat certain fields: a board for general questions, an employment agency, an economic commission, a patents commission, a commission for professional ranking, a lecture commission and an education commission.[14] But these permanent commissions were dissolved in 1899 and replaced by commissions that were established when needed.

In 1906 a committee relating to trade policy was elected. It had to inform the plenary assembly about questions of tax laws and customs policy. In 1907 a committee was installed to formulate a draft agreement for chemists working in the private industry. At that time chemists were usually working without a contract, which was unfavourable on the termination of employment and in case of the utilization of inventions made during employment. In 1909 this committee published a proposal for an employment contract and in 1910 this proposal was complemented by the general assembly.[15] This model contract was not obligatory but was used in many cases as a basis for employment. In 1914 a commission for the advancement of the chemical industry in war time was established.[16]

1.3.4   Various Activities

The main concern of the society was the professional recognition and the status of Austrian chemists. Other goals of the association were to further chemistry and chemists in all areas of science and economy, and to sustain research and education in Austria. Any political influence should be excluded from the activity of the association. VÖCH supported its members by arranging positions for chemists free of charge. A commission for that was established in 1897, as mentioned previously. The nine members of that commission came from Brno (Brünn), Prague, Vienna and Graz.[14] Fifteen to twenty jobs were offered each year with the help of VÖCH. Efforts to establish a legal title for

[14] Hazura (1898c).
[15] *ÖCHZ* (1910a).
[16] *ÖCHZ* (1915).

chemists from the Technical University (chemists from the university had a Ph.D.) were time consuming.[17] Finally in 1917 the government approved the title "Ing." (engineer).

Because chemists with academic degrees working in public service were paid less than public officers with a law degree, VÖCH tried to convince the government that this was unjust.[18] Until 1918 these efforts were only partly successful. Most letters were not answered by the ministry. In some cases improvements were reached.

At the end of the nineteenth century a new law concerning food quality was passed.[19] To control its regulations new types of food experts had to be instructed and afterwards examined. VÖCH was involved in the improvement of the examinations which were necessary to become a food expert.[20]

In 1910 the VÖCH sent a letter to the Ministry of Public Labour complaining that chemists did not have the possibility to become professional consultants.[21] In consequence, in 1913 the Ministry approved the profession "civil engineer for technical chemistry" for persons who were allowed to design and examine devices and machinery in the field of technical chemistry.[22] They also could deliver certificates for apparatuses and processes. Problems with this new profession were reported in 1914.[23] VÖCH also made efforts to improve the patent laws but without success.

The engagement in chemical education started in 1900, because in 1898 new curricula reduced the number of chemistry lessons in secondary schools. VÖCH suggested a new type of secondary school with living languages instead of Greek and Latin and enhanced teaching in science. In 1907 Wilhelm Ostwald gave a talk in Vienna[24] suggesting installing science, instead of languages, as the basis of secondary school education. In 1911 the VÖCH wrote to the Ministry[25] to support a memorandum from the *Česká společnost chemická pro vědu a průmysl* (Czech Chemical Society for Science and Industry) concerning secondary chemical education. Because the society of the Monarchy preferred the humanistic education, these forward thinking plans were not accepted.

Plans for new chemistry curricula at the universities and technical universities were formulated by VÖCH in 1901.[26] The training in both institutions should have become similar and comparable. Since 1901 it was possible to write a thesis at the *Technische Hochschule* in order to obtain the degree "doctor of technical science".[27] But this was difficult and time consuming, so most students of the *Technische Hochschule* finished their studies with the title "Ing."

---

[17] *ÖCHZ* (1899a), (1899b), (1900a), (1900b).
[18] *ÖCHZ* (1911c).
[19] Hazura (1898a).
[20] Hazura and Meissl (1898), (1899).
[21] Hazura and Wegscheider (1911a).
[22] *ÖCHZ* (1913).
[23] *ÖCHZ* (1914a), (1914b), (1914c).
[24] *ÖCHZ* (1907).
[25] Hazura and Wegscheider (1911b).
[26] Stiassny and Meissl (1901).
[27] Hartel (1901).

The VÖCH organized a great number of lectures held at the regular meetings of the association. Most speakers came from Austrian institutions and only a few from abroad. The reason was probably the limited financial sources of the association. Famous speakers in most cases gave lectures organized by the *Österreichischer Ingenieur- und Architektenverein* (ÖIAV, Austrian Association of Engineers and Architects, founded in 1848), which had a section for chemists. This association had its own building in Vienna and had a larger budget than the VÖCH.

A few examples of lectures held at meetings of VÖCH in the years 1898 to 1899 show which themes were important for the new association:

"About x-rays and more recent instruments for their production and use"[28]
"About Patent Legislation and the Austrian Patent Laws"[29]
"About progress in the production of nitric acid"[30]
"About methods to label margarine"[31]
"Chemistry of the usual techniques for reproduction"[32]
"Which chemical reactions occur spontaneously?"[33]
"Chemical speed"[34]
"Experience from the forensic-chemical practice"[35]

This list shows that practical and industrial chemistry had priority for the members of VÖCH and that physical chemistry became an important branch of research and education at Austrian universities around 1900.

1.3.5 International and National Contacts

There existed well-established relations of VÖCH with other professional societies in Vienna, in other parts of the monarchy and abroad.

One of the most important scientific congresses, the III. International Congress for Applied Chemistry (III. Internationaler Congress für angewandte Chemie), took place in Vienna in 1898, soon after the foundation of the VÖCH. Several members of VÖCH worked in the organizing committee in 1897 (Emerich Meißl, president of VÖCH, Ernst Ludwig, the first vice-president of VÖCH, Hans Heger, editor of the ÖCHZ, Josef Maria Eder, professor of photochemistry, Georg Vortmann, professor of analytical chemistry, Alexander Bauer, professor of general chemistry, all at the Technical College Vienna) together with high officials from Austrian ministries. At the opening ceremony the mayor of Vienna, the Minister of Commerce and high representatives from several institutions addressed the meeting. 765 participants attended lectures in 12 sections. The *American Chemical Society* was represented by Harvey W. Wiley from

---

[28] Valenta (1898).
[29] Kuzel (1898).
[30] Hölbling (1898).
[31] Ulzer (1898).
[32] Hrdliczka (1899).
[33] Wegscheider (1899).
[34] Klaudy (1899).
[35] Ludwig (1899).

Washington D.C. (president of the American Society 1893 and 1894).[36] Dr Wiley had been the chairman of the organizing committee of the World Congress of Chemists in Chicago in 1893. Honorary chairmen for nine divisions came from Germany, England and Switzerland. 76 papers were presented. In Vienna about 160 lectures were given, among them the lecture of Eduard Buchner "About cell-free fermentation", given at the opening of the congress, which raised general interest.[37] Buchner's studies, of which he spoke in Vienna, were honoured with the Nobel Prize for chemistry in 1907. Henri Moissan, who received the Nobel Prize for chemistry in 1906 for his investigation and isolation of the element fluorine, talked about "Production and properties of pure and crystallized calcium" in the section on electrochemistry.[38] He was asked in Vienna to take over the presidency of the IV. International Congress of Applied Chemistry in Paris in 1900 and he accepted. A decision taken at that congress in Vienna was the definition of the unit of the amount of electricity: the amount of electricity 96 500 coulombs, which results in the deposition of one gram-equivalent of any element, was defined as "1 Faraday (F)" in remembrance of Michael Faraday, following a suggestion of the German Electrochemical Society.[39]

The VÖCH had good relations with other national and international societies. VÖCH members attended meetings of the *Österreichischer Ingenieur- und Architekten Verein* (ÖIAV). Its chemistry section also organized lectures of famous chemists from abroad, which the VÖCH could not afford. Examples of lectures given in 1906 in the lecture-hall of the ÖIAV are:

Jacobus Henricus van 't Hoff, Berlin, "Thermo-Chemistry" (Thermo-Chemie).
Walter Nernst, Berlin, "Electrochemistry" (Die Elektrochemie).
Georg Lunge, Zürich, "Cooperation of chemistry and engineering in technical sciences" (Das Zusammenwirken von Chemie und Ingenieurwesen in der Technik).
Also Austrian chemists with international reputations lectured for ÖIAV in 1906:
Josef Maria Eder, Vienna, "Photochemistry" (Die Photochemie).
Zdenko Hans Skraup, Graz, "Constitution and synthesis of chemical compounds" (Konstitution und Synthese chemischer Verbindungen).

Good relations existed between VÖCH and the *Chemisch-physikalische Gesellschaft* (Chemical-Physical Society), which has been described above.

The VÖCH also kept friendly relations with the *Verein Deutscher Chemiker* (VDCh, Association of German chemists). ÖCHZ published reviews of lectures presented at sessions of VDCh regularly. Also summaries of lectures organized by the *Deutsche Chemische Gesellschaft* (German Chemical Society) were reviewed in ÖCHZ. This will be described in more detail in the following section on publications.

---

[36] Browne and Weeks (1952).
[37] Buchner (1898).
[38] Moissan (1898).
[39] Strohmer and others (1898).

1.3.6 Publications

As already mentioned, the *Sitzungsberichte der kaiserlichen Wiener Akademie der Wissenschaften* (Proceedings of the Imperial Viennese Academy of Sciences) were founded in 1848. Whereas the proceedings published papers from all natural sciences and mathematics, the papers dealing only with chemistry were also reprinted after 1880 in the new journal *Monatshefte für Chemie und Verwandte Wissenschaften* (Chemical Monthly).[4] Papers of industrial interest were also published in the *Berichte der Österreichischen Gesellschaft zur Förderung der chemischen Industrie* (Proceedings of the Austrian Society for the Advancement of the Chemical Industry), which appeared in Prague from 1879 to 1898. Starting in 1887 the journal *Zeitschrift für Nahrungsmittel-Untersuchung und Hygiene* was edited by Hans Heger in Vienna. When the VÖCH was founded in 1898, Heger suggested widening the scope of his journal and giving it a new name *Österreichische Chemiker-Zeitung"* (Austrian Chemists' Journal, *ÖCHZ*).

With a second editor, Eduard Stiassny, the *ÖCHZ* became the official organ of the VÖCH. The aim of the journal was to inform the members of VÖCH about the progress in the field of chemistry in the Habsburg monarchy and abroad. Paying members received the *ÖCHZ* without charge. Printing costs of the *ÖCHZ* were always a problem (they were about 2100 Gulden in 1898). In 1899 *ÖCHZ* also became the official journal of the *Österreichische Gesellschaft zur Förderung der chemischen Industrie in Prag* (Austrian Society for the Advancement of the Chemical Industry in Prague). This society had relatively few members (61 ordinary and 96 extraordinary members in 1908), but was quite influential.[40] It had issued its own journal from 1879 to 1898 as mentioned above. Its affiliation to *ÖCHZ* from 1899 to 1914 is shown on the front page.

The *ÖCHZ* was published by the two private owners until 1914, when VÖCH bought half of the journal. When the first number of *ÖCHZ* appeared in 1898[41] a large number of co-workers, including most established Austrian chemistry professors (what is now called the editorial board), was shown on the front page (Figure 1.3). There were 34 chemists from Vienna, 8 from provinces of Austria (Bohemia, Galicia and Bukowina) and 24 from abroad (Germany, Hungary, Italy and Greece) mentioned. No one from the German-speaking Switzerland was on the board. We may find among the co-workers personalities such as Alexander Bauer, Josef Maria Eder, Paul Friedlaender, Adolf von Lieben, Eduard Valenta, Georg Vortmann, Hugo Weidel, Hans Molisch (at that time in Prague) and Zdenko Hans Skraup (at that time in Graz). Biographical notes on most of these persons are given in Robert Rosner's book *Chemie in Österreich 1740–1914.*[42]

---

[40] Kornfeld (1909).
[41] *Österreichische Chemiker-Zeitung* erster Jahrgang (neue Folge) 1898 und *Zeitschrift für Nahrungsmittel-Untersuchung, Hygiene und Warenkunde* (12. Jahrgang), Officielles Organ des Vereines Österreichischer Chemiker in Wien, Verlag der Österreichischen Chemiker-Zeitung, Wien.
[42] Rosner (2004).

Figure 1.3   Title page of the first issue of the first volume of the *Österreichische Chemiker-Zeitung*, May 1, 1898.

The *ÖCHZ* published various original papers, summaries of articles from other journals, reviews, communications about patents, short reviews of recently edited books, questions from readers to be answered by experts, reports from industry and trade and announcements and reports from VÖCH. Also the *Chemisch-Physikalische Gesellschaft* decided to publish its minutes in the *ÖCHZ*. However, we cannot find in *ÖCHZ* original research papers in contrast to the publications of other chemical societies (like *Deutsche Chemische Gesellschaft* or *American Chemical Society*) as there had existed since 1880 the *Monatshefte für Chemie* (Chemical Monthly),[4] where most Austrian chemists published their scientific papers. The intended readers of *ÖCHZ* were mainly chemists from industry (chemical and food) and administration, who wanted information about industrial developments in Austria and abroad, about new institutions, biographical notes about famous chemists and reports summarizing new developments in various fields of chemistry.

Statistics of the sections of *ÖCHZ* show the proportion of articles about different branches of chemistry published from 1899 to 1914: inorganic chemistry 21.1%, organic chemistry 22%, physical chemistry 7%, analytical chemistry 14.6%, chemical technology 17.1% and "various subjects" (like obituaries, reports about buildings, congresses, new curricula of chemical education, jurisdiction concerning trade and patents, *etc.*) 18.2%.

The journal appeared twice a month. Its 24 issues per year had about 500–600 pages, half of which were filled with advertisements. The volume of 1915 had only half the normal size, demonstrating the shock of the starting war.

Although the *ÖCHZ* generally dealt only with subjects that had some relationship to chemistry or chemical industry, it also published several reports dealing with events concerning the imperial family, like the assassination of the Empress Elisabeth,[43] or the assassination of the successor to the throne Franz Ferdinand,[44] an incident which initiated the First World War. On the occasion of the 50th anniversary of the accession to the throne of Franz Joseph, an article was published glorifying the emperor[45] as a ruler, who had contributed to the development of chemistry.

Several articles discussed achievements of Austrian chemists, often in connection with obituaries or when a chemist was awarded the Lieben Prize. One of the obituaries was for the chemist Adolf Lieben, who had initiated the *Ignaz Lieben Preis* in 1863 from money inherited from his father Ignaz Lieben. It was at that time a completely new idea to award a prize for excellence in science. The Lieben Foundation was controlled by the Imperial Academy of Sciences in Vienna. *ÖCHZ* always reported on the sessions of the mathematical-scientific class of the academy and also on the winners of the Lieben Prize.[46] For example *ÖCHZ* reported on the Lieben Prize for Rudolf Wegscheider, professor at the University of Vienna and president of VÖCH since 1905, and Hans Meyer, professor at the University of Prague, who later died in the concentration camp Theresienstadt in 1942.[47] Wegscheider had published papers on the esterification of various acids and Meyer about the use of thionyl chloride in organic synthesis.

On another occasion we find a notice about the planned closing of the chemistry section of the *Technologisches Gewerbemuseum* (TGM), where at that time Paul Friedlaender was doing research into dye chemistry as head of the chemistry section. Friedlaender, who was also on the editorial board of *ÖCHZ*, was awarded the Lieben Prize in 1909 for his work on the antique purple and left Vienna in 1911 for Darmstadt when the chemistry section of TGM was finally closed and a new chemistry school without research was opened instead.[48]

An inspection of the articles published in the *ÖCHZ* shows that the Association tried hard to stay in contact with the rapid development of chemistry in other European countries (mainly Germany). In the years before the First World War the research of several important chemists in Berlin, like Emil Fischer,

[43] *ÖCHZ* (1898a).
[44] *ÖCHZ* (1914d).
[45] *ÖCHZ* (1898b).
[46] Biographies of all Lieben Prize winners have been collected by Werner Soukup. See Soukup (2004).
[47] *ÖCHZ* (1905a).
[48] *ÖCHZ* (1911a), (1910b).

Table 1.4 The employment of Austrian chemists in 1911 (including the Czech Lands). From Grünwald (1913), 135.

Number of chemists in certain professions (examples)

In public service
1603 (38%)
    118 university professors
    40 university lecturers (*Dozenten*)
    134 university assistants (*Assistenten*)
    280 teachers in secondary schools
    238 teachers in professional secondary schools (Gewerbeschulen)
    367 in public technical control
    222 in public research institutes

Privately employed
2567 (62%)
    308 chemical heavy industry
    817 sugar industry
    285 dye industry
    157 mineral oil industry
    142 pharmaceutical industry
    113 mines and metallurgical industry
    111 private research institutes

Total 4170 (from this number 525 people were working in Hungary and 270 in other countries outside the monarchy).

Richard Willstätter, Otto N. Witt, Paul Ehrlich, Hans Landolt, Walther Nernst and Fritz Haber, is described in the *ÖCHZ*. From Austria the work of organic chemists like Zdenko H. Skraup, Hans Molisch, Paul Friedlaender, of analytical chemists like Friedrich Emich (Graz) and Fritz Pregl (Innsbruck) and of physical chemists like Rudolf Wegscheider, Anton Skrabal (Graz) and Emil Abel (Vienna) was presented in the *ÖCHZ*. Some articles discussed lectures about important developments, such as a lecture given by Fritz Haber about the synthesis of ammonia and Friedrich Bergius about the hardening of fats. In both cases catalysis, which was detected as the most important factor for chemical kinetics by Wilhelm Ostwald (Nobel Prize 1909), played a central role.

The *Österreichischer Chemiker-Schematismus*, a collection of the names, addresses and professional positions of about 2600 chemists working in Austria, was first published by VÖCH in 1898.[49] The booklet, which also contains useful statistical information, had 142 pages and was sold for 60 kr. In 1905 the 4th edition appeared with enumeration of 2700 chemists working in the Austro-Hungarian Monarchy. The 5th edition in 1908 was subdivided into three parts. In the first part as many as 4500 names were given in alphabetical order. In parts 2 and 3 the names were arranged according to professional fields and places of residence. These parts were sold for 2.50 K each. In 1911 the 6th edition was printed. Information from that booklet is given in a paper which appeared in VÖCH with a table summarizing the numbers of chemists in public and private employment.[50] Part of these data is given in Table 1.4.

---

[49] Hazura (1898b).
[50] Grünwald (1913), Tabelle 1, 135.

The number of experts by VÖCH who were allowed to deliver certificates about industrial enterprises in different fields of chemistry was 124.[51] The 7th edition of the *Chemiker-Schematismus* appeared in 1914.

An address register of all industrial companies connected to chemistry was edited by VÖCH in 1897 and its second edition appeared in 1898.

1.4   Concluding Remarks

The Association of Austrian Chemists was founded at a time when the number of chemical companies, in which all sorts of chemicals were made, grew rapidly, with the consequence that more and more chemists were required to work in the industry. The industrial production shows clearly that this growth started around 1890 (Table 1.5).[52]

The textile and sugar industries were growing extremely rapidly. Engineering for the transport and armaments industries was also developing rapidly. In addition, the production of the chemical industry increased steadily during these years. The chemical productions that used electricity moved from Bohemia to the Alpine regions. Parallel to the growth of the chemical industry the number of students at universities and technical universities of the Monarchy and the number of graduated chemists increased steadily during the last years of the nineteenth century and the beginning of the twentieth century.

The new association had as its primary aim the representation of chemists in industry or in administration both legally and economically but also supported the employers in the chemical industry in their dealings with the authorities. A further aim was for an improvement of chemical education in order to keep the industry competitive.

In its activities the VÖCH made a great effort to keep its members informed about the rapid developments of chemistry in every field and therefore often invited prominent scientists to lecture at the meetings of the association. But the exchange of information in scientific meetings was not a major activity of the VÖCH in contrast to other chemical societies. This can be seen in the *ÖCHZ*, the journal of the VÖCH, which contained articles discussing various new developments in chemistry, biographies of important Austrian chemists and articles dealing with a variety of legal and economic questions, but no articles in

Table 1.5   Industrial production in Austria 1870–1913. The production of 1913 is taken as 100%. From Rosner (2004), p. 292.

1870	1875	1880	1885	1890	1895	1900	1905	1910	1913
33.7	35.4	35.8	44.0	54.2	65.0	70.6	78.7	93.5	100

---

[51] *ÖCHZ* (1911b).
[52] Rosner (2004), 292, taken from Sylla and Toniolo (1991), 227.

which research results were discussed. The journal in which Austrian chemists could report their research results was the *Monatshefte für Chemie und Verwandte Wissenschaften* (Chemical Monthly), which was published by the Imperial Academy of Sciences.

Although the primary aim of the VÖCH was the representation of chemists working in industry, many of the elected officials were chemists with an academic background, probably because of the high prestige of these persons. As a result, more attention was paid to the research activities of Austrian chemists in later years. Compared to basic research, the Austrian chemical industry was quite successful during the second half of the nineteenth century. Practical application of scientific research resulted in an increasing gross national product. Before the First World War, Austria's rate of economic growth was second in Europe and it appears now that the activities of the VÖCH have contributed to the positive development in the area of chemical industry in Austria during the period from the 1890s to the beginning of the First World War.[53]

Finally, the author of this chapter wants to report an observation, which in various ways induces associations concerning the past and present of chemistry and the *ÖCHZ*. The paper of the original *ÖCHZ* journals that I could study in two libraries of Linz is often very brittle. This is probably caused by the acidic reactions of the paper. My speculation is that this paper was made from cellulose prepared from wood using the new successful "Kellner" process with magnesium sulphite. At the end of the nineteenth century many paper factories using this process were built and the rivers were extremely contaminated with wastes from the factories. Karl Kellner, the inventor, founded the Kellner-Partington Paper Pulp Co. Ltd, one of the most important industrial companies of the world. An obituary of Kellner, who also invented a new method of chlorine-alkali-electrolysis (Castner-Kellner-process), appeared in the *ÖCHZ* 1905.[54] In 1912, Benjamin Reinitzer, professor at the Technical College Graz, the brother of Friedrich Reinitzer (discoverer of liquid crystals), wrote in the *ÖCHZ* about contamination of Austrian rivers by paper factories.[55] He developed a method to oxidize sulfite by chlorine and precipitate it as calcium sulfate. This method improved the situation in many cases. But the final cure for the rivers came only in the second half of the twentieth century and is still a problem now.

Acknowledgements

I want to thank Soňa Štrbáňová and Robert Rosner for their comments and suggestions which helped me to improve this chapter.

---

[53] Narbeshuber in Markl (1997), 74.
[54] *ÖCHZ* (1905b).
[55] Reinitzer (1912).

Abbreviations

CPG   *Chemisch-physikalische Gesellschaft* (Chemical-Physical Society)
fl   Gulden (old Austrian currency, divided into 100 Kreuzer, abbr. kr)
GÖCH   *Gesellschaft Österreichischer Chemiker* (Austrian Chemical Society, before 1982 the name was VÖCH)
K   Krone (Austrian crown, currency after 1900 by a law of 1892, divided into 100 Heller abbr. H, 1 fl = 2 K)
K.k.   Kaiserlich königlich (Imperial and royal)
ÖCHZ   *Österreichische Chemiker-Zeitung* (Austrian Chemists Journal)
ÖIAV   *Österreichischer Ingenieur- und Architektenverein* (Austrian Association of Engineers and Architects)
TGM   Technologisches Gewerbemuseum (Technological Industrial Museum, a new type of higher technical school)
VDCh   *Verein Deutscher Chemiker* (Association of German Chemists)
VÖCH   *Verein Österreichischer Chemiker in Wien* (Austrian Chemists Association in Vienna; in 1901 the supplement "in Wien" was omitted from the name, in 1982 the name was changed into GÖCH)

References[56]

Browne, C. A. and Weeks, M. E. (1952), *A History of the American Chemical Society*, American Chemical Society, Washington D.C.

Bolton, Henry Carrington (1902), *Chemical Societies of the Nineteenth Century*, Smithsonian Institution, Washington D.C.

Buchner, E. (1898), Ueber zellfreie Gärung, *ÖCHZ* **1**, 229–232.

Grünwald, J. (1913), Die Stellung des Chemikers in der modernen Industrie, im Staatsdienst und seine soziale Stellung, *ÖCHZ* **16**, 133–138.

Hartel, M. P. (1901), Die Verleihung des Promotionsrechtes an die technischen Hochschulen der im Reichsrathe vertretenen Königreiche und Länder, Rigorosenordnung", *ÖCHZ* **4**, 183–184.

Hazura, K. (1898a), Das Lebensmittelgesetz vom 16. Jänner 1896 und seine Durchführung, *ÖCHZ* **1**, 83–84 and 117–120.

Hazura, K. (1898b), Der Chemiker-Schematismus, *ÖCHZ* **1**, 326 and 368.

Hazura, K. (1898c), Verzeichnis der Ausschuss- und Commissions-Mitglieder des Vereines Österreichischer Chemiker in Wien, *ÖCHZ* **1**, 434.

Hazura, K. and Meissl, E. (1898), Eingabe des Vereines Österr. Chemiker in Wien an das k.k. Ministerium des Inneren in Angelegenheit der Prüfungsordnung für dipl. Lebensmittelexperten vom 13. October 1897, *ÖCHZ* **1**, 80–82.

Hazura, K. and Meissl, E. (1899), Eingabe des Vereines Österreichischer Chemiker in Wien an das k.k. Ministerium des Inneren und das k.k. Ministerium für Cultus und Unterricht, *ÖCHZ* **2**, 348–349.

---

[56] Some articles in *ÖCHZ* do not show any author. These are referred to as *ÖCHZ*.

Hazura, K. and Wegscheider, R. (1911a), Eingabe des Vereines in Angelegenheit der Ziviltechniker-Verordnung, *ÖCHZ* **14**, 12–14.

Hazura, K. and Wegscheider, R. (1911b), Eingabe des Vereines in Angelegenheit der Ausgestaltung des Chemie-Unterrichtes an Mittelschulen, *ÖCHZ* **14**, 40.

Hölbling, V. (1898), Ueber Fortschritte in der Fabrication von Salpetersäure, *ÖCHZ* **1**, 141–143 and 172–173.

Hrdliczka, F. (1899), Chemismus der gebräuchlichen Reproductionsverfahren, *ÖCHZ* **2**, 38–40.

Kernbauer, A. (1997), Chemical Education in the Habsburg Monarchy's Universities and Technical Colleges around 1861, in W. Fleischhacker and T. Schönfeld ed., *Pioneering Ideas for the Physical and Chemical Sciences*, Plenum Press, New York & London, 289–296.

Klaudy, J. (1899), Die chemische Geschwindigkeit, *ÖCHZ* **2**, 364–366 and 433–437.

Kořistka, C. (1863), *Der höhere polytechnische Unterricht in Deutschland, in der Schweiz, in Frankreich, Belgien und England*, in H. Sequenz, *150 Jahre Technische Hochschule Wien*, Springer, Wien & New York, 31–33, 151.

Kornfeld, F. (1909), Geschäfts- und Rechenschaftsbericht für das Jahr 1908, *ÖCHZ* **12**,137–138.

Kuzel, H. (1898), Ueber Erfindungsschutz, Patentsysteme und moderne Patent-Gesetzgebung, *ÖCHZ* **1**, 25–28 and 65–66.

Kuzel, H. (1899), Ueber die Aufgaben der Patentgesetzgebung und über österreichisches Patentrecht, *ÖCHZ* **2**, 226–231.

Liebig, J. (1838), Der Zustand der Chemie in Österreich, *Annalen der Pharmacie* **25**, 339–347.

Ludwig, E. (1899), Erfahrungen aus der gerichtlich-chemischen Praxis, *ÖCHZ* **2**, 634–636.

Markl, P. ed. (1997), *Chemie in Österreich, 100 Jahre Gesellschaft Österreichischer Chemiker 1897–1997*, Gesellschaft Österreichischer Chemiker, Wien.

Meißl, E. (1898), "Protokoll der Generalversammlung des, Vereines österreichischer Chemiker in Wien' am 25. Mai 1898", *ÖCHZ* **1**, 185–188 and 219–220.

Moissan, H. (1898), Herstellung und Eigenschaften des reinen und krystallisierten Calciums, *ÖCHZ* **1**, 255–256.

ÖCHZ (1898a), Kaiserin Elisabeth, *ÖCHZ* **1**, 309.

ÖCHZ (1898b), Zum 2. Dezember 1898, *ÖCHZ* **1**, 445–446.

ÖCHZ (1899a), Zur Erlangung der Doctorwürde an technischen Hochschulen, *ÖCHZ* **2**, 563.

ÖCHZ (1899b), Die gesetzliche Regelung des Ingenieurtitels, *ÖCHZ* **2**, 614.

ÖCHZ (1900a), Ingenieurtitelausschuss, *ÖCHZ* **3**, 136.

ÖCHZ (1900b), Standesbezeichnung Ingenieur, *ÖCHZ* **3**, 550.

ÖCHZ (1905a), Preisverleihung der kaiserl. Akademie der Wissenschaften, *ÖCHZ* **8**, 258.

ÖCHZ (1905b), Dr. Karl Kellner, *ÖCHZ* **8**, 279.

ÖCHZ (1907), Naturwissenschaftliche Forderungen zur Mittelschulreform, *ÖCHZ* **10**, 342–343.

ÖCHZ (1910a), Jahresbericht über das Vereinsjahr 1909, *ÖCHZ* **13**, 51.

ÖCHZ (1910b), K.k. höhere Staatsgewerbeschule chemisch-technischer Richtung in Wien 17, *ÖCHZ* **13**, 230.

ÖCHZ (1911a), Prof. Dr. Paul Friedlaender, *ÖCHZ* **14**, 61.

ÖCHZ (1911b), Die neuernannten handelsgerichtlich beeideten Sachverständigen und Schätzmeister der chemischen Branchen (Gruppe 61), *ÖCHZ* **14**, 131–133.

ÖCHZ (1911c), Die Zurücksetzung der Techniker im Staatsdienst, *ÖCHZ* **14**, 258.

ÖCHZ (1913), Zivilingenieure für technische Chemie, *ÖCHZ* **16**, 141–142.

ÖCHZ (1914a), Die neue Ziviltechniker-Verordnung, *ÖCHZ* **17**, 124.

ÖCHZ (1914b), Titelführung der Ziviltechniker, *ÖCHZ* **17**, 139.

ÖCHZ (1914c), Die Prüfungsordnung für Ziviltechniker, *ÖCHZ* **17**, 156.

ÖCHZ (1914c), Erzherzog Franz Ferdinand, *ÖCHZ* **17**, 163.

ÖCHZ (1915), Kommission zur Förderung der chemischen Industrie in Kriegszeiten, *ÖCHZ* **18**, 31.

Pohl, G. W. and Soukup, W. (2005), 125 years 'Monatshefte für Chemie/ Chemical Monthly', *Monatshefte für Chemie* **136**, 5–14.

Reinitzer, B. (1912), Ueber die Verunreinigung der Gewässer durch die Abläufe der Sulfitzellulosefabriken und ihre Bekämpfung durch chemische Mittel, *ÖCHZ* **15**, 61.

Rosner, R. W. (2004), *Chemie in Österreich 1740–1914, Lehre-Forschung-Industrie*, Böhlau Verlag, Wien-Köln-Weimar.

Soukup, R. W. (2004), *Die wissenschaftliche Welt von gestern*, Böhlau Verlag, Wien-Köln-Weimar.

Stiassny, E. and Meissl, E. (1901), Eingabe des 'Vereines österr. Chemiker' an das k.k. Ministerium für Cultus und Unterricht, in Angelegenheit der Reform der chemischen Studien an den österreichischen Hochschulen, *ÖCHZ* **4**, 288–289.

Strohmer, F. *et al.* (1898), III. Internationaler Congress für angewandte Chemie, 27.Juli bis 2. August 1898, *ÖCHZ* **1**, 234–240 and 258–274.

Sylla, R. and Toniolo, G. ed. (1991), *Patterns of European Industrialisation. The Nineteenth Century*, Routledge, London & New York.

Tumpel, R. (1995), *Die Geschichte des Vereines Österreichischer Chemiker* Diplomarbeit Kennzahl J 150, Matrikel-Nr. 8203922, Wirtschaftsuniversität Wien.

Ulzer, F. (1898), Ueber Methoden zur Kennzeichnung der Margarine, *ÖCHZ* **1**, 452–453.

Valenta, E. (1898), Ueber Röntgenstrahlen und neuere Apparate zur Erzeugung und Verwendung derselben, *ÖCHZ* **1**, 19–22 and 99–101.

Wegscheider, R. (1899), Welche chemischen Reactionen verlaufen von selbst?, *ÖCHZ* **2**, 274–279.

CHAPTER 2

BELGIUM: From Industry to Academia: The Belgian Chemical Society, 1887–1914

BRIGITTE VAN TIGGELEN AND HENDRIK DEELSTRA

Today, the *Société Royale de Chimie* (French-speaking part of Belgium) and the *Koninklijke Vlaamse Chemische Vereniging* (Dutch-speaking part of the country) consider themselves as typical chemical societies, holding or sponsoring conferences and meetings, editing scientific journals and sustaining a local community connected to the international federations and scientific unions. The members would be most surprised though if they were to receive either the agenda or the minutes of the early meetings of the *Association des Chimistes Belges*, as it was first suggested to be called in 1887, as they would be astonished to discover for which practical aims the Belgian chemical society was founded. Chemical problems, finding consensus about the best and most reliable analytical procedures, and laws and rules pertaining to chemical expertise were then at the core of the debates and discussions among the members. Chemical problems? What was considered as chemistry at the time of the birth of the Belgian chemical society? Why did the first name of this *Association des Chimistes Belges* very soon change to *Association Belge des Chimistes*? And what was at stake when the general assembly decided to relabel itself *Société Chimique de Belgique*?[1] By focusing on a limited number of questions in the frame of this collective volume

---

[1] In this chapter, we will not discuss the later changes in name that occurred after the First World War. The split of the *Société Chimique de Belgique* into two chemical societies according to language occurred in 1939. It is furthermore the tradition in Belgium that associations that exist for many years may apply for the royal patronage and use "royal" in their denomination. This happened in 1987, for the centennial anniversary, for both societies, hence the actual names *Société Royale de Chimie* and *Koninklijke Vlaamse Chemische Vereniging*.

Creating Networks in Chemistry: The Founding and Early History of Chemical Societies in Europe
Edited by Anita Kildebæk Nielsen and Soňa Štrbáňová
© The Royal Society of Chemistry 2008

devoted to the emergence of chemical societies across Europe, this chapter aims to provide some insights on these questions that would deserve a much closer study; a fine grain analysis has not yet been achieved for the Belgian case.[2]

2.1   Chemical Education: A Kaleidoscope[3]

When, in 1797, the Université de Louvain (University of Louvain) was closed by the French republican authorities, chemistry had been taught as a subject in the medical faculty for more than two centuries. Of course, chemistry was first thought of as being a division of medicine, but in the course of the eighteenth century the teaching evolved to match the development of this science and one of the last royal professors of chemistry, Karel van Bochaute, even included Lavoisier's revolutionary theories in his teaching and research. By that time some chemistry, viewed this time as the science of matter, was also present in the classrooms at the faculty of arts, while youngsters were learning "*physica*". Switching from the French to the Dutch authority in 1815, three State universities were founded in the Belgian provinces in 1816–1817. Science became a subject, and students could earn a candidate in science after a two-year course, although this stage was still seen as a preparation for medical studies, except for secondary school teachers who could immediately apply for a teaching position. After the Belgian state came into being in 1830, the distinction was made between the mathematical and physical sciences on the one hand and the natural sciences on the other. The latter directly led to further medical studies and consisted of botany, zoology, geology, mineralogy and chemistry. Only in 1890 did the specialization in chemistry appear in the legal titles of the diploma: at the end of their four years, the students were trained chemists, having learned to manipulate and also to pursue and produce original research.

This evolution was met earlier in some "free universities" – those that are not state founded and subsidized – in Leuven, and then in Brussels, and was prompted by two different demands. Several natural sciences university professors, among them the organic chemist Louis Henry, strongly insisted on research as a professional duty for university teachers; along the same line, he insisted on providing schooling in research through practical laboratories and original dissertation.[4] He was successful in training numerous Ph.D. students in chemistry from 1876 onwards. The specialization process in pure chemistry education was enhanced in 1929: the licentiate level was created, granted after four successful years at the university, before a non-compulsory Ph.D. level that was by then exclusively dedicated to research. But chemical education was put under another constraint during the nineteenth century, as a subject in the curricula of pharmacists and engineers.

---

[2] Since the *Société Chimique de Belgique*'s archives have not survived, the main source for a detailed investigation of its history is the *Bulletin*, of which a complete set is not easily found. H. Deelstra owns such a set for the early years and has also conducted most of his research in the unclassified archives of the sugar factory at Tienen, Belgium.

[3] For a more detailed description, see Deelstra (2001).

[4] Van Tiggelen (1993).

In 1836, two "Ecoles spéciales" (special schools) were attached to the two surviving state universities from the Dutch period, Ghent and Liège. Another "Ecole des mines" (school for mining) was established in Mons, at the heart of the coal-mining region of the Borinage, independently from any university a year later, in 1837. Along with the traditional curriculum devoted to civil construction ("Génie civil"), two specializations appeared at Ghent and Liège where courses of applied chemistry were part of the teaching: "Ecole des mines" and "arts et manufactures". To mark the difference, students from the latter specialization would receive a diploma as "ingénieur de l'industrie" (industrial engineer) instead of the title "ingénieur civil" (civil engineer). The formation in chemistry was extended in 1867 and especially in 1869 with a course on analytical chemistry, for which laboratories were to be organized by the science faculty teachers.

The free universities were less keen on opening a special section for engineers since most of the positions would be in public services and they were not delivering a diploma sanctioned by the state. The only reason for founding an engineering school in a free university was proselytism, both catholic and free-thinking; for instance, in 1865 a special school was founded at the University of Louvain, attached to the science faculty. The situation changed completely in 1890 when the monopoly of the state universities was abolished and in the case of the engineering curricula in the universities five legal titles were created, among them one for "ingénieur chimiste" (chemical engineer), specifically dedicated to training engineers in the realm of chemistry.

The Belgian State was slower to organize lawfully the education of other professionals, the pharmacists, sticking to the will of freedom of education stated in the Constitution and relying on the traditional organization of pharmacy training by local authorities and professional corporations. In 1849, the academic degree in pharmacy, attached to the medical faculty, was introduced. As has been stated above, the students in pharmacy followed the natural science curriculum. But only in 1876 were the courses in analytical chemistry, toxicology and falsification of drugs part of the legal curriculum. These courses again were taught by professors from the science faculty and forced students to have laboratory training associated with the theoretical teaching. Thus academic chemists, passing on their knowledge as a service for different professional education, were forced to develop practical training, which eventually led to inclusion and justification of practical training in the science faculty itself.

As well as engineering and pharmacy, a third "university" curriculum was also strongly reliant on chemistry: agronomy. The first school for this subject was established by the State in Gembloux in 1860, and this Institut agricole de l'Etat (State Agricultural Institute) very soon attracted even foreign students. During their three-year curriculum, students were mostly trained in practical agronomy, and the chemistry teaching was given by only one teacher. In 1896, three chemistry chairs were legally organized: one for general and organic chemistry, one for analytical chemistry and the last one for "technologie agricole" (agricultural technology). The following year, the curriculum was extended to four years. In the meantime, the agriculture crisis of 1870 also led the Université Catholique de Louvain (University of Louvain) to open an Ecole supérieure d'agriculture (Higher School for Agriculture). The goals were to use the discoveries and ideas of Liebig, Boussingault, Schloesing and Georges Ville, and to promote intensive farming. The beginnings were difficult but in 1883 the reformed "Institut Agronomique" (Agronomic institute) was attached to the science faculty very much in the same way as the engineering school. Among other industries derived from agriculture, brewing has always been very important to Belgium. In 1887 two brewing schools were created, one in Ghent, the Ecole professionnelle de Brasserie (Professional School for Brewing), and the other in Louvain, the Ecole supérieure de Brasserie (Higher School for Brewing), attached to the agronomic institute mentioned earlier.

In addition to these academic schools devoted to more practical or professional aims, several technical and industrial schools also provided training loosely linked with the chemical industry or the use of chemical knowledge in industrial processes. One of the most successful was the Institut Meurice de Chimie (Meurice Institute for Chemistry), founded in the industrial landscape of Charleroi by Charles Meurice, and then moved to Brussels in 1897 by his son Albert, who worked as an assistant for two semesters in Remigius Fresenius's laboratory. A technical school for brewing was also founded in Ghent in 1892, out of the shadow of academic teaching.

2.2   First Step to Union: Facing Disagreement

The first gathering that ultimately led to the foundation of a Belgian chemical society had only one point on the agenda: solving a distinct technical problem, in a specific field of expertise inside the practice of chemistry. The problem, measuring the actual content of sugar in sugar beets, was not a new one but had forged controversies between producers of beet sugar, sugar sales associations and, last but not least, the farmers, for years before 1887, since this piece of information was crucial to settle the price of these goods. To solve this issue a specific commission was created appointed by the *Société Belge des Fabricants de Sucre* (Belgian Association of Sugar Manufacturers) and other commercial associations, among them the powerful *Société Commerciale, Industrielle et Maritime d'Anvers* (Commercial, Industrial and Maritime Society of Antwerp) and the Commission mixte des départages (Joint Commission for Arbitrage). The latter was a joint commission of experts bringing together chemists employed by sugar factories and chemists representing the interests of the commercial associations. In 1903 the Minister of Agriculture created another "Commission mixte des départages" bringing together the official chemists, nominated by the government and defending the interests of the farmers, and the private chemists, the chemists of the sugar industry.

However, the method for the determination of sugar used by then seemed to be controversial, and François Sachs along with others thought that the best and most definite way to find a consensus on the method of determination was to discuss the problem thoroughly and on free grounds among all "chemists" interested. The meeting was called in a letter dated 30th March 1887, and published on 1st April in *La Sucrerie Belge* (The Belgian sugar factory), the journal of the *Société Belge des Fabricants de Sucre* (Belgian Society of Sugar Makers).[5] The first assembly, devoted to the "determination of the best method for the commercial analysis of sugar beets", was held in a private room of *La Brasserie Belge* in Brussels in the early afternoon of April 14th. The signers of the prospectus were joined by thirteen other "chimistes", and together they took the initiative to found the *Association des Chimistes Belges*.[6] Most of them again were sugar chemists and among those present at the birth of the *Association* were a manager of a sugar factory and an influential landowner.[7]

During the first year of its establishment, the association changed its name from *Association des Chimistes Belges* to *Association Belge des Chimistes*. Euphonically the two locutions arc almost the same, but a closer look at the international membership does provide a clue to this shift: starting from a very local and specific problem, the association very soon reached out to international experts. *La Sucrerie Belge* had a French counterpart, *La Sucrerie Française* (The French sugar factory) and the problems reported there were by no means different from or less crucial than those set on the agenda of the first meeting. Actually, before the *Association* came into being, a suggestion was even made to affiliate the associations to be founded as a kind of Belgian section of *l'Association des Chimistes Sucriers et de Distillerie de France* (The French Association of Sugar and Distillery Chemists), but, as reported 25 years later by François Sachs, "the majority resolved to found an exclusively Belgian society".[8] Hence, probably, the emphasis on "chimistes Belges" (Belgian chemists) during the first year of existence. This French temptation was also thematically too narrow a perspective since Sachs insists that the goal was to create "a society that would gather not only all sugar chemists but all chemists in general". From its early beginnings though, one notices that the association did have ordinary members residing outside of Belgium along with designated "foreign" members.[9]

Surprisingly little is known about the founding fathers of the Belgian chemical society and the later history of the *Association* can probably account for this lack of information. For instance, Edouard Hanuise was not even granted an obituary, even though he had a leading role inside the *Association* for eight years.[10] Even less is known about some others; for instance, Delville, Lembourg and Gaillard, who were to remain faithful members of the society

---

[5] The letter is reprinted in the very first *Bulletin* **1**, 1887/1888, 1–2.

[6] F. Sachs, Historique de la société, *Bulletin* **26**, 1912, 141.

[7] *Bulletin* **1**, 1887/1888, 22, provides also the list of the 13 other names.

[8] *Bulletin* **26**, 1912, 141.

[9] Later, in 1900, when the section of Luxembourg is created, very interestingly, to legitimise the inauguration of a local section outside of Belgium's border, the president refers to the existence of a Belgian section in the *Verein deutscher Chemiker*; see *Bulletin* **15**, 1901, 56 (annual report of the president L. L. De Koninck). This extra-territorial section though did not last more than two or three years.

[10] Deelstra and Van Tiggelen (2003).

through the years. The scant information gathered on the background and function of the founding fathers does, however, reflect the frame in which the first encounter was placed:[11] six of them are "chimistes départageurs" sitting at the "Commission mixte des départages", the three others are known as "chimistes privés" (private chemists), who are chemists employed directly by sugar industries. More striking is the diversity in backgrounds and actual professions: obviously this chemical expertise in the specific problem of sugar chemistry did not provide a (full) living or status at that time. Out of five engineers, one of them also held an academic position at a higher technical school (Ecole des Mines de Mons), two were "chimistes", one a "professor" – actually a secondary school teacher – and one was a "docteur en sciences" (Ph.D. in science) who gained his degree after graduating in pharmacy. The training in chemistry of most of the founding fathers was achieved at the Ecole des Arts et des Manufactures de Gand, the polytechnic school attached to the University of Ghent, or at the Ecole des Mines de Mons, where Edouard Hanuise, who was to be the first president of the forthcoming chemical society, was teaching. Among them there is even a mechanical engineer, Julien de Puydt. Only the secondary school teacher and the pharmacist have a degree genuinely from a science faculty, and for three out of nine of the men we have no information (M. Gaillard, Armand Le Docte and Edmond Lembourg). Out of these three, Le Docte probably trained in chemistry but what is more interesting is his job as a maker of chemical instruments: there are a few remaining aerometers and densitometers that came from his workshop.[12] His expertise is also attached to a method of analysis of the content of sugar in beets, developed in collaboration with Sachs, the method known as "Sachs–Le Docte".[13]

2.3   Chemists, Among Other Things

On the 4th August 1887, at the founding of the *Association* meeting, there were 32 members and fourteen "members honoraires" – members that were not actually labelled as chemists, *e.g.* director of a sugar factory, landowner, *etc.* "Honorary members" are described in the statutes as "industrials, makers, engineers, and more generally all people who are interested in chemical and industrial issues". In 1888–1889, half of the members were "honoraires".[14] By the following year, however, the membership list did not make any distinction between those two categories.[15] And, indeed, the honoraires held the same rights as the effective members, "members effectifs", and paid the same fees. In the first approximation, this blended association seemed to be a success, as two years after its foundation the *Association* had 292 members in 1890, to be

---

[11] See Table 2.1.
[12] Private collection of Brigitte Van Tiggelen.
[13] Mélard (1999 and 2001).
[14] *Bulletin* **2**, 1888/1889, 92–95.
[15] *Bulletin* **3**, 1889/1890, 119–127.

Table 2.1 The signatories of the first call to found a Belgian chemical society (1887).

Name	Role 1887	Education	Profession
Delville, Edouard	Chimiste départageur	Ecole Normale des Sciences, Ghent 1864	Teacher at the Athénée Royal, Tournai
Depuydt, Julien (1842–1919)	Chimiste départageur	Mechanical engineer, Ecole des Mines, Mons 1864	Chemist at the *Société Commerciale, Industrie et Maritime*, Antwerp
Francken, Victor (1842–1892)	Chimiste départageur	Engineer, Ecole des Mines, Department Arts et Manufactures, Liège 1862	Teaching assistant in the course on the analysis of industrial products, University of Liège
Gaillard, M.	Chimiste privé		Chemist at a sugar factory in Tienen
Hanuise, Edouard (1842–1913)	Chimiste départageur	Engineer, Ecole des Mines, Mons 1864	Professor of geology and mineralogy at the Ecole des Mines, Mons
Le Docte, Armand (?–1935)	Chimiste privé	Did "nice studies of chemistry" (=?)	Constructor of laboratory equipment, Brussels
Lembourg, Edmond	Chimiste départageur	Engineer, Ecole Spéciale du Génie Civil, Dept. Arts et Manufactures, Ghent 1871	Chemist at a sugar factory in Quiévrain
Sachs, François (1849–1919)	Chimiste privé	Pharmacist, University of Louvain 1864	Chemical engineer at Gembloux (agricultural school, or sugar industry)
Van Melckebeke, Edouard (1843–1915)	Chimiste départageur	Ph.D. in natural science, University of Louvain, 1870	Chemist at the *Société Commerciale, Industrie et Maritime*, Antwerp

Table 2.2a Evolution of membership for the *Association Belge des Chimistes* (1887–1903). Table compiled with the member lists and with the annual reports printed in the *Bulletin*.

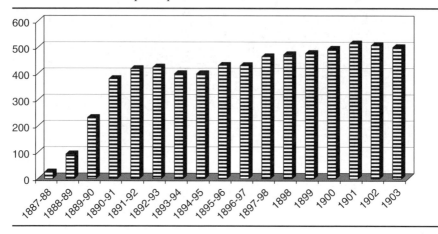

Table 2.2b Evolution of membership for the *Société Chimique de Belgique* (1904–1914). Table compiled with the annual reports printed in the *Bulletin*.

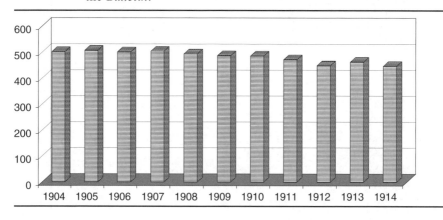

compared to the peak value of 507 attained in 1901.[16] The rules of admission were quite typical of the way associations perform in Belgium: two members submitted the name of the newcomer to the central committee, where the final decision was taken. Since no remark appears in the journal of the *Association* about the reasons why new members joined and which criteria applied to the admission, one can not know whether there was ever a refusal of admission.

---

[16] See Table 2.2.

Table 2.3   Thematic sections of the *Association Belge des Chimistes* (1887–1896).

Year	Name	Example of topics
1887	Section sucrière	Sugar chemistry
1889	Section des denrées alimentaires (et d'hygiène)	Food adulteration
1889	Section agricole	Fertilizers
1890	Section des industries de fermentation (et des industries connexes)	Vinegar production
1894	[later: Section de chimie biologique]	
1896	Section industrielle	Sulfuric acid plants
1896	Section de métallurgie et docimasie	Metallurgical chemistry

To trace the educational background or the position on the labour market is very difficult since, for most members, we only have information provided by the *Bulletin* itself. Often the names are followed by "chimiste", which carries no definite depiction of education or profession: it could also mean private experts, chemical experts appointed by the State, chemists working in a specific industrial plant, or even private manufacturers. Along with these active members, two other categories were created: foreign members, "membres correspondants", and "membres d'honneur", members of honour. A smaller group, the central committee, was keeping the management of the association and was elected by the general assembly according to the statutes.

The founding members, though mostly sugar chemists, felt the need to invite all "chemists" at an early stage of the association. Several thematic sections were founded from 1889 onwards (see Table 2.3): less than two years after the foundation, a section was created for food control and hygiene, along with an agricultural section devoted to chemical fertilizers and other issues related to farming. The chemists especially interested in fermentation (brewery, alcohol making, vinegar, *etc.*) opened their own section in 1890, and a more general "industrial chemistry section" and a metallurgical section were founded in 1896 to gather the chemists involved in expertise other than that discussed in the already existing sections. The organization was clearly, as it was from the very first gathering, more problem oriented than actually thematic. Some themes do indeed overlap – or could belong to different sections. Very symptomatic is the failed attempt to have a "pure chemistry" section: for a section to be proposed to the central committee, at least ten members were needed to create the section; the pure chemistry section never made it because in 1896 only eight members showed interest. Inside the sections, there was a strong will to achieve consensus: even the themes to be studied were submitted to votes and, to be sure that the results were shared, some samples were distributed among the members to review the results of analysis at the next meeting.[17] Meetings were

---

[17] For instance, *Bulletin* **10**, 1896/1897, 396. Samples are distributed during the meeting of the agricultural section, and one of the first topics of the industrial chemistry section is voted at the first meeting, *Bulletin* **10**, 1896/1897, 337.

held at least twice a year; after 1895, monthly meetings were held. In addition to these meetings, the different sections met separately three or four times a year, and of course joint meetings of two or three sections were not excluded: for instance, joint meetings of the sugar section and the agricultural section on the trade conditions of sugar beets, or meetings of the food section and the biological section on topics such as the analysis and legislation of vinegar and beer.

During this stage, each section was allowed to suggest a speaker or a topic to the central committee, who in turn decided who should be invited to address the assembly during the annual meeting. All papers given on this occasion were in the field of applied chemistry, with, of course, a strong presence of analysis of sugar, falsification of food and composition of fertilizers.

2.4   Extending to All Chemical Experts

The creation of two new thematic sections in 1889 provoked the first substantial change in the statutes. The *Association* had been given two aims: "establish between its members longstanding relations and to use the relationships hence created to (1) study chemical and technical questions, and (2) discuss professional interests".[18] The words "intérêts professionnels" (professional interests) were changed into "intérêts généraux" (general interests) in 1889 when two new sections were created. These professional interests seemed indeed so closely related to the sugar chemists that other chemists only agreed to join when the statutes were modified.[19] The opening of the thematic section on food control, for instance, attracted a substantial number of pharmacists to become an active part of the association (see Table 2.4). Pharmacists, even if they had training and often recognized expertise in chemistry, were at first representatives of a very specific profession, whose organization had begun much earlier. One could think, therefore, that their "professional interests" were taken care of elsewhere. This is only true for their traditional roles, making and selling medicines, but surely not for their other expertise often used by the local or public authorities: toxicology, legal expertise, public hygiene, food control. For instance, in the matter of food control, pharmacists were still fighting to gain the monopoly for fraud detection against other "chemists" (*i.e.* the official agricultural chemists), who would also have been able to demonstrate expertise in that area. Becoming members of the association or, better stated, the association inviting them to become members, is very much the follow up of the strategy first used among "sugar chemists": to provide a place where the conflicting specialists would discuss the problems between themselves, instead of putting the debate in the public arena. It is also in this way that the emphasis put on the scientific debate should be understood: the expectation was that the scientific debate would provide common ground to achieve consensus and

---

[18] *Bulletin* **1**, 1887/1888, 6 (proposition of the central committee), *Bulletin* **1**, 1887/1888, 35 (decision by the general assembly of 4th August 1887); see also *Bulletin* **2**, 1888/1889, 45.

[19] This is at least how Sachs remembers the early history, *Bulletin* **26**, 1912, 142.

Table 2.4 Pharmacists at the *Association Belge des Chimistes* (1889–1895). Table compiled by H. Deelstra with the member list in the *Bulletin*.

Year	Total number of members	Number of members of sections	Total number of pharmacists in sections	Members in the food control section	Pharmacists in the food control section
1889–1890	232	181	49 (27%)	60	48 (80%)
1890–1891	380	323	60 (18.5%)	83	60 (72%)
1891–1892	418	351	68 (19.5%)	89	67 (75%)
1893–1894	424	360	63 (17.5%)	89	61 (68.5%)
1894–1895	399	336	57 (17.5%)	82	55 (67%)

eventually fortify the community of "chemists". It was not unusual that when a harsh debate was taking place about prices, methods or lawfully recognized expertise, tenants of both sides, experts and counter-experts, private and official, pharmacists and analytical chemists, were actually members of the same society.

A very surprising achievement of the young association was the organization of the first international conference for applied chemistry in 1894, planned only seven years after the *Association* had come into being. Here again, the interplay between the sugar beet connection and the rest of the chemists inside the society was crucial in the dynamics of this project. The secretary, Sachs, a prominent member of the sugar section, first thought of a congress devoted to his favourite subject, but the chairman of the food section, Désiré Van Bastelaer, insisted he had good experience of such an organization since he was president of the very successful Conference on Pharmacy in Brussels in 1885. From then on the four sections were associated with the project and it occupied the *Association* until 1894 when the Congrès International de Chimie Appliquée (International Congress of Applied Chemistry) was held in Brussels and Antwerp between 4th and 11th August 1894.[20] This gathering marked the first of many such meetings. Interesting to note is the way the meetings were planned: it was very much the same system that we have already outlined for the beginning of the *Association*. This was a community of experts at work, together trying to solve discrepancies or controversies about methods and measures. In 1904, a list of questions set up by the delegates at the 1903 Berlin meeting became the list of themes to be discussed in Rome two years later, and again the meeting was all about "formuler des méthode uniformes" (to state standardized methods) – achieving consensus.[21] Delegates of the Belgian Chemical Society were, of course, to be sent to all successive meetings in applied chemistry,[22] as well as, for instance, to the Congress on Hygiene and Demography (1903), to the Flemish Congress on Medical and Natural Science (1903) and many other specialized meetings on the sugar industry, food chemistry, brewing, *etc.*

This consensus inside the association was not always easy to achieve, however, and early clashes between the "scientists" and "practitioners" resulted. A very good example is the heated debate around vinegar: on the one hand a "practitioner", a vinegar producer, is harshly criticized for not using the appropriate scientific terms and more than once denounced for his ungrounded knowledge of the chemistry behind vinegar making. On the other, a "scientist" is accused by the former of lacking the experience acquired by vinegar making and hence underestimating its importance.[23] Since the *Bulletin* reported less and less the actual discussions taking place during the thematic section gatherings, it

[20] On this congress, see Deelstra and Fuks (1995).

[21] *Bulletin* **18**, 1904, 145.

[22] Paris (1896), Vienna (1898), Paris (1900), Berlin (1903), Rome (1906), London (1909) and New York (1912).

[23] Aulard even states that "as members of a chemists' association, we can not deny our studies, our beliefs and our trials", *Bulletin* **4**, 1890/1891, 258–265. It is interesting to note that the early issues of the *Bulletin* reproduced the details of the discussion, even though the atmosphere was very heated.

is, of course, hard to know whether these debates did or did not affect the later history of the association. However, the emphasis on the chemical understanding of the processes and the necessity of a good chemical education therefore, did not weaken after the fundamental reorganization that took place in 1898.

2.5   From Expert Chemists to Chemically Educated

In 1898, following the initiative of L. L. De Koninck, professor of analytical chemistry at the Université d'Etat de Liège (State University of Liège), the society was reorganized in local sections, thus abandoning the problem oriented structure that shaped the association from its early beginnings.[24] There were, of course, practical advantages to doing so: not all members were able to go to Brussels once a month, especially those coming from the "far-away" provinces. But the shift from problem oriented sections to local sections had deeper implications. Honorary members did not appear any more, and a new category was opened, the "members associés" (associated members) to welcome students or "chemists-to-be". Paying half the fees, they were not allowed to vote but were invited to participate in all scientific activities, attend meetings, give talks or even publish.[25] With this new opportunity, De Koninck hoped very much to attract students who were pursuing "studies more or less specialized in chemistry", whatever diploma they were pursuing: chemists, but also engineers or pharmacists. Since the local sections were mostly organized in localities with at least either a university or a good technical school (Mons, Liège, Leuven, Brussels, Ghent and Gembloux), except for Antwerp and Charleroi, these sections could also better include, and be useful to, a higher number of student members, binding together more closely the younger members.

As for other matters, it was then left to local sections to decide for themselves how many meetings were to be planned each year and how to organize them. The reorganization also had an impact on the general meetings of the association: the general gatherings were from now on held each year in January instead of the traditional month of April, in remembrance of the first meeting in 1887, and in different places, offering to the chemical community the opportunity to travel around Belgium. This was more interesting for the members, since visiting one or more industrial plants or chemically related institutions (state laboratories, *etc.*) was added to the ordinary agenda of these meetings: society business and lectures. The industrial or institutional visits were even extended to neighbouring countries, such as northern France, The Netherlands and Germany.[26] Inside these sections, themes were to be chosen by the leading members, and for the general gathering, a wider variety of topics was observed.

---

[24] Deelstra and Fuks (1998).

[25] *Bulletin* **11**, 1897/1898, 7.

[26] The *Association* went to Lille to visit the Pasteur Institute with Calmette, a soap factory and a distillery, *Bulletin* **14**, 1900. In Germany, the memorable part was the visit to the Bayer plant in Leverkusen, *Bulletin* **26**, 1912, 136.

At the turn of the century, L. L. De Koninck explicitly referred to a background in chemistry as the common point between all members, and opening the *Association* to students is the most obvious evidence of this new feature. The list of members in 1905 shows the following trends.[27] The vast majority of the members were trained as pharmacists (69),[28] followed by two equal groups: engineers (66) and holders of a Ph.D. in sciences (66, among them thirteen labelled as "docteur en chimie"). One finds also 48 "ingénieur-agronome" (specialized in agronomy) and sixteen "ingénieurs chimistes" (but this category is not well defined: it could as well be engineers with a specialization in chemistry (arts and manufacture) as holders of a degree from technical schools highly valued at that time). This leaves 85 "chimistes", about whose chemical education we know nothing and, even more frustrating, 33 members about whom we do not even have any mention of training or profession. Noteworthy, however, is the fact that out of 453 members, 33 were teachers at the higher level: in 1887–1889, 35% of the chemistry "breeders" at higher level were members of the society. Even if the absence of pure chemistry is striking, most of those holding a chair in pharmacy or engineering schools, by 1900 the presence of the "breeders" had doubled to 73.5%.

The number of members increased but, in reality, some left and were replaced by more new members. The change was acknowledged as a success, and the increase of members was parallel with the local section activity, but the trend soon reversed, except for some well-established local sections such as Brussels and Liège (see Tables 2.2 and 2.5). At the annual meeting of 31st January 1904, the title of the association was changed to *Société Chimique de Belgique*,[29] very much at the persistence of De Koninck who insisted in 1900 that the imprint of "le péché originel" (the original sin – [sic]!) should be overcome at last. But what must have seemed more unexpected to the founding fathers was the selection of such a topic as "Le fluor dans ses combinaisons organiques" (Fluorine in its Organic Compounds) by Frédéric Swarts, chosen for the general assembly of 1906. If pure chemistry and research had seemed to be underrepresented during the early years of the *Association Belge des Chimistes*, this was not the case for the *Societe Chimique de Belgique* after the turn of the century. It should be noted though that all these activities were only attended by a few members.[30] Hence the crucial role of the *Bulletin* for linking together the members.

---

[27] Since there is no list of members available for 1900 or 1914, the analysis of the membership is based on the 1905 list of members.

[28] Pharmacists were considered as trained chemists and were often asked to be experts for lawsuits or food control and hygiene. Their presence increased even more when a section for the study of food control and hygiene was created in the *Association*. See Table 2.4.

[29] Deelstra and Van Tiggelen (2004); *Bulletin* **14**, 1900, 102.

[30] For instance in 1900, while the association had more than 450 members, the visit to northern France only gathered 40 members. The number was not much higher at the second general assembly: 50 members; the meeting was organized by the Charleroi section, which had by that time 44 members, *Bulletin* **14**, 1900, 103.

Table 2.5 Member repartition in local sections and in brackets numbers of meetings. Table compiled with the data found in the annual reports printed in the *Bulletin*.

Year	1898	1899	1900	1901	1902	1903	1904	1905	1907	1909	1911	1912	1913
Local sections													
Anvers	34 (4)	36 (3)	36	37 (10)	28	29 (3)	33	30 (?)	27 (10)	22 (2)	? (1)	(6)	(6)
Bruxelles	120 (6)	133 (8)	137	143 (10)	151	165 (10)	171	173 (9)	189 (11)	161 (10)	150 (9)	(10)	(9)
Charleroi	44 (6)	44 (6)	46	42 (10)	44	38 (4)	38	38 (2)	35 (1)	? (0)	? (0)	(0)	(1)
Gand	50 (2)	46 (3)	45	43 (2)	43	41 (3)	41	? (?)	46 (7)	47 (7)	45 (8)	(8)	(8)
Gembloux	30 (2)	28 (2)	30	29 (1)	29	25 (0)	25	21 (1)	16 (0)	? (0)	? (0)	(0)	(?)
Liège	90 (5)	98 (8)	102	94 (8)	92	86 (8)	84	94 (9)	95 (8)	110 (6)	101 (7)	(6)	(9)
Louvain	21 (2)	21 (3)	23	22 (3)	20	20 (2)	17	18 (1)	17 (1)	? (0)	? (0)	(0)	(?)
Mons	61 (3)	60 (3)	58	58 (6)	56	50 (3)	49	? (3)	51 (4)	41 (3)	49 (6)	(6)	(9)
Luxembourg			10	12	—	? (0)							

2.6   The *Bulletin*, the Organic Link

The call for the first meeting appeared, as said at the beginning of this contribution, in the professional journal, *La Sucrerie Belge*. And so did the minutes of the early meetings during the first year (1887–1888). Even the revised texts of the conferences and debates were published in this journal, until the decision was taken, in 1888, to publish an autonomous journal, entitled *Bulletin de l'Association Belge des Chimistes*. The first issue came out on 1st May almost one year after the *Association* had been founded and François Sachs, already secretary general, naturally became editor-in-chief (see Figure 2.1). In 1893, an editorial committee was officially set up, composed of one member per section, joining the secretary general and in 1895 the editorial board was brought under

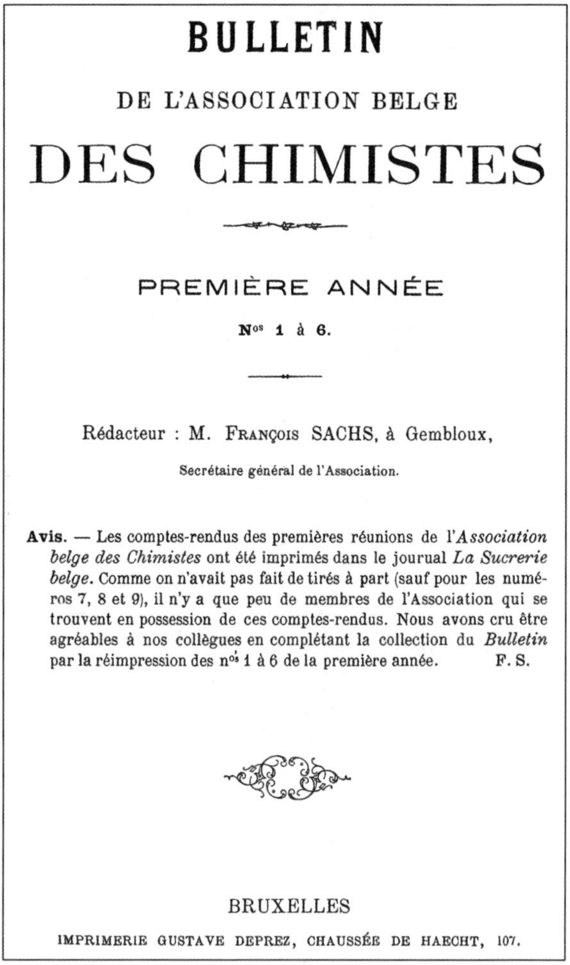

BULLETIN

DE L'ASSOCIATION BELGE

DES CHIMISTES

PREMIÈRE ANNÉE

Nᵒˢ 1 à 6.

Rédacteur : M. Fʀᴀɴçᴏɪꜱ SACHS, à Gembloux,

Secrétaire général de l'Association.

Avis. — Les comptes-rendus des premières réunions de l'*Association belge des Chimistes* ont été imprimés dans le journal *La Sucrerie belge*. Comme on n'avait pas fait de tirés à part (sauf pour les numéros 7, 8 et 9), il n'y a que peu de membres de l'Association qui se trouvent en possession de ces comptes-rendus. Nous avons cru être agréables à nos collègues en complétant la collection du *Bulletin* par la réimpression des nᵒˢ 1 à 6 de la première année.   F. S.

BRUXELLES

ɪᴍᴘʀɪᴍᴇʀɪᴇ ɢᴜꜱᴛᴀᴠᴇ ᴅᴇᴘʀᴇᴢ, ᴄʜᴀᴜꜱꜱéᴇ ᴅᴇ ʜᴀᴇᴄʜᴛ, 107.

Figure 2.1   Title page of the first issue of the *Bulletin de l'Association Belge des Chimistes*, 1887.

the direct guidance of the central committee. This editorial board thus reflected the core structure of the association. When the distribution of members was changed into local sections, each section was to send its representatives. This makes sense as long as the *Bulletin* was originally aimed at reflecting the life of the *Association*, reporting not only the minutes of the meetings but also the contributions in a definite form. But, parallel to this very democratic system, the *Bulletin* also mentions collaborators, half of whom are from the academic teaching realm. At a later stage, the editorial committee was designated by the central committee and it seems to have become more of a permanent job, looking at the stability of the names listed.

The frequency of the issues also reflects very much the rhythm of life inside the *Association*. Meetings were supposed to be held every month (except for the summer months considered as academic holidays): there were thus ten issues per year. During the first year, the content was almost a written version of the associative life led by the most active members, most probably intended to keep the members who did not have the opportunity to gather in Brussels every month posted about what was discussed and decided. During the first six months, readers could find the minutes of the central committee, as well as the minutes of the meetings of the sections and the general meetings, and also news about the profession (necrology, appointments, contests, meetings, *etc.*). The conferences held during these gatherings were also systematically published.

But very soon, in 1888-1889, some new features appeared that transformed the journal into a mixed blend of scientific journal and the former associative bulletin, keeping the members in close proximity to each other. Short book reviews, abstracts of national and international journals, tables of contents of publications and periodicals received at the association appeared as well as, most importantly, original contributions categorized into the seven thematic disciplines relating more or less to the thematic structure of the *Association*: General Chemistry, Analytical Chemistry, Food Chemistry, Agricultural Chemistry, Sugar Chemistry, Biological Chemistry and Industrial Chemistry. One can take for granted that the number of submitted contributions exceeded the time available during the meetings. Interestingly, this evolution occurred before the profound reorganization of 1898.

To encourage high standards, a prize was founded in 1897 to reward the three best original contributions published every year. Since the archives of the *Société* have been lost, it is difficult to gain more insight about how publications were selected and reviewed, but it is obvious from the creation of this distinction that the editors were looking forward to original contributions and wanting to meet scientific standards. What is also very much to be noted is the will to associate students or chemists with the life of the association through the journal: it became a tradition that students were invited to submit their first publication to the *Bulletin*. The complete change into a scientific journal took some time, however, and was achieved long after the First World War, when the minutes of the meetings and other routine matters disappeared completely from the table of contents.

2.7   Consensus *versus* Demarcation

Though timidly, the *Association* started with the expectation of being able to define and protect professional autonomy. The community was consciously created in order both to solve technical problems by achieving consensus on the best available methods, be they instrumental, experimental or theoretical, and to discuss professional interests. Controversies were seen as a real hindrance to achieving an expert status. One solution, as Auguste Aulard stated it in the early days, was thus to reach an agreement in-between themselves before going public. There lies the necessity of a scientific discussion: "As soon as we agree on the scientific part, we should apply the same method that we will have ascertained as the fairest ... We have to achieve our duty as chemists and we don't have to let us be imposed with whatever method ..."[31] However scientific the consensus is meant to be, the mention of fairness refers to the embedding of chemical expertise in socio-economic settings and thus also relates to the professional status to be reached in civil society through the power of chemistry as an exact science.

Now, what was meant by the term "professional interests" is not fully clear, especially when referring to the variety of backgrounds and professions of the members in the first phase of the *Association*. It is true that the "private and independent chemists" who took the initiative of the first gathering did quickly attract the official state chemists to become integrand partners inside the association – what did professional interest mean to both of these categories that were sharing the same kind of expertise in different markets? A kind of workers' union the *Association* was certainly not, and would not have to become one, since several already existed and had, such as was the case for sugar chemists, their own syndicates and journals and lobbying forces (cement, brewery, chemical products, *etc.*).

In this way the early union of the community was achieved as an amalgam rather than as a pure product for aspects that related to educational background, profession or public function. This was even the case for the conflicting professional or commercial interests. By opening the doors to other chemical experts, sugar chemists knew they would have to enlarge the scope of their "professional interest" in order to better ground their status as experts in chemistry. The pharmacists, the teachers at every level, the mechanical or agricultural engineers and the official analysts (official, independent and private) only shared one thing in common: chemistry itself as training, as an expertise, not as a discipline. But as the years passed by, the structure of teaching chemistry in Belgium evolved, and the discipline was more and more shaped as an academic pursuit, showing a very strong difference between "pure" and "applied" science. Therefore, the fact that the association changed between 1898 and 1904 from an association of *chemists* into a society of *chemistry* reflects the real nature of the union achieved before the First World War: the amalgam was progressively purified from its more technical slag.

---

[31] *Bulletin* **2**, 1889/1890, 69.

Instead of the emphasis on methods and instruments, the focus shifted to modern chemical theories and fundamental research conveyed by academic teaching. The early history of the society explains the difficulties for the *Société Chimique de Belgique* to appear both as a fully scientific society and as an attractive place for chemists working in both the industrial and the academic worlds to meet each other. Long after these first steps, one can hear the call for more involvement from the industrials, and the incompatible yet parallel claim to be, as any other foreign chemical society, a learned society.

References

Bulletin = *Bulletin de l'Association Belge des Chimistes*, 1887/1888–1903 and *Bulletin de la Société Chimique de Belgique*, sequel to *Bulletin de l'Association Belge des Chimistes*, 1904–1919.

Bulletin de l'Association Belge des Chimistes, 1887/1888–1903.

Deelstra, Hendrik (1995), La professionnalisation de la Chimie en Belgique, *Sartonia* **8**, 85–101.

Deelstra, Hendrik (1999), La contribution des pharmaciens au développement de l'Association Belge des Chimistes (1887–1906), *Chimie Nouvelle* **17**, 3003–3007.

Deelstra, Hendrik (2000), Contribution de l'Association Belge de Chimistes (1887–1898) à la chimie alimentaire, *L'Actualité Chimique* 45–46.

Deelstra, Hendrik (2000), Het 150 jarig bestaan van het diploma van apotheker in België, *Farmaceutisch Tijdschrift van Belgïe* **1**, 25–30.

Deelstra, Hendrik (2001), La chimie dans les universités et les écoles supérieures, in A. Despy-Meyer, G. Vanpaemel and J. Vandersmissen, ed., *Histoire des Sciences en Belgique*, La Renaissance du Livre, Dexia, Brussels, 159–178.

Deelstra, Hendrik (2003), A. Meurice, *Chemie Magazine* **1**, 47.

Deelstra, Hendrik (2004), De fermentatiescheikunde in België als voorloper van de biochemie (1890–1900), *Cerevisia, Belgian Journal for Brewing and Biotechnology* **29**, 211–215.

Deelstra, Hendrik (2005), De afdeling "Landbouwscheikunde" van de Association Belge des Chimistes, *Scientiarum Historia* **31(2)**, 51–60.

Deelstra, Hendrik and Fuks, Robert (1995), La Belgique organise en 1894 le premier congrès international de chimie appliquée, *Chimie Nouvelle* **13**, 1443–1447.

Deelstra, Hendrik and Fuks, Robert (1998), La réorganisation fondamentale de l'Association belge des chimistes (1898), *Chimie Nouvelle* **16**, 1971–1977.

Deelstra, Hendrik and Van Tiggelen, Brigitte (2003), Edouard Hanuise, Président-fondateur de l'Association des Chimistes en Belgique, *Chimie Nouvelle* **21**, 21–24.

Deelstra, Hendrik and Van Tiggelen, Brigitte (2004), L'Association Belge des Chimistes devient la Société Chimique de Belgique (1904), *Chimie Nouvelle* **22**, 112–114.

Développement de la chimie en Belgique. Numéro spécial à l'occasion de l'année internationale de la Chimie, B. Van Tiggelen, ed., *Chimie Nouvelle* **17**, 1999, p. 60 (Special issue on the history of chemistry in Belgium).

Mélard, François (1999), Les ingénieurs chimistes et leurs instruments: le cas de l'analyse polarimétrique dans l'industrie sucrière, *Chimie Nouvelle* **17**, 3008–3011.

Mélard, François (2001), *L'autorité des instruments dans la production du lien social. Le cas de l'analyse polarimétrique dans l'industrie sucrière belge*, doctoral thesis from l'Ecole des Mines, Paris (promotor: B. Latour).

Vanpaemel, Geert and Van Tiggelen, Brigitte (1998), The profession of chemistry in nineteenth century Belgium, in D. Knight and H. Kragh, ed., *The Making of the Chemist: The Social History of Chemistry in Europe, 1789–1914*, Cambridge University Press, Cambridge, 191–206.

Van Tiggelen, Brigitte (1993), De la chaire aux laboratoires: Louis Henry et la professionnalisation de la recherche en sciences naturelles en Belgique, *De toga om de wetenschap. Ontwikkelingen in het hoger onderwijs in de Geneeskunde, Natuurwetenschappen en Techniek in België en Nederland (1850–1940)* **16**, 192–203.

Van Tiggelen, Brigitte (1999), Les premiers Conseil de chimie Solvay (1922–1928). Entre ingérence et collaboration, les nouvelles relations de la physique et de la chimie, *Chimie Nouvelle* **17**, 3015–3018.

Van Tiggelen, Brigitte (2001), Les théories chimiques, in A. Despy-Meyer, G. Vanpaemel and J. Vandersmissen, ed., *Histoire des Sciences en Belgique*, La Renaissance du Livre, Dexia, Brussels, 179–194.

Van Tiggelen, Brigitte (2004), *Chimie et Chimistes de Belgique*, Labor, Bruxelles.

CHAPTER 3

CZECH LANDS: Chemical Societies as Multifunctional Social Elements in the Czech Lands, 1866–1919[1]

SOŇA ŠTRBÁŇOVÁ

3.1 General Historical Background

In order to understand the essence of the establishment and functioning of the Czech chemical societies in the given time period, it is important to characterize first of all the landscape of society, science and education in the nineteenth century in the Czech Lands.

The Czech Lands were inhabited since the early Middle Ages by several ethnic groups among which the Czechs and Germans prevailed. Since the beginning of the nineteenth century, the movement of the Czech National Revival set an ambitious objective: to create a modern Czech language which would be able to become a tool of the formation of the modern Czech nation and a language of science and higher education. We may consider the Czech National Revival the main factor of the nineteenth-century development of science and education in the Czech lands.[2]

The new political and social situation demarcated by the 1848 revolution, abolition of the so-called Bach Absolutism[3] in Austria in 1860 and the rising

---

[1] The author wishes to express her thanks to Dr Robert Rosner, Dr Anita Kildebæk Nielsen and Dr Jiří Jindra for their suggestions and advice, which helped in the preparation of this chapter.

[2] The role of the Czech National Revival in the scientific developments in the Czech Lands was described in detail in Janko and Štrbáňová (1988). The book refers to additional literature.

[3] Bach Absolutism, named after Alexander Bach (1813–1893), is a historical term indicating the era of a powerful police control in the Monarchy covering the years 1849–1859 when Bach was the Austrian Minister of Interior.

Creating Networks in Chemistry: The Founding and Early History of Chemical Societies in Europe
Edited by Anita Kildebæk Nielsen and Soňa Štrbáňová
© The Royal Society of Chemistry 2008

Czech national movement evoked basic changes in high education and not only there. The Czech National Revival in science reached its culmination: the Czech scientific terminology was completed,[4] and the Prague Polytechnic (at that time already called Polytechnische Landes-Institut) and the Prague Charles-Ferdinand University divided into their Czech and German counterparts in 1869 and 1882, respectively. Since the 1860s numerous Czech scientific associations were founded with their own specialized journals, which not only looked after the advancement of their fields but also became vehicles of further emancipation of the linguistically Czech science from the German one. In a stepwise fashion a separate Czech intellectual community evolved, led by the *Česká akademie věd a umění* (Czech Academy of Sciences and Arts) founded in 1890. The cultural and economic emancipation of the Czech nation, separation of the Czech scientific institutions from the German ones and the widening gap between the German and Czech scientific communities were also accompanied by rising nationalism,[5] which also found its reflection in the scientific associations of that period.

In the intertwined network of business, basic and applied science, education, social, political and national life, the various associations, corporations and scientific societies played irreplaceable roles. Here could meet the various interest groups, scientists, entrepreneurs, university and high-school teachers, technologists, politicians and the lay public. They could discuss new findings, influence political decisions, exchange experience and disseminate new knowledge. The societies and associations became catalysts of intellectual progress in the Czech Lands and in the complicated transfer between science and practice. As the majority of scientific and other professional societies founded after 1860 almost exclusively used the Czech language for communication, they actually demarcated the border between the German and Czech professional and scientific communities. It is important to note that in spite of this most societies, unlike the Czech Academy of Sciences and Arts and some cultural associations, seldom expressed extreme nationalistic attitudes.

3.2   Chemical Industry and Chemical Education in the Czech Lands

The dominant chemical enterprises in the Czech Lands in the nineteenth century were sugar, fermentation, food, agricultural and textile industries and, since the second half of the nineteenth century, production of inorganic chemicals, while industry of organic chemicals, especially synthetic dyestuffs, only started on a large scale at the beginning of the twentieth century.[6] Ownership in industry was nationally divided. While the fermentation, food and sugar industries were mostly in Czech hands, inorganic and organic

---

[4] The process of the formation of the Czech scientific terminology, including the chemical one, is described in Štrbáňová and Janko (2003).

[5] See Janko and Štrbáňová (1988), 74–121 and 215–246; more detailed discussion on ethnicity and ethnic identifications in the Czech Lands is presented in Kořalka (1996).

[6] For some of these issues and further reading see Štrbáňová (1992).

chemicals and textiles were made in plants belonging predominantly to German owners; however, this partition was far from being exclusive.

The university chemistry education at Prague University (founded in 1348) was focused in the nineteenth century on the theoretical chemical fields (organic, inorganic, analytical chemistry), medical chemistry and chemistry of plants. The more technically oriented chemistry was pursued at the Technical University, originally the Prague Ständisches Polytechnisches Institut (Prague Polytechnic) founded in 1806.[7] This school had undergone in the nineteenth century a number of important transformations exerting its influence not only on chemistry teaching, but also on the formation and further development of the Czech chemical societies. In 1863, the Czech and German languages became equal in instruction; in 1868 a reform separated chemistry instruction into two chairs – general chemistry and chemical technology; in 1869 the Polytechnic, providing at that time university-type education, divided into its Czech and German counterparts and since then the Czech and German chemical communities developed to a large extent independently. The two nationally separated polytechnics were officially legitimized into independent Czech and German technical universities in 1876. In the second largest Czech city Brno (Brünn) technical education evolved in a similar way but the Czech language was only introduced in 1899.[8] The gap between the Czech and German scientific communities widened even more after the division of the Prague University in 1882 into Czech and German counterparts. A survey of these events is given in Table 3.1.

3.3 The Changing Framework of the Czech Chemical Societies, 1866–1919[9]

Recounting the history of the Czech Chemical Society represents for the historian a quite complicated task because the society's history is full of twists and turns: it has undergone various splits, re-unifications and programmatic changes and has also accordingly changed its name several times (see Table 3.2). In this chapter, the history of these interrelated Societies will be discussed until 1919 because this was the year when the "Czech Society" became "Czechoslovak Society", bringing together the Czech and Slovak chemists in the young Czechoslovak Republic created in 1918.

---

[7] For the history of chemical education at the Polytechnic and the Technical University in Prague, see Quadrat (1966)) and Schätz (2002). For the early development of chemistry in Bohemia, see Wraný (1902) and chapters on chemistry in Nový *et al.* (1961). Chapters on chemistry instruction at the Prague Charles University can be found in Havránek (1997). Schätz's book contains many biographies to which we refer, although there exist many other sources as well.

[8] For the history of the Technical University in Brno see Franěk (1969).

[9] The archives of the Czech Chemical Society have been inaccessible for several years, therefore the basic facts on the Czech chemical societies were taken from the following sources: Neumann (1906); Hanč (1966) and (1966a); *Československá společnost chemická* (1980); Jindra (2003), esp. 409–416; Jindra (2003a); Bolton (1902), esp. 3–4. Details can be found in the volumes of the journals issued by the Czech chemical societies, listed in Table 3.2, where annual and other reports on the societies' activities appeared regularly.

Table 3.1   Universities in the Czech Lands with chemistry education before
1914 – a short historical overview.

1806	Prague Polytechnic (Ständisches polytechnisches Institut) founded
1849	Technical College (K.k. Technische Lehranstalt) in Brno (Brünn) founded
1863	The Prague Polytechnic reorganized to bilingual German and Czech Polytechnical Land Institution (Polytechnische Landes-Institut)
1869	Prague Polytechnic divides into two independent Czech and German counterparts
1873	The Technical College in Brünn is changed into the Technical University in Brünn – K.k. Technische Hochschule in Brünn
1876	The Czech and German technical universities legitimized as: - Czech Technical University (České vysoké učení technické) and - German Technical University (K.k. Deutsche technische Hochschule)
1882	Charles-Ferdinand University in Prague (Karl-Ferdinands-Universität) divides into independent Czech and German Universities: - Czech Charles-Ferdinand University (Česká universita Karlo-Ferdinandova) - German Charles-Ferdinand University (Deutsche Karl-Ferdinands-Universität)
1899	Czech Technical University (Česká vysoká škola technická) founded in Brno (Brünn)

Although the chapter focuses mostly on the history of what is today the Czech Chemical Society, other professional bodies will be mentioned where necessary. These deserve attention either because their activities were closely linked to the Chemical Society or because they offered a communication platform to various specialized groups of chemists and/or pharmacists (Table 3.3).

3.3.1   *Isis*, 1866–1872

The year of birth of the present *Česká společnost chemická* (Czech Chemical Society) is officially considered to be 1866, although in that year only *Isis* – an embryonic form of the society – was established.[10] In 1863, the Prague Polytechnic was reorganized into a university-type institution under the name Polytechnische Landes-Institut (Polytechnical Land Institution) and teaching became officially bilingual – Czech and German. This started its transitory period, which ended in 1869 with the division into the Czech and German Polytechnics. During this transitory period in 1865, the first attempts to establish something like a chemical society were made. The initiative came from the Czech students of technical chemistry, who established in 1866 the *Přírodovědecký spolek Isis* (Isis Association for Natural Sciences), a self-help group that attempted to enhance studies with additional activities: collecting books for a library and organizing scientific lectures, debates and excursions to factories. By the statutes accredited in 1866, only matriculated

---

[10] Hanč (1966), 23–27.

Table 3.2   Survey of the development of the Czech chemical societies 1866–1920.

Title	Years of existence	Periodical(s)	Historical background
Přírodovědecký spolek Isis (Isis, Association for Natural Sciences)	1866–1872	*Kritické listy*, 1869–? (Critical Letters)	Abolition of Bach absolutism, culmination of the Czech National Enlightenment; division of the Prague Polytechnic into Czech and German counterparts
Spolek chemiků českých (Society of Czech Chemists)	1872–1906	*Zprávy vědecké o činnosti Spolku*, 1871–? (Scientific Reports about the Association's activities) *Zprávy Spolku chemiků českých*, 1872–1875? (Bulletin of the Society of Czech Chemists) *Listy chemické*, 1876[a] (Chemical Letters)	Division of the Czech and German scientific communities; division of the Prague university into Czech and German counterparts
Společnost pro průmysl chemický v Království českém (Society for Chemical Industry in the Czech Kingdom)	1893–1906	*Časopis pro průmysl chemický*, 1891–1906 (Journal for Chemical Industry)	Rise and modernization of chemical industry
Česká společnost chemická pro vědu a průmysl (Czech Chemical Society for Science and Industry)	1907–1919	*Chemické listy pro vědu a průmysl*, 1907– (Chemical Letters for Science and Industry)	First World War, founding of Czechoslovakia in 1918

[a] The first volume of the journal was issued on 1st October 1876; therefore the first volume covers both 1876 and 1877, and the second volume covers 1878.

students of technical and analytical chemistry at the Polytechnic could be members, while students of other universities or graduates could participate as guests. This way, the association used to have around 60–70 regular members (see Table 3.4).

Isis helped needy students who could borrow textbooks from its voluminous library, but otherwise it did not have much money; its main resource was support from the so-called "founding members" who should actually be

Table 3.3 Other selected associations in the Czech Lands, where chemists were allied before 1914, and their journals.

Title	Periodical	Notes
Verein zur Hebung der Zuckerfabrikation im Königreich Böhmen, 1868–1874 (Association for Improvement of the Sugar Industry in Bohemia)	*Zeitschrift für Zucker-Industrie*, 1872–1874 (Journal for the Sugar Industry)	
Spolek cukrovarníků východních Čech, before 1869 (Association of Sugar Industrialists of East Bohemia)	*Časopis chemiků českých*, 1869–1871 (Journal of the Czech Chemists); since July1870 the journal became supplement of the journal *Průmyslník* (Industrialist)	The journal served since vol. 1, No. 10 also as organ of the Czech sugar industrialists. News about this and other sugar industrialists' associations was regularly published later in the *Listy Chemické*
Spolek farmaceutů, 1875 (Association of Pharmacists)	*Časopis českého lékárnictva*, 1882 (Journal of the Czech Pharmacists)	News about the Association was often published in *Listy Chemické*
Spolek pro průmysl cukrovarnický, 1876 (Association for Sugar Industry)	*Listy cukrovarnické*, 1883 (Sugar Industry Letters)	The Association financed for some time the journal *Listy Chemické* and published there its official proceedings
Spolek pro průmysl pivovarnický v Království českém, 1873 (Society for the Brewing Industry in the Czech Kingdom)	*Kvas*, 1873 (Brew)	News about these Associations was often published in *Listy Chemické*
Spolek pro průmysl lihovarnický v král. českém (Society for the Distilling Industry in the Czech Kingdom)		
Österreichische Gesellschaft zur Förderung der chemischen Industrie, 1878 (Austrian Society for the Advancement of Chemical Industry)	*Bericht der Österreichischen Gesellschaft zur Förderung der chemischen Industrie*, 1879–1898 (Report of the...)	Located in Prague, associated both chemists and industrialists, Czechs and Germans. Its significance was all-Austrian. Its main goal was to support the advancement of the chemical industry
Exkursionsfond der deutschen Chemiker in Prag, 1885 (Excursion Fund of the German Chemists in Prague)		Supported chemistry students at the German Technical University
Chemiker-Verband in Brünn, 1896–1900 (Union of Chemists in Brünn)		Short-lived attempt to establish a German chemical society in the Czech Lands

Table 3.4 Selected membership of the chemical societies in the Czech Lands 1866–1919.[a]

Title of the society	Years	No. of members
Isis	1866–1872	1866: 63
		1868: 75
Spolek chemiků českých (Society of Czech Chemists)	1872–1906	1872: 172
		1877: 211
		1883: 191
		1892: 304
		1893: 373
		1894: 434
		1897: 443
		1902: 407[b]
		1906: 376
Společnost pro průmysl chemický v Království českém (Society for Chemical Industry in the Czech Kingdom)	1893–1906	1893: 340
		1894: 430
		1897: 570
		1905: 514[b]
		1906: 519
Česká společnost chemická pro vědu a průmysl[c] (Czech Chemical Society for Science and Industry)	1907–1920	1907: 630[b]
		1915: 615
		1917: 580
		1919: 636
Česká společnost chemická (Czech Chemical Society)	2006	about 4000

[a] Not all years are shown, only those that reflect some general tendencies. For comparison the membership of the present society is shown.
[b] These years are used as samples in Tables 3.6, 3.7, 3.8 and 3.9.
[c] When the societies re-united, the number of members was not equal to the sum of members of both societies because during the split some people were members of both societies.

considered "supporting members". Among these benefactors were companies, like sugar factories, and individuals – mostly graduates and teaching staff of the Polytechnic, managers of enterprises, and even representatives of the Czech gentry. This diverse circle of benefactors demonstrates the importance attributed to a mere students' corporation, which was to become the core of the future powerful Czech chemists' societies. In spite of the expression "natural sciences" in its title, Isis focused almost exclusively on chemistry. It attempted to supplement the regular university education with regular lectures from the Czech chemistry professors, especially Vojtěch Šafařík and Fratišek Štolba,[11] about whom will be written in more detail below. Industrial specialists took the students to factories where they became acquainted with the manufacturing process. The students were encouraged to give speeches themselves and to contribute to the handwritten bulletin Kritické listy (Critical Letters).

In 1871 the Czech chemists started to consider the possibility of establishing a regular chemical society; Isis became its crystallizing centre.

---

[11] For Šafařík's biography see Schätz (2004), 47–49; for Štolba's biography see Schätz (2004), 44–46.

3.3.2   *Spolek chemiků českých* – Society of Czech Chemists, 1872–1906[12]

In the environment where the Czech professionals in the Czech Lands constituted numerous specialized associations and chemical societies appeared in several European countries, the founding of a Czech chemical society seemed inevitable.[13] Its first general assembly on 28th January 1872 transformed *Isis* into the *Spolek chemiků českých* (Society of Czech Chemists – SCHC) by a simple change of the *Isis* statutes.[14] In the first annual report of the year 1872 we may read that "the main purpose of the society was and still is to join the scattered chemical forces to a common goal and cultivate chemical science especially by regular lectures and publishing a chemical journal".[15] This goal was defined very modestly and the actual activities of the society transcended it stepwise to a large extent. The emphasis on scientific chemistry without mentioning technical chemistry or chemical technology was probably motivated by the mostly theoretical alignment of the two principal "godfathers" of the society, the professors of the Polytechnic Vojtěch Šafařík and František Štolba. Šafařík, the son of the notable Czech archaeologist and linguist P. J. Šafařík, was a pupil of Liebig's disciple Josef Redtenbacher at Prague University, and of the German chemist Friedrich Wöhler.[16] His linguistic talent predestined him to pioneering work on the modern Czech chemical terminology and to writing textbooks, but he was more of an organizer than a researcher. Štolba, pupil of the fermentation chemist K. N. Balling, was oriented more experimentally, especially in inorganic preparations and analytical chemistry.[17] This "scientific" tendency of the society was soon complemented with interests in chemical technology and chemical industry; nevertheless "science" remained somehow fixed in its principles and later caused some problems leading even to a temporary split.

3.3.2.1   *The Executive*

The executive of the SCHC usually consisted of its president, vice-president, secretary and other posts, such as treasurer, recorder and librarian. In accordance with the prevailing scientific inclination of the society, they were mostly from the faculty of the Czech Technical University or Prague University or specialists from the sugar and fermentation industries. The latter were in a

---

[12] Neumann (1906); Hanč (1966), 28–33. Much of what is said about the society in this period applies to the whole time of its existence, including the separatist SPCH and the unified CSCH, therefore this paragraph also presents a number of facts related to the following periods.

[13] This movement is described in Janko and Štrbáňová (1988), see esp. 233–263.

[14] Until now the author of this chapter could not find the statutory text of the statutes and had to rely on secondary sources. In fact, even in 1878 the statutes were not officially approved by the Land Governor's office; compare Zprávy spolkové (1879).

[15] Quoted from Hanč (1966), 28.

[16] Since 1840, when J. Redtenbacher came to Prague, most Czech chemists have belonged to the Liebig "family tree". See *e.g.* Štrbáňová (1992) and Štrbáňová (2005).

[17] K. J. N. Balling, professor of chemistry at the Prague Polytechnic since 1835, called in Europe "Grand Master of Zymotechnology", educated dozens of professionals, especially in fermentation and sugar-refining chemistry and ironmongery. For his biography see Schätz (2004), 32–33.

minority in spite of the fact that the society's membership was dominated by people who came mostly from the practical sphere. As already hinted, the imbalance between the membership and leadership could have been one of the reasons of the break-up to come many years later. Karel Preis became the first president, later professor of analytical and inorganic chemistry at the Czech Technical University, active also in sugar industrial research.[18]

3.3.2.2   The Members

The SCHC aimed at embracing among its members practically all persons who considered themselves chemists and Czechs, including faculty and teachers of all types of schools, students, industrial chemists, medical and agricultural chemists, mineralogists, chemical technologists and technicians, and even lay sympathizers. By the statutes there existed several types of membership differentiated mainly by their yearly membership fees.[19] Founding members could be any chemists, people interested in chemistry and chemical enterprises or associations that paid 100 K (Austrian Crowns) per year. Companies or associations (but not individuals) could also become contributing members and only paid 10 K per year. The rest were acting members – individuals – chemists of any specialization, students or anybody interested in chemistry and paid four K per year. In the years 1878–1892 active members were divided into two groups: ordinary members paying four K per year and extraordinary members – students who paid only half. Later the membership fees slightly increased, but all members were entitled to receive the society's journal for free.

The founding members of the SCHC were in reality wealthy sponsors of the society who wanted to support the rise of a modern scientific discipline that promised prosperity in terms of flourishing industry. Among them were businessmen, directors of sugar factories or the factories themselves, but remarkably also several municipalities. Contributing members were only businesses, among which sugar factories prevailed. The structure of the active or ordinary members was very heterogeneous: university faculty, grammar school teachers, owners, directors and employees of chemistry-related enterprises, researchers, shopkeepers, pharmacists, *etc.* Such a structure probably reflected the allocation of the Czech chemists in different domains within the Czech Lands and the fact that the society did not exclude any applicants. Since 1880 the society also elected honorary members – both Czech and foreign.

If we consider professional stratification of the members (also compare Table 3.5), one peculiarity is quite conspicuous: practitioners from the sugar industry regularly outnumbered the other members.[20] The strength of this group can be exemplified by some facts: in 1892 the sugar specialists formed a separate sugar industry section, whose main goal was to organize conventions of sugar

---

[18] For biographical data of Preis see Schätz (2002), 52–54.
[19] Neuman (1906). In characterizing the various types of membership I relied on the statutes of the separatistic SPCH whose basic regulations were similar to those of the SCHC; see Stanovy (1892).
[20] Compare Protokol (1892).

Table 3.5　*Spolek chemiků českých* (Society of Czech Chemists). Profile of the membership in 1902.[a]

Group of members	Type of membership				
	Honorary	Founding	Contributing	Acting	Total
University teachers	7	1	–	16	24
Teachers in higher educational establishments, researchers in non-university institutions, owners and workers of private laboratories	–	–	–	50	50
Industrialists, landowners, business- and tradesmen, business and industry employees	4	5	–	201 (54 + 147[b])	210
Corporations	–	16	70 (9 + 61[b])	–	86
Secondary school teachers		–		14	14
Government employees		–		8	8
Pharmacists		–		2	2
Others		3		10	13
Total	11	25	70	301	407

[a] The statistics were compiled from Baur (1903); this list indicates sugar factories or workers of sugar factories, while in the other lists such information is missing.
[b] Sugar factories or workers of sugar factories.

industrialists; the first one in 1892 was attended by 150 participants.[21] The same year the society had a total of 304 members, hence it follows that the share of sugar specialists among the members reached as much as 50% and this number even increased later on.[22]

The number of members steadily increased (see Table 3.4). At the end of 1872, when the society was founded, it had a total of 172 members, in 1906, 376, in 1914, 615 members and in 1919, the last year of its existence, 636.[23] Naturally, the temporary split of the society in 1893, which will be discussed below, affected the numbers considerably.

---

[21] At the beginning the conventions took place every year, later every two years; the last one in 1899. Neumann (1906), 302.
[22] Only a year later, in 1893, out of 373 members of the SCHC the sugar-industry section had 242 members; see Zpráva (1894).
[23] Since 1907 the numbers belong to the *Česká společnost chemická pro vědu a průmysl* (Czech Chemical Society for Science and Industry).

3.3.2.3   Financing

The SCHC was financed largely from membership dues and apparently only a little from donations; for instance the *Spolek pro průmysl cukrovarnický v Čechách* (Association for Sugar Industry in Bohemia) and K. Preis regularly subsidized the society's organ *Listy Chemické*[24] (Chemical Letters), while the state did not provide any assistance. Nevertheless, after some initial monetary difficulties, in the 1890s the economic situation of the SCHC improved, and it did not report financial problems, especially thanks to a substantial rise of the membership, which reached its peak – 448 members in 1896; then it could even afford a paid administrative secretary.

3.3.2.4   Survey of Activities

The regular activities of the SCHC included in particular the following:

- Publishing journals, monographs, handbooks and textbooks (will be treated separately);
- Running the library;
- Regular monthly meetings of members with lectures to which non-members also used to be invited;
- Organizing excursions;
- Dissemination of knowledge about the progress of domestic and foreign chemistry and chemical industry;
- Discussing matters of chemical industry with the aim of helping in its modernizing and introducing new productions;
- Supervision of chemistry instruction in schools of various levels with serious attempts to influence curricula and structure of the university chemical subjects;
- Further adjustment of the Czech chemical nomenclature;
- Participation in domestic and international meetings;
- Co-operation with other Czech scientific and technical associations;
- Professional counselling at various levels;
- Economic and legal counselling;
- Fostering of employment of graduates in chemistry and their position on the labour market;
- Facilitation and promotion of international contacts;
- Financial support of publishing scholarly literature;
- Awarding prizes for special professional accomplishments, mostly to students but also some lecturers;
- Election of domestic and foreign Honorary Members.

Some of these activities deserve more detailed explanation.

---

[24] For instance, in 1877 the Association for Sugar Industry donated 100 Florins; see Zprávy (1878).

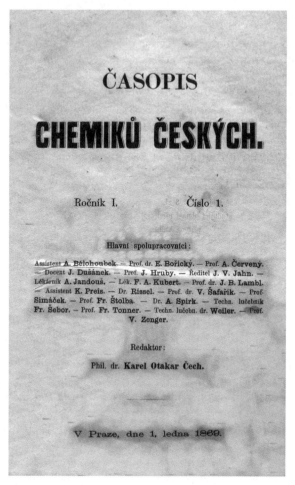

Figure 3.1   Title page of the first short-lived chemical journal in the Czech language
started in 1869.

3.3.2.5   Publications, Lecturing and Excursions

Since 1869, still in the period of *Isis*, deliberations about making a regular
association of Czech chemists were going on hand in hand with attempts to
publish a specialized journal (see also Tables 3.2 and 3.3).

   In 1869 a group of Czech chemistry leaders, mostly teaching staff of the Czech
Technical University and some pharmacists, made their first attempt to start a
periodical that would publish both theoretical and technical chemical papers.
However, the *Časopis chemikŭ českých* (Journal of the Czech Chemists,
1869–1871 Figure 3.1) soon lost its independent character. Probably for financial
reasons, it changed into the organ of the *Spolek cukrovarníkŭ východních Čech*
(Association of Sugar Industrialists of East Bohemia) and since July 1870 it
became only a supplement of the journal *Prŭmyslník* (Industrialist).

The next attempts were directly associated with the endeavour to establish a society of the Czech chemists. The *Zprávy vědecké o činnosti Spolku* (Scientific Reports about the Association's Activities, 1871–?) and the *Zprávy Spolku chemiků českých* (Bulletin of the Society of Czech Chemists, 1872–1875?), issued irregularly by V. Šafařík, mostly reported about events in scientific life or printed some popular or scientific articles.

The first volume of the regularly issued journal *Listy Chemické* (Chemical Letters),[25] appeared on 1st October 1876 as the organ of the Society of Czech Chemists (Figure 3.2).[26] It has been uninterruptedly published until today under slightly changing names, and has always been widely read. Its first editors were the already-mentioned Karel Preis and Antonín Bělohoubek, professor of fermentation chemistry and textile chemistry specialist at the Polytechnic.[27] The journal was dedicated in its beginnings more to technical chemistry than to scientific objectives, and it also served until 1883 as a publication and information base for sugar industrialists.[28] In a stepwise manner, the journal turned into the principal publication base of Czech chemists with its original scientific papers, reports on advances in chemistry at home and abroad, book reviews, translations of essential foreign articles, regular reports of the society's activities (including annual reports), reports on other local and foreign associations, commemorative articles and advertisements.

The journal was mainly financed by the SCHC; nevertheless in an incredibly long 15 years' interval of 1882–1896 the money necessary for publishing was provided from the pocket of K. Preis himself.[29] Since 1891, however, the members received the journal for free.

The improving financial situation also enabled the SCHC to create the *Literární fond* (Literary Fund) in 1889, which got its money especially from the wealthy sugar-making community.[30] Through the fund the series *Chemická knihovna* (Chemical Library) came into being, which issued text- and handbooks intended to replenish the shortage of chemical literature in the Czech language.[31] The society's members and students could buy the books with a discount, in spite of which the series was making a profit reused for further editions.

---

[25] The 100th anniversary of the journal *Listy chemické* was commemorated in 1976 by two volumes of *Chemické listy* (successor of *Listy chemické*) with several articles related to the history of the journal and Czech chemistry. (*Chemické listy* **70**, 1976, No. 9 and 10). See especially Gut (1976) and Koštíř (1976).

[26] The name of the journal was apparently inspired by Liebig's *Chemische Briefe*. Liebig (1844).

[27] For Bělohoubek's biographical data see Schätz (2002), 58–59.

[28] Karel Preis was the principal organizer of the community of Czech sugar manufacturers. In 1883 he founded a separate journal *Listy cukrovarnické* (Letters of Sugar Industry) solely for the needs of the sugar-refining community. Preis edited the *Listy chemické* until 1896, and in 1882–1896 he paid all the expenses of the journal.

[29] Neumann (1906), 304.

[30] The fund was originally started on the initiative of the Association of Sugar Industrialists of East Bohemia.

[31] Among the eleven titles issued in the years 1890–1906 there were, for instance, handbooks such as Neumann (1890); monographs, such as Raýman (1893); university textbooks, such as Preis and Votoček (1902); and dictionaries such as Šetlik and Votoček (1906).

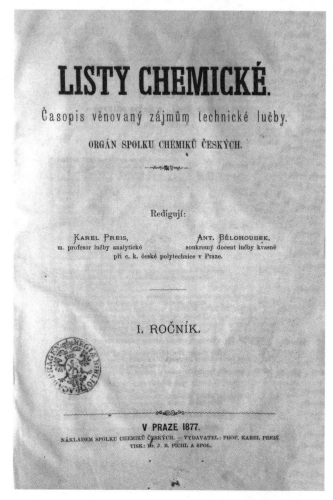

Figure 3.2   The title page of the first issue of *Listy Chemické*, 1877, a journal which has
been published under various titles up to the present day.

Lectures and excursions were frequently provided by the SCHC.[32] Until 1906
the members and guests delivered at the monthly meetings 420 lectures covering
most domains of theoretical and practical chemistry. The society awarded to
the best of them financial prizes, but these were given, for obvious reasons,
mostly to students and to external guests. The 62 excursions involved visits to
various factories and plants related to chemical, metallurgical and food
industries, and also exhibitions. A special excursion fund enabled even the
neediest students to participate.

---

[32] Neumann (1906), 302.

3.3.2.6   Professional, Economic and Legal Counselling

The members of the SCHC, beside regular conducts characteristic of most professional associations, were also involved in various commissions solving communal or public problems, mostly at the initiative of the SCHC. Among them were commissions charged by instituting unified standards of analyses of mineral fertilisers (1888), drinking water (1894) and others. The Customs Commission (1898) intervened with legal matters. Of special importance was the society's care for modernizing chemistry education. In 1893 a new well-subsidized laboratory for organic chemistry was created at the Prague Czech Technical University, on account of a petition submitted by the SCHC to the Austrian Ministry of Education. This success was followed in 1901 by formation of a commission for reform of teaching chemistry at the Technical University.

3.3.2.7   Cooperation with Other Corporations

The close contacts with the sugar makers' associations have already been described. The SCHC was also represented in various decision-making professional corporations such as the Zemědělská rada pro království české (Agricultural Council for the Bohemian Kingdom), Stálá zdravotní komise (Permanent Health Commission) and Obchodní a živnostenská komora (Commercial and Trading Chamber).

It should be noted that the SCHC did not yield fully to the ever-strengthening atmosphere of the Czech-German animosity of those times and to a certain extent cooperated also with the all-Austrian *Österreichische Gesellschaft zur Förderung der chemischen Industrie* (Austrian Society for the Advancement of the Chemical Industry). The SCHC also never lost its ties to the separatist *Společnost pro průmysl chemický v Království českém* (Society for Chemical Industry in the Czech Kingdom), which will be mentioned further. For instance, in 1894–1895, these three major chemical societies in the Czech Lands exerted joint efforts to abolish discrimination of technical chemists who were not authorized to perform official analyses of foodstuffs. Their memorandum to the Government and Parliament eventually led to changes in legislation, which granted the graduates of the technical university the rights of foodstuff adjudication and even enabled studies of food technology at the technical universities. The three societies also formed together an independent representation of chemists from Bohemia in several commissions of the six International Congresses of Applied Chemistry between 1894 and 1906.[33]

3.3.2.8   International Contacts

The SCHC cooperated from the very beginning especially with its Russian and Polish counterparts, and with the British, French, German and several other chemical societies, with whom journals were regularly exchanged.

---

[33] Neumann (1906), 303.

Table 3.6   Foreign honorary members of the Czech Chemical Society (SCHC) and the Czech Society for Chemical Industry (SPCH).

Name, place, country	Elected in SCHC	Elected in SPCH
A. M. Butleroff (St. Petersburg, Russia)	1880	–
D. I. Mendeleev (St. Petersburg, Russia)	1880	1894
N. A. Menshutkin (St. Petersburg, Russia)	1880	–
B. Radziszewski (Lwow, Poland)	1880	1894
L. Pasteur (Paris, France)		1893
E. Ch. Hansen (Carlsberg, Denmark)		1893
K. Lintner (Weihenstephan, Germany)		1894
M. Maercker (Halle/Saale, Germany)		1894
V. D. Lambl (Warsaw, Poland)		1895
E. Nölting (Mühlhausen, Germany)		1896

Particularly intense were the contacts with the Slavic chemists. Official delegations of the Czech Chemical Society used to be present at most Polish and Russian chemical congresses, and the journal *Listy chemické* published detailed accounts of their proceedings. The Slavic scientific collaboration culminated in the years 1880–1914 when five Congresses of Czech Scientists and Physicians (1880, 1882, 1901, 1908 and 1914) with international participation were organized in Prague with close involvement of the societies of the Czech chemists. Since 1882 the Congresses became unofficial international meetings of Slavic scholars as a result of the idea initiated by the Czechs and Poles, who planned to transform them into all-Slavic scientific meetings as a point of departure of versatile scientific cooperation of the Slavic nations. The idea behind this initiative was to use this teamwork as a political tool that would improve the position of the Slavic scientists and their national institutions in Austria-Hungary and render them more autonomous. Most eager to establish permanent bonds, especially between the Czechs and the Poles, were the chemists and physicians. From the Czech side the chemists V. Šafařík, B. Raýman and K. Chodounský were engaged in these activities and from the Polish side B. Radziszewski from Lemberg (Lwow). Although the Congresses became in reality for some time an informal platform of communication between Slavic scientists, all attempts to formalize this cooperation eventually failed due to nationalistic antagonism, particularly between the Russians and the Poles.

On the pages of the journal *Listy chemické*, most other European chemical meetings and congresses were also acknowledged, and the official representatives of the society attended all European Congresses of Applied Chemistry.

Electing foreign chemists as honorary members of the SCHC was to become the expression of highest esteem of their scientific achievements. As shown in Table 3.6, however, only four foreign honorary members were elected in 1880: the Russian chemists D. I. Mendeleev, A. M. Butlerov and N. A. Menshutkin, and the Polish B. Radziszewski. It is not clear why the next round of elections only happened in 1927, and many important chemists, who would have deserved this honour, are missing from the list.[34]

The society's international contacts were also apparently affected by the Czech-German antagonism in the second part of the nineteenth century, which led, especially since the 1880s, to some manifestations of extreme narrow-minded nationalism on both sides infiltrating all strata of the society in the Czech Lands. The more enlightened representatives of the Czech scientific community, however, understood that nationalism could have disastrous consequences for the development of Czech science and on its international reception. The young Czech intellectuals, including chemists, went frequently for study trips abroad, mainly to Germany, France and England, published in foreign languages and made strong personal ties to topmost foreign institutions and leading scientists without any visible national distinction. This way the majority of Czech chemists identified themselves with the European scientific community and refrained from false nationalistic patriotism.[35] In spite of that, regular contacts of the Czech scientific societies with those located in Austria or Germany were almost non-existent.[36] Nevertheless we may find surprising instances when the SCHS refrained from the governing nationalistic tendencies: in 1880 it became a founding member of the Society of Naturalists and Physicians in Vienna and in 1905 contributed 50 K to the *Verein Österreichischer Chemiker in Wien* (Association of Austrian Chemists in Vienna) to support building of a monument for the Bohemian-born Austrian chemist H. Hlasiwetz.[37,38]

---

[34] For instance Pasteur was only elected by the SPCH and Marie Curie became honorary member as late as 1932, in spite of her "eligibility" due to her scientific accomplishments and close ties to the Czech Lands (the ore from which M. Curie isolated radium came from Jáchymov – Joachimsthal – in Bohemia). Absence of M. Curie's name in the list of honorary members of the CSCH is also peculiar because she was elected honorary member of the *Česká akademie věd a umění* (Czech Academy of Sciences and Arts) in 1908.

[35] For instance Bohuslav Rayman, a prominent member of the Chemical Society, pupil of the German chemist August Kekulé and the secretary general of the Czech Academy, pushed through in 1893 the first foreign language journal published by a Czech scientific institution; the *Bulletin International* of the Czech Academy enabled communication of Czech scientific papers abroad in understandable languages, mostly in French, without being accused of "lukewarm patriotism". For Rayman's biography see Štrbáňová (2005).

[36] The oddity of this situation becomes even more striking in the light of the strong Liebig School tradition in the Czech Lands and the leading role of German chemistry in Europe. Compare Štrbáňová (1992).

[37] *Listy chemické* **4**, 1880, 128.

[38] Hlasiwetz, Heinrich, a prominent Austrian organic chemist, was born in Liberec (Reichenberg) in Bohemia, studied in Prague and Vienna, was professor of chemistry at the University of Innsbruck since 1854 and at the Technische Hochschule in Vienna since 1867. He was one of the founders of the *Chemisch-physikalische Gesellschaft* (Chemical-Physical Society). For his biography see *e.g.* Lexikon (1988), 206; see also Rosner (2004), 250–253.

3.3.3 Temporary Split: *Společnost pro průmysl chemický v Království českém* (Society for Chemical Industry in the Czech Kingdom), 1893–1906[39]

3.3.3.1 Founding a New Society

In 1891, professors of chemical technology of the Czech Technical University, Antonín Bělohoubek and František Štolba, started a new journal *Časopis pro průmysl chemický* (Journal for Chemical Industry) as an expression of their discontent with the insufficient representation of matters of chemical industry in the *Listy Chemické*.[40] In the first volume of the new journal they disclosed their endeavour to cultivate "all branches of the chemical industry and also cultivate agricultural chemistry, the science of foodstuffs..., statistics and industrial hygiene".[41] Among the reasons for founding a new journal they mentioned the lag of the Czech chemical industry behind the German, English and French plants, insufficient organization of the Czech technological chemists and lack of a platform that would promote the interests of the chemical industry and industrial chemists.[42] In fact, there was some rationale in founding the new journal as the *Listy Chemické* mainly focused on scientific articles and to some extent neglected the problems of the chemical industry.

Industrial chemists and entrepreneurs started to gather around the new journal and to organize, since the beginning of 1892, informal gatherings called "*dýchánky*" (in free translation tea-parties), where various matters scientific and technological were discussed. By the end of 1892 the statutes of a new association were prepared and its Founding General Assembly was called on 17th December 1892. The foundation year of the *Společnost pro průmysl chemický v Království českém* (Society for Chemical Industry in the Czech Kingdom, SPCH) is considered to be 1893.[43]

3.3.3.2 The Executive, Membership and Activities

The "defector" SPCH proclaimed at its December 1892 Founding Assembly its main objective to be promoting the stagnating chemical industry in the Czech Lands, and elected its first executive president, Jan Baptista Lambl, professor of the Prague Czech Technical University, but among the 12 members, as many as 10 were directors or owners of factories, one was a high-school director and one a pharmacist; nevertheless Bělohoubek remained the driving power of SPCH with his contacts both in the academy and in industry.[44] Such division of

---

[39] Vondráček (1906); Hanč (1966), 34–36.

[40] According to some historians, the main reasons for Bělohoubek's attempt to challenge the old Society were of personal character. Compare Hanč (1966), 35.

[41] Náš program (1891).

[42] Osvědčení redakce (1891).

[43] Compare Ze Společnosti (1893).

[44] J. B. Lambl's background was pharmacy, but his main subject was economy of agriculture. According to Hanč (1966), 62–63, he was the driving force of the SPCH and in 1906 exerted an immense effort for reunification of the SCHC and SPCH.

power reflected the main focus on industrial matters with some interest in education and trade. The statutes did not exclude anyone from membership and thus several persons were members of both societies.[45]

The statutes and the scope of activities of the SPCH did not differ too much from those of the mother society; only the agenda, like regular lectures or excursions, was somewhat more related to the chemical industry. Members were divided into acting, founding, corresponding and honorary. Acting members could be "chemists, theoretical or practical or specialists working in any branch of chemical industry or personalities interested in chemistry or chemical industry".[46] Founding members differed from the acting ones only by contribution of 50 Fl in two consecutive years; also enterprises, schools or various institutions could fall into this category. Honorary and corresponding membership was awarded as an expression of the person's merits.

In 1893 the Society for Chemical Industry had in total 340 members compared to its rival, the Society of the Czech Chemists, whose membership was only slightly higher – 373. However, the fresh opportunities quickly attracted new people, mostly with some relation to industrial circles and of less academic affiliation, so that in 1897, at its peak, the membership of the SPCH totalled 570, while its competitor SCHC had markedly fewer members – 443. The last statistics of the year 1906 enumerate 21 honorary members, 67 corresponding members, 57 founding members and 372 acting members, a total of 519 (compare with the slightly differing numbers for 1905 in Table 3.7).[47]

It seems from the columns of the journal that the SPCH was somewhat more focused on domestic affairs than SCHC and its members participated less in international activities. Nevertheless in the era of SPCH the number of foreign honorary members increased by 8. These awards (compare Table 3.6) were apparently bestowed quire randomly, particularly to those chemists who had closer ties to prominent members of the society (Mendeleev, Radziszewski, Lambl, Maercker, Nölting) or whose specialty markedly influenced the fermentation industry in the Czech Lands (Pasteur, Lintner, Hansen).[48]

The journal *Časopis pro průmysl chemický* (Journal for Chemical Industry) became the publishing base of the SPCH officially in 1894. As early as 1893 the SPCH also started a new programme of publishing chemical books called *Chemická knihovna technologická* (Chemical Technological Library). Although the main project in view was a several volume chemical technology in the Czech language, it only materialized about 30 years later. Those books which appeared in the years 1893–1905 had surprisingly little to do with chemical technology. We may find among them such divergent titles as biographies of historical characters, one handbook of saccharimetry and even Ostwald's popular *Die Schule der Chemie*.[49–51]

---

[45] Stanovy (1892).
[46] Stanovy (1892), 439.
[47] Vondráček (1906), 243.
[48] Short biographies of honorary members are in Hanč (1966), 60–75.
[49] For instance, Jahn (1894).
[50] Diviš (1897).
[51] Ostwald (1905).

Table 3.7   *Společnost pro průmysl chemický v Království českém* (Society for Chemical Industry in the Czech Kingdom). Profile of the membership in 1905.[a]

Group of members	Type of membership				
	Honorary	*Corresponding*	*Founding*	*Acting*	*Total*
University teachers	10	8	5	9	32
Teachers in higher educational establishments, researchers in non-university institutions, owners and workers of private laboratories	3	17	4	42	66
Industrialists, landowners, business- and tradesmen, business and industry employees	7	32	28	167	234
Corporations	–	–	12	74	86
Secondary school teachers	1	2	–	15	18
Government employees	–	4	4	44	52
Pharmacists	–	4	2	6	12
Others	–	–	2	12	14
Total	21	67	57	369	514

[a] The statistics were compiled from Seznam (1906).

The SPCH, and similarly the SCHC, intervened in economic and legislation matters (sometimes even jointly with the "old" chemical society as has been shown) and offered counselling services. Perhaps the most important accomplishment of SPCH was its considerable influence on modernizing the chemical industry in Bohemia.

3.3.3.3   Plea for Reunification

It is believed that the split also had personal motivation and served some chemists, especially Antonín Bělohoubek, as a vehicle for gaining a more influential position in the Czech community of chemists. Nevertheless, surprisingly, the existence of two simultaneously existing societies did not evoke animosities, rather some kind of healthy competition that helped to raise the professional standards in both and also the number of chemists organized in specialized associations. If we compare (Table 3.4) the increase in the total number of members in both societies, we realize how steeply the Czech chemical community grew in this period. Even so, the advent of the new century brought forward a young generation of chemists in both societies who came to realize that the split decreased the potency and lowered the prestige of the strong professional community of chemists. At the same time the representatives of industry in the SPCH, who were its main sponsors, felt somewhat disappointed that the society did not fully meet their expectations and started to leave; thus the society got into financial difficulties. For these and other reasons the voices in SPCH pleading to rejoin the mother society were becoming ever stronger. Many members called for the creation of a strong society which would build on the positive features of both societies and attract chemists from all directions. Eventually, at the end of 1905, the Presidium of the SPCH obtained a letter signed by 12 representatives of the younger generation, proposing reunification of the two split societies and a subsequent poll among the members proved that the absolute majority wanted a unified chemical society.[52]

3.4 Reunification: *Česká společnost chemická pro vědu a průmysl* (Czech Chemical Society for Science and Industry), 1907–1919[53]

If we browse through the volumes of the *Listy Chemické* journal issued in the years of the split, we may find there a few indirect reactions to the schism in the Czech chemists' community. The editorial of the 1894 volume emphasizes the necessity to maintain a scientific journal, but at the same time promises to pay more attention to technical disciplines.[54] This and similar feedback indicate that even some chemists who stayed loyally with the "old" society realized the necessity of paying more attention to practical domains – chemical industry and chemistry-related businesses like private research laboratories, pharmacies, trading with chemicals and others. At the turn of the nineteenth century, likewise as in the SPCH, the younger members of the SCHC also became conscious of the negative consequences of the schism and appealed for reunification. The double pressure from both societies brought their representatives to the negotiating table: a joint committee appointed by both societies first met on 18th January 1906.[55]

After one year of preparations the reunited *Česká společnost chemická pro vědu a průmysl* (Czech Chemical Society for Science and Industry, CSCH) was constituted, officially on 26th January 1907. Quite characteristically, František V. Goller, one of the older generation sugar industry leaders, was elected the new president and was expected to bridge the gap and get the support of the industrial circles.[56]

Karel C. Neumann, chemist and sugar-making specialist, vice-president of the reunited society, greeted the establishment of the Czech Chemical Society for Science and Industry in 1907 with words reminding members of the rich membership stratification of the society: "We should not forget that that we are not a monolithic body consisting only of graduated technologists

---

[52] Hanč (1966), 36.
[53] Vondráček (1906); Hanč (1966), 37–41.
[54] Redakce (1894).
[55] Ředitelstvo (1906).
[56] Hanč (1966), 64.

or academics, but we should realize that among us are also graduates from high schools, lower economic- and professional schools, industrialists and other personalities who were attracted to us by their interest in scientific and industrial chemistry. These differences in educational background should be respected in our lectures which should be interesting in general and comprehensible to all participants."[57]

The new/old society attempted to cover all activities and programmatic goals of the two maternal societies; therefore it constituted two divisions: the Scientific Division (Sekce vědecká) and the Industrial Division (Sekce průmyslová), administered by a common Executive. Which direction was stronger is obvious; the membership of the Industrial Division was more than three times higher than that of the Scientific Division (see Table 3.8), although the more theoretically oriented chemists from Prague University and Technical University represented a powerful minority on account of their academic positions and international reputation.[58] The membership structure followed the pattern of the initial SPCH, which means that the members were divided into honorary, founding, contributing and acting.[59] However, the members in the last group belonged to either the Scientific or the Industrial Division.

The two journals also united as the associated organ of the society under the name *Chemické listy pro vědu a průmysl* (1907, Chemical Letters for Science and Industry, Figure 3.3). The title expressed both the industrial and scientific trends of the journal, but its structure remained traditional. The other publishing activities, which included university textbooks and some technological monographs, took up the projects of both societies.[60] During the First World War the long-planned voluminous series *Chemická technologie* (Chemical Technology) was launched in cooperation with the *Česká matice technická* (Czech Technical Foundation) and the *Jednota českých matematiků a fysiků* (Union of Czech Mathematicians and Physicists).[61] This and other initiatives indicate that even during the First World War the society continued working, although to lesser extent. Among the most significant events was the establishment of the nomenclature commission, which after many years of discussions pushed through the modernization of the Czech chemical nomenclature in 1918.

In 1918, after the end of the First World War and the disintegration of Austria-Hungary, the new independent state of Czechoslovakia was founded and the society immediately reacted with a memorandum that proposed establishment of a chemical section in the Ministry of Industry, which would act as a supporting factor for chemical industry development and a counselling

---

[57] Quoted from Hanč (1966), 37–38.

[58] Among them were, for instance, Bohuslav Raýman (see note 35); Bohuslav Brauner, inorganic chemist, personal friend of Mendeleev, specialist in atomic weights who made several modifications of the Periodic System, for a biography see Štrbáňová (2003); Emil Votoček, organic chemist, known for his pioneering research into the chemistry of carbohydrates, for a biography see Schätz (2002), 77–79.

[59] The honorary members included those still living from both SCHC and SPCH.

[60] Probably most important was the first edition of Votoček's, at that time very modern, Organic Chemistry, see Votoček (1912–1916).

[61] Hanč (1966), 39.

Table 3.8 Česká společnost chemická pro vědu a průmysl (Czech Chemical Society for Science and Industry). Profile of the membership in 1907.[a]

Group of members	Type of membership			Acting in section		Total
	Honorary	Founding	Contribut.	Scientific	Industrial	
University teachers	7	3	–	27	3	40
Teachers in higher educational establishments, researchers in non-university institutions, owners and workers of private laboratories	3	3	–	27	46	79
Industrialists, landowners, business- and tradesmen, business and industry employees	6	29	17	3	217	272
Corporations	–	26	80	–	4	110
Secondary school teachers	1	1	1	35	7	45
Government employees	–	–	–	3	35	38
Pharmacists	–	2	–	1	1	4
Others	–	8	2	4	28	42
Total	17	72	100	100	341	630

[a] The statistics were compiled from Výroční zpráva (1908), 48–56.

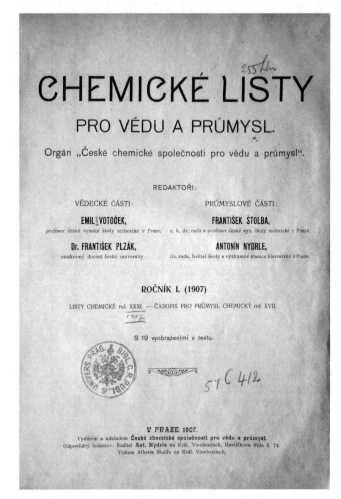

Figure 3.3   *Chemické listy pro vědu a průmysl* issued since 1907 after the reunification of the Czech Chemical Society.

organ in all matters related to chemistry. In 1919 the society was transformed into the *Československá společnost chemická* (Czechoslovak Chemical Society) with its new members – the Slovak chemists – and extended territorial reach: Slovakia and the Sub-Carpathian Ukraine. But this is an entirely different story.

3.5   Other Professional Associations of Chemists in the Czech Lands

In the Czech Lands there existed a variety of strong professional groups of chemists which founded their associations in the second half of the nineteenth century (see Table 3.3), serving the interests of more demarcated circles of specialists. Some of them have already been mentioned; nevertheless they deserve a few additional remarks.

The sugar industry represented a traditionally strong field in the Czech Lands. It was estimated that most graduates of the Prague Polytechnic and later Technical University found their employment in the sugar and fermentation industry. In 1868 the sugar industrialists and chemists employed in the sugar industry founded the bilingual *Verein zur Hebung der Zuckerfabrikation im Königreich Böhmen* (Association for Improvement of the Sugar Industry in Bohemia), which was disbanded in 1874. Its journal *Zeitschrift für Zucker-Industrie in Böhmen* (Journal for Sugar Industry in Bohemia) was only issued in three volumes (1872–1874).[62] This not quite successful attempt was followed in 1876 by the flourishing bilingual but Czech-dominated *Spolek pro průmysl cukrovarnický v Čechách*, (Association for Sugar Industry in Bohemia) closely collaborating with the Czech Chemical Society. Similar professional societies were established by the brewing and distilling industries, which were mostly in Czech hands. Chemists were acting in all these associations, and the SCHC, SPCH and CSCH became a unifying element, a place of interaction across fields and professions.

Last but not least we should not omit the powerful all-Austrian *Österreichische Gesellschaft zur Förderung der chemischen Industrie* (Austrian Society for the Advancement of the Chemical Industry), founded in Prague in 1878.[63] It was one of the rare bases where the Czech and German chemists still communicated, in spite of the rising nationalism.[64] The membership included not only chemical enterprises and representatives of the chemical industry, but also university teachers of chemistry; in the first year of the *Gesellschaft*'s existence it amounted to 82. The journal *Berichte der Österreichischen Gesellschaft zur Förderung der chemischen Industrie* (Reports of the Austrian Society for the Advancement of the Chemical Industry) appeared between 1879 and 1898.

3.6 Membership Profile and Demarcation of the Czech Chemical Societies

To learn about the membership profiles and compare them in the three interrelated Czech chemical societies – the original SCHC, the dissident SPCH and the unified CSCH – some statistical probes were necessary (see Tables 3.5, 3.7, 3.8 and 3.9). To get comparable results, lists of members of the three societies from years close to each other were chosen. For SCHC it was the year 1902, in the time of the schism, when the total number of members was nearing its maximum. The statistics of the SPCH come from 1905, which is the year

---

[62] Bolton (1902), 3.
[63] Its establishment is described in Geschichte (1879). For the statutes see Statuten (1879). Compare also Rosner (2004), 298.
[64] Among the founders was also the chemist F. Šebor, a leading Czech sugar specialist; see Geschichte (1879).

Table 3.9 Proportion (percentage) of various groups of members of the Czech chemical societies.

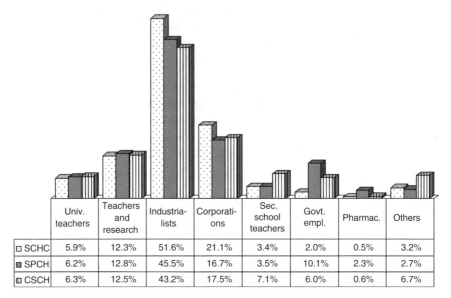

	Univ. teachers	Teachers and research	Industria-lists	Corporati-ons	Sec. school teachers	Govt. empl.	Pharmac.	Others
☐ SCHC	5.9%	12.3%	51.6%	21.1%	3.4%	2.0%	0.5%	3.2%
▨ SPCH	6.2%	12.8%	45.5%	16.7%	3.5%	10.1%	2.3%	2.7%
▥ CSCH	6.3%	12.5%	43.2%	17.5%	7.1%	6.0%	0.6%	6.7%

SCHC – *Spolek chemiků českých* 1902 – total membership 407 (Society of Czech Chemists).
SPCH – *Společnost pro průmysl chemický v Království českém* 1905 – total membership 514 (Society for Chemical Industry in the Czech Kingdom).
CSCH – *Česká společnost chemická pro vědu a průmysl* 1907 – total membership 630 (Czech Chemical Society for Science and Industry).
The calculations are based on numbers taken from Tables 3.6, 3.7 and 3.8.

before the unification also when the membership was almost at its highest. The membership of the CSCH was taken from the year 1907, the first year that a list of members of the unified society was available.

Before making comparisons and drawing some conclusions from the tables, let us give some explanation of the groups of members and categories of membership.

The item "University teachers" includes all those members who were involved in university education, that is not only full and extraordinary professors, but also other educators at universities: "dozents", "private dozents" and "assistants".[65] The next group is formed by the wide range of teachers in higher educational establishments, non-university researchers, mostly in private laboratories, and a few museum workers.[66] The item "Corporations" means mostly manufacturing works and businesses of all

---

[65] The rank of "dozent" conformed with reader or associate professor standing; "private dozent" had the right to teach, but was not in the "dozent" position at the university. The standing of "Assistant" is comparable to lecturer or assistant professor.
[66] These were various vocational schools specializing in all possible kinds of studies, such as agricultural, business or commerce, or even ceramics, fishing, *etc.*

possible types, not only those connected with chemistry, with the obvious prevalence of sugar and fermentation plants.[67] Nevertheless we may find among the corporations even municipalities, a few vocational or secondary schools and the Prague Technological museum. Not all secondary school teachers were teaching chemistry. Government employees were represented in the societies by members of state and municipal administration, such as revenue officers, technical financial inspectors, school inspectors, *etc*. Among the "Others" were placed those whose affiliation it was not possible to determine; some are identified as "chemists" or "engineers".

If we compare the lists of members with their affiliations and distributions in the individual societies (Tables 3.5, 3.7 and 3.8) we may detect numerous peculiarities. Those who supported the societies with high membership fees, that is the founding and contributing members, were represented in the SCHC by various corporations, especially the sugar companies; in the SPCH and CSCH also the industrialists, landowners, business- and tradesmen and business and industry employees. Nevertheless, the three societies had a very similar share of members from the industrial and business circles so in this respect we cannot see any special difference between the societies, although the SPCH was originally founded to attract more people and corporations from industry. This fact also suggests that the representatives of industry and business probably attached great importance to the society as a social centre of contact and an institution serving as counsel and supporter of their interests. Unanticipated are also the overall low differences in the distribution of the individual professional groups of members (Table 3.9) whose percentage share is nearly the same in all three societies.

If we look for differences, we may note the slightly increasing participation of university teachers over the course in time, which is understandable in the light of extending university education in chemistry. The same applies to the increase in the membership of secondary school teachers who were interested in science and in deepening their knowledge through the activities of the chemical societies. The low percentage of pharmacists among the members was not due to any marginalization of this profession but rather by the fact that they assembled in the *Spolek farmaceutů* (Association of Pharmacists) founded in 1875. Another noticeable distinction is the relatively high proportion of government employees in the SPCH, which is not so easy to explain.

Worth mentioning is the fact that there were chemists from abroad among not only the honorary but also the corresponding members, such as the university professors Gustav Janeček from Zagreb, Sima Lozanič from Belgrade and Bronislaw Pawlewski from Lemberg (Lwów), or industrialists such as J. Tourtel, owner of the well-known Tourtel brewery in Tantonville near Nancy.[68,69]

---

[67] Such as machine works, brickworks, warehouses, *etc*.

[68] As already mentioned, honorary and corresponding membership was awarded as an expression of the person's merits, and was not related to nationality.

[69] Compare http://perso.orange.fr/magicam/bierfr1.htm

The richly structured membership demonstrates that all three chemical societies had no pronounced policy of demarcation whether in statutes or in practice and were open to all those interested in joining. Although the lists of members do not distinguish chemists from non-chemists, there were evidently quite a few members who had no specialized chemical education. This shows that the societies were open to any professionals not only by their statutes but also in reality. Probably the only firm and almost insurmountable demarcation was affiliation to the linguistically Czech community. The SPCH membership of Carl Zulkowski, the only professor from the German Technical University (*Deutsche technische Hochschule*) in Prague, was the exception that confirms the rule.

3.7   Conclusion

The present *Česká společnost chemická* (Czech Chemical Society) takes 1866 as the year of its foundation; thus it can be considered one of the oldest chemical societies in the world, which has undergone since its nuclear form – the students' association *Isis* – several institutional transformations.

In the period in question most of the Austro-Hungarian chemical (or chemistry-related) industry was in the territory of Bohemia, especially the manufacture of glass and porcelain, the sugar industry, the food industry (including the fermentation industry) and metallurgy.[70] A great number of chemists found their worth not only in the plants and factories but also in teaching and research institutions, pharmacies, hospitals, *etc.* Most of them were members of at least one of the three Czech chemical societies that emerged successively in the years 1872, 1893 and 1907, considered in hindsight transitory forms of the same scientific society.

It follows that none of the three societies in question had any policy of demarcation. They were open not only to chemists of all possible vocations and branches, but also to the lay public. The notion of "chemist" was nevertheless very wide. Chemists were those who graduated in chemistry from the universities or specialized grammar schools, but also physicians who worked in some chemistry-related laboratories, university or high-school teachers, students, technicians, factory chemists and managers, people who owned shops with chemicals, as well as pharmacists. We may only consider two groups that did not join the Societies, although their statutes did not prohibit their membership – women and Germans. The absence of women members can be explained by the traditional cultural environment of Czech society, whereby women were only just starting to gain access to higher education.[71] The absence of Germans was due to the nationally divided society, although we may say that the societies remained free from the extreme nationalism of the period. Perhaps a specific phenomenon was the prevalence of specialists from the sugar industry

---

[70] Compare Redakce (1928). According to this source 75% of the Austro-Hungarian chemical industry was located in the territory of what was to become Czechoslovakia.
[71] Štrbáňová (2004).

in the Society of Czech Chemists and people from the practical sphere in the transitory Society for Chemical Industry.

In spite of some discrepancies that led to a temporary split in 1893–1906, the activities of the societies must be considered of crucial significance for the formation of the Czech chemical community and all other aspects of chemistry in the Czech Lands not only in the given period, but also in the long run. All three societies played a multifunctional role in developing the field itself, its teaching (both at the universities and the secondary schools), nomenclature, publishing activities, circulating new knowledge, affecting the structure of the chemical industry, businesses, legislation, *etc.* The fact that their members included not only chemists, but also people who were just interested in chemistry or attracted by a strong social group, enhanced the general social and cultural character of these societies, which besides stimulating scientific and technological progress in the Czech Lands also played, along with other professional associations, a significant role in the cultural, political and intellectual life of Czech society in general.

Abbreviations

SCHC   *Spolek chemiků českých* (Society of Czech Chemists)
SPCH   *Společnost pro průmysl chemický v Kríálovství českém* (Society for Chemical Industry in the Czech Kingdom)
CSCH   *Česká společnost chemická pro vědu a průmysl* (Czech Chemical Society for Science and Industry)

References

Baur, V. (1903), III. Zpráva o stavu členstva r. 1902 [Report on the state of membership in the year 1902], *Listy chemické* **27**, 57–64.

Bolton, H. C. (1902), *Chemical Societies of the Nineteenth Century*, Smithsonian Institution, Washington D.C.

Československá společnost chemická – Slovenská chemická spoločnost' 1966–1975 [Czechoslovak and Slovak chemical societies 1966–1975], Academia, Praha.

Diviš, J. V. (1897), *Saccharimetrie na základě fysikalném a chemickém* [Saccharimetry based on chemistry and physics], Chemická knihovna technologická 2, Šimáček, Praha.

Franěk, O. (1969), *Dějiny české vysoké školy technické v Brně* [History of the Czech Technical University in Brünn], **1**, VUT, Brno.

Geschichte der Bildung der Gesellschaft (1879), *Berichte der Österreichischen Gesellschaft zur Förderung der chemischen Industrie*, Nos. 1 and 2, 1–2.

Gut, J. (1976), 100 let Chemických listů [100 years of the Chemické listy journal], *Chemické listy* **70** (100), 897–906.

Hanč, O., ed. (1966), *100 let Československé společnosti chemické, její dějiny a vývoj 1866–1966* [100 years of the Czechoslovak Chemical Society, its history and development], Academia, Praha.

Hanč, O., ed. (1966a), *100 Years of the Czechoslovak Chemical Society, Its History and Development*, Academia, Prague.

Havránek, J., ed. (1997), *Dějiny Univerzity Karlovy III, 1802–1918* [History of the Charles University], Karolinum, Praha.

Jahn, J. V. (1894), *A.V. Lavoisier, jeho život a práce* [Life and work of Lavoisier], Šimáček Praha.

Janko, J. and Štrbáňová, S. (1988), *Věda Purkyňovy doby* [Science in Purkinje's time], Academia, Praha.

Jindra, J. (2003), České vědecké společnosti exaktních věd v 19. století [Czech scientific societies for exact sciences], in H. Binder, B. Křivohlavá and L. Velek, ed., *Místo národních jazyků ve výchově, školství a vědě v habsburské monarchii 1867–1918* [Position of National Languages in Education, Educational System and Science of the Habsburg Monarchy 1867–1918], *Práce z dějin vědy* **11**, 401–420.

Jindra, J. (2003a), Počátky České chemické společnosti [The beginnings of the Czech Chemical Society], *Chemické listy* **97**, 220–222.

Kořalka, J. (1966), *Češi v habsburské říši a v Evropě* [The Czechs in the Habsburg Empire and in Europe], Argo, Praha.

Koštíř, J. (1976), O redaktorech a zajímavých autorech našeho časopisu [About the editors and interesting authors of our journal], *Chemické listy* **70** (100), 907–932.

Lexikon bedeutender Chemiker (1988), Bibliographisches Institut, Leipzig.

Liebig, Justus von (1844), *Chemische Briefe, 1. Auflage*, Akademische Verlagshandlung von CF Winter, Heidelberg.

Náš program (1891) [Our Programme], *Časopis pro průmysl chemický* **1**, 1–2.

Neumann, K. C. (1890), *Příruční kniha ku rozborům cukrovarnickým* [Handbook for analyses in sugar industry], Knihovna chemická, Praha.

Neumann, K. C. (1906), Dějinný přehled o činnosti Spolku českých chemiků za prošlých 35 let [Historical survey of the activities of the Society of Czech Chemists in the last 35 years], *Listy chemické* **30**, 301–326.

Nový, L. *et al.* (1961), *Dějiny exaktních věd v českých zemích do konce 19. století* [History of exact sciences in the Czech Lands until the end of the nineteenth century], Nakladatelství Československé akademie věd, Praha.

Ostwald, W. (1905), *Škola chemie* [School of chemistry], Weinfurter, Praha.

Osvědčení redakce (1891) [Statement of the editors], *Časopis pro průmysl chemický* **1**, 2–5.

Preis, K. and Votoček, E. (1902), Anorganická chemie [Inorganic chemistry], Knihovna chemická, Praha.

Protokol prvního sjezdu chemicko-cukrovarnického (1892) [Protocol of the First Chemical and Sugar Industry Convention], *Listy chemické* **16**, 234–237.

Quadrat, O. (1966), *Nástin historického vývoje Vysoké školy chemicko-technologické v Praze (do roku 1945)* [Outline of the historical development of the Chemical Technological University in Prague until 1945], Státní pedagogické nakladatelství, Praha.

Raýman, B. (1893), *Cukry a sloučeniny příbuzné* [Carbohydrates and related compounds], Knihovna chemická, Praha.

Redakce (1928), 1918–1928, *Chemický obzor* **3**, 325–328.

Redakce Listů chemických (1894), Našim čtenářům [To our readers], *Listy chemické* **18**, 1.

Ředitelstvo Společnosti pro průmysl chemický v království českém (1906), Ke sloučení obou českých spolků chemických [To the unification of both Czech chemical societies], *Časopis pro průmysl chemický* **16**, 64.

Rosner, R. (2004), *Chemie in Österreich 1740–1914*, Böhlau Verlag, Wien-Köln-Weimar.

Seznam pp. členů (1906) [The list of members], *Časopis pro průmysl chemický* **16**, 9–15.

Schätz, M. (2002), *Historie výuky chemie* [History of chemistry instruction], Vysoká škola chemicko-technologická, Praha.

Stanovy Společnosti pro průmysl chemický v Království českém (1892) [Statutes of the Society for Chemical Industry in the Czech Kingdom], *Časopis pro průmysl chemický* **2**, 439–447.

Statuten der österreichischen Gesellschaft zur Förderung der chemischen Industrie (1879), *Berichte der Österreichischen Gesellschaft zur Förderung der chemischen Industrie*, No. 1 and 2, 2–5.

Šetlík, B. and Votoček, E. (1906), *Příruční slovník chemicko-technický* [Concise chemical technical dictionary], Chemická knihovna, Praha.

Štrbáňová, S. (1992), The Liebig and Hofmann schools at the Prague universities and the Development of Chemistry in Bohemia, in C. Meinel and H. Scholz, eds., *Die Allianz von Wissenschaft und Industrie - August Wilhelm Hofmann (1818 – 1892)*, Verlag Chemie, Weinheim, New York, Cambridge, Basel, 211–220.

Štrbáňová, S. (2003), Brauner, Bohuslav, in: D. Hoffmann, H. Laitko, S. Müller-Ville and Ilse Jahn , ed., *Lexikon der bedeutenden Naturwissenschaftler* 1. Band, Spektrum Akademischer Verlag, Heidelberg-Berlin, 249–251.

Štrbáňová, S. (2005), Correspondence Strengthening the Network of a Scientific School: Unknown Letters of the French Chemists C. Friedel and C. A. Wurtz to the Czech Chemist B. Rayman, in: H. Kant and A. Vogt, ed., *Aus Wissenschaftsgeschichte und – Theorie*, Verlag für Wissenschafts- und Regionalgeschichte, Dr. Michael Engel, Berlin, 257–276.

Štrbáňová, S. and Janko, J. (2003), Uplatnění nového českého přírodovědného názvosloví na českých vysokých školách v průběhu 19. století [Assertion of the new Czech scientific terminology at the Czech universities in the course of the nineteenth century], in H. Binder, B. Křivohlavá and L. Velek, eds., *Místo národních jazyků ve výchově, školství a vědě v habsburské monarchii 1867–1918* [Position of National Languages in Education, Educational System and Science of the Habsburg Monarchy 1867–1918], *Práce z dějin vědy* **11**, 297–311.

Štrbáňová, S. (2004), The Institutional Position of Czech Women in Bohemia 1860–1938, in S. Štrbáňová, I. H. Stamhuis and K. Mojsejová, ed., *Women Scholars and Institutions, Proceedings of the International Conference (Prague, June 8–11, 2003)*, Studies in the history of sciences and humanities, **13A**, **13B**, Research Centre for the History of Sciences and Humanities, Prague, 69–97.

Verzeichnis der Mitglieder der Österr. Gesellschaft zur Förderung der chemischen Industrie (1879), *Berichte der Österreichischen Gesellschaft zur Förderung der chemischen Industrie*, No. 4, Beilage zu den Berichten der Österreichischen Gesellschaft zur Förderung der chemischen Industrie, no pages given.

Vondráček, R. (1906), Retrospektiva Společnosti pro průmysl chemický v království Českém [Retrospective of the Society for Chemical Industry in the Czech Kingdom], *Časopis pro průmysl chemický* **16**, 224–244.

Votoček, Emil (1912–1916), *Chemie organická* [Organic chemistry], 2 volumes, Politika, Praha.

Výroční zpráva o činnosti 'České chemické společnosti pro vědu a průmysl v Praze za správní rok 1907 (1908) [Annual report of the activities of the Czech Chemical Society for Science and Industry], *Chemické listy pro vědu a průmysl* **2**, 43–56.

Wraný, A. (1902), *Geschichte der Chemie und der auf chemischer Grundlage beruhende Betriebe in Böhmen bis zur Mitte des 19. Jahrhunderts*, Řivnáč, Prag.

Ze Společnosti pro průmysl chemický v království českém (1893) [From the Society for Chemical Industry in the Czech Kingdom], *Časopis pro průmysl chemický* **3**, 76–78.

Zpráva výboru o činnosti 'Spolku českých chemiků' (1894) [Report of the Board on the activities of the Society of Czech Chemists], *Listy chemické* **17**, 21–28.

Zprávy spolkové (1878) [Reports of the Society], *Listy chemické* **2**, 139–140.

Zprávy spolkové (1879) [Reports of the Society], *Listy chemické* **3**, 131.

CHAPTER 4

DENMARK: *Creating a Danish Identity in Chemistry between Pharmacy and Engineering, 1879–1914[1]*

ANITA KILDEBÆK NIELSEN

4.1 Introduction

On 13th October 1879, the Danish daily *Berlingske Tidende* carried the following announcement: "The [Danish] Chemical Society. At a meeting held in the spring for a smaller circle [of people], a society was founded. Under the above mentioned name it has as its object **through lectures and discussions, minor communications and social gatherings** to establish a tie between men who have an interest in chemistry. The annual fee is three DKK collected once a year annually. It is the intention during the winter term to hold four to six ordinary meetings, the first meeting will take place in the first week of November. Enrolment is possible by signing one of the lists that are available at the porters at the Polytechnic College and the chemical laboratory at the university, or to the undersigned, who at the first meeting are authorized to be in charge of the provisional arrangements".[2]

The announcement was signed by four young men, who will be introduced below. As stated in the quotation, the background of the proclamation was a meeting in the spring of 1879, which took place as a result of a self-study club in chemistry established by young and newly graduated chemists in Copenhagen

---

[1] This chapter is primarily based on Nielsen (2000), **1**, sections 2.9 and 4.1.
[2] *Berlingske Tidende*, 13th October 1879. Translated by the author. The present name for the Polyteknisk Læreanstalt (Polytechnic College) is Denmark's Technical University.

Creating Networks in Chemistry: The Founding and Early History of Chemical Societies in Europe
Edited by Anita Kildebæk Nielsen and Soňa Štrbáňová
© The Royal Society of Chemistry 2008

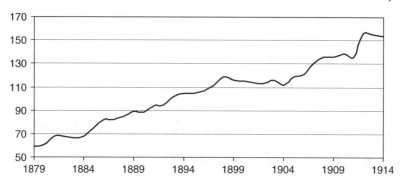

Figure 4.1   Number of members in the Danish Chemical Society, 1879–1914.[4]

around 1877.[3] The meeting in the spring of 1879 was initiated by some of the members of the club in order to learn whether a chemical society should and could be established, and – judging from the announcement quoted above – the assembly came to a positive conclusion. By the end of 1879, 59 men joined *Kemisk Forening* (The Danish Chemical Society), and the number rose by almost three times in 1914 (see Figure 4.1).

This paper analyses the establishment and early history of the Danish Chemical Society with particular focus on the relationship between the different groups of members. Furthermore, it will discuss how the definition of a "chemist" in Denmark in the analysed period was influenced by the structure of the Danish Chemical Society and how this affected the balance between pure and applied chemistry, and between the producers and applicants of chemical knowledge. But before further introduction of the chemical society, it is necessary to provide some background information on the Danish chemical community around 1880.

4.2   A Short History of Chemistry in Denmark in the Nineteenth Century

In 1880, Denmark had but one geographical centre of university-level learning, namely the capital Copenhagen.[5] It comprised two centres of chemical education: Københavns Universitet (the University of Copenhagen) and Polyteknisk Læreanstalt (the Polytechnic College).

The University of Copenhagen was founded in 1479. The first professorship of chemistry was primarily a chair of physics and it was established in 1753 at the faculty of medicine. From 1806, the chair was transferred to the faculty of arts simultaneously with the appointment of Hans Christian Ørsted as professor. Ørsted was instrumental in the establishment of the first full professorship of chemistry in 1822, to which William Christopher Zeise was appointed. When

---

[3] According to Christensen (1902), nothing further is known about the study club.
[4] Nielsen (2000), **2**, 79–80.
[5] For further details on the history of Danish chemistry in the nineteenth century see Kragh (1998) and Nielsen (2000), **2**, chapter 2 and references therein.

in 1850 a faculty of mathematics and science was established, this chair and second in chemistry established in 1840 was transferred to the new faculty. The faculty had but seven chairs at the point of establishment. Moreover, the faculty could not offer its students any qualifying examination, which resulted in a very low number of students.[6] In the first 30 years, only about 20 students graduated from the whole faculty. When a qualifying examination was granted in 1883, the number of students increased slightly and in 1913 the faculty had about 150 students in total, but still produced only 10 graduates annually. Consequently, the number of chemistry students at the university was very low. Only about 20 students completed university degrees in chemistry during the second half of the nineteenth century. Between 1900 and 1914 another eight men and two women graduated in chemistry as their major subject –with focus on either pursuing an academic career (the majority) or obtaining a position as a secondary school teacher. In the period 1850 to 1914, 13 doctorates in chemistry were awarded.

At the Polytechnic College, the students could obtain a degree in chemical engineering. The curriculum put much emphasis on pure chemistry and, until a reform in the 1890s, less on applied, engineering chemistry. The prioritizing of academic chemistry should be understood on the background that the science teachers at the college were the university professors of science. In the nineteenth century, Denmark was still an agrarian country with hardly any natural resources. The industrialization only slowly developed towards the end of the century, and there was little chemical industry. The Polytechnic College was established in 1829, but its primary aim was not to train engineers to the industry but to provide the King's administration with scientifically skilled civil servants who could be employed to build and run a growing number of public works.

The profile of the graduates of the Polytechnic College made it difficult for the arising chemical industry to fit them into the lines of production. The graduates lacked craftsman skills; they were considered academics with primarily theoretical abilities. Furthermore, the self-taught engineers and other subordinate leaders of the factories were in no hurry to give up competence to the new man. Consequently, many factory owners chose to employ pharmacists rather than graduates of the Polytechnic. Only from 1867 did the Polytechnic College start to consider this problem by appointing a teacher of technical chemistry, but the chair was not turned into a permanent professorship before the 1890s. Its first occupant was Carl August Thomsen. But the pharmacists kept the upper hand in the competition with the polytechnic graduates in industrial positions for the decades up to the First World War.[7]

As the number of science students at the university was negligible, the lectures for those and for the students at the Polytechnic College were practically one and the same thing, and the lectures were oriented towards the scientific, pure chemistry.[8] In the period 1829 to 1890, the Polytechnic College

[6] Nielsen (2000), **1**, 47–48.
[7] Petersen (1996), 42.
[8] Wagner (1993), 153. It is worth noting that the industrialization of Denmark only took off in the 1870s and first gained full speed in the 1890s; see Hansen (1970).

produced a little less than 500 graduates, but the number exploded around the turn of the century, and between 1890 and 1930 approximately 3500 graduated.[9] About one quarter of the total number of graduates between 1832 and 1914 became chemical engineers.[10] In the last decades of the nineteenth century, the chemical lectures were given by the professors Julius Thomsen, brother of C. A. Thomsen, and the slightly younger Sophus Mads Jørgensen. Both preferred to do scientific research without collaborators, and neither formed research schools. Jørgensen, however, gave room to assistants and he was appreciated as a lecturer.[11]

Chemistry, of course, also played a central role in the education and training of Danish pharmacists and pharmaceutical assistants. Although future graduates of pharmacology have been state-examined since 1672, a separate college of pharmacology (Farmaceutisk Læreanstalt (The Pharmaceutical College)) was not established until the early 1890s. Until then, the pharmaceutical students often sat in on the same lectures and participated in the same laboratory courses as the chemical and medical university students and the students at the Polytechnic College. In fact, the group of pharmacists far outnumbered the groups of university and polytechnic graduates, as some 1000 graduated in pharmacology in the second half of the nineteenth century.[12] Even though these three groups had different priorities and different examination requirements, their educational backgrounds were quite similar, owing to a lack of a separate college of pharmacology and to the peculiar relationship between the university and the Polytechnic College.

Helge Kragh has carried out a bibliometrical analysis of Danish chemical authors and their publications in the nineteenth century, which shows a structure in accordance with the educational pattern.[13] The majority of the authors had medical, physiological, pharmaceutical, agricultural or polytechnic educational backgrounds. In the second half of the century, the pharmacists alone accounted for one third of the publications, while the engineers constituted about 20% of the authors, which taken together wrote about one quarter of the papers. The few university-trained chemists, however, contributed a disproportionately large part of the publications. They only constituted 2.5% of the authors (a total of six men) but they wrote 15% (164) of the publications. The group thus put a finger mark on the development of Danish scientific chemistry in spite of their small number. The exact same situation became prevalent in the Danish Chemical Society.

The archival and printed sources do not provide clues as to why the Danish Chemical Society was founded exactly in 1879 and not earlier or later. The establishment may have been motivated by the foundation of similar societies abroad, but nothing certain is known. The establishment first and foremost rested on the personal initiative of the members of the study group mentioned

[9] Harnow (1993), 169.
[10] *Dansk Civilingeniørstat 1942*, Copenhagen, 10–11.
[11] Nielsen (2000), **1**, 40.
[12] Kofod (1967), 28.
[13] Kragh (1998), 247.

in the beginning of the chapter. Furthermore, it is not even known if the actual profile and activities of the society were quite in accordance with what the initiators visualized the society would be. None of the initiators ever provided such information. From the historian's point of view, it is clear that the steady, albeit not overwhelming, number of graduates of chemistry from the university, the Polytechnic College and in pharmacology, together with the dramatic developments of chemistry as a scientific discipline throughout the nineteenth century, provides one of the main reasons why the Danish Chemical Society was founded – and at approximately the same time as chemical societies in many other European countries.

4.3   Initiation and Early Years of the Danish Chemical Society[14]

As mentioned at the start of the chapter, the Danish Chemical Society was unofficially founded in the spring of 1879 by a small group of men. At this spring meeting predating the official establishment in October, the five initiators were elected to the first, provisional executive committee.[15] In hindsight, the most prominent of them was probably Odin Tidemand Christensen, who became an associate (and later full) professor of chemistry at Den kgl. Veterinær- og Landbohøjskole (the Royal Veterinary and Agricultural College) in 1892. He was one of the rare university graduates of chemistry (1877) and he obtained a doctor's degree in 1886. The second man to sign the release was the polytechnic graduate Niels Georg Steenberg, who from 1865 had a position as factory manager in a cordage- and wire-producing company. From 1895, his job was combined with a professorship in technical chemistry at the Polytechnic College. The third, Christian Frederik August Tuxen, graduated in agriculture in 1874 and became associated with the Royal Veterinary and Agricultural College shortly after, from 1903 as professor of agricultural chemistry. The last two signers were pharmacists. Erik Christian Verlauff Steenbuch later worked as an apothecary in Copenhagen and Sigfred Frederik Edvard Valdemar Stein bought a commercial chemical laboratory in 1865. Behind this group of five was another chemical engineer, Thomas Thomsen, a younger brother of the professors August and Julius Thomsen. Together with Christensen and Steenbuch, he had sent out the invitation for the spring meeting of 1879. At this time, Thomsen, as Steenbuch, worked at the university's chemical laboratory, but soon after (in 1884) he gave up the career as chemist to become a vicar.

The official founding meeting of the society took place on 22nd November 1879. Here, the first executive committee was elected, but the first set of rules had already been passed on an earlier date, probably in early October. The first set of rules, which only comprised three paragraphs, was identical to the newspaper announcement, namely the purpose of the society, the number of

---

[14] The only primary source of the history of the Danish Chemical Society in the nineteenth century is Christensen (1902).
[15] Christensen (1902), 1–2.

annual meetings and the annual fee. Paragraph two furthermore stated the size of the executive committee (five, to be elected or re-elected each year), and the admission rule (recommendation of one existing member). The first committee included only Stein and Thomsen from the provisional committee. It is not known why the founding fathers didn't want seats in the committee or even if they gave up the seats without struggle. A conscious strategy might have been instead to select chemists for the committee who were already established in academia or in private companies in order to give the new society more stability, prestige and first-hand connections to chemical institutions. The elected members were Sophus Mads Jørgensen, Johan Gustav Christoffer Thorsager Kjeldahl director of the private Carlsberg Laboratory's chemical department, and Frederik Engelbreth Holm, who had a managing position at the Royal Copenhagen Porcelain Manufactory. He was also the former editor of a journal of applied chemistry that was published in the 1870s.[16] Throughout the century, Jørgensen remained in presidency. His reign was later denoted "enlightened despotism" by one of the later presidents.[17] When, at the turn of century, the statutes of the society were altered, one of the new paragraphs stated that a person could only be a member of the executive committee (and thus president) for three years in succession. Jørgensen was, however, re-elected in 1903 for another three-year period. Already in 1889, the size of the board had been reduced to three people. I shall discuss the constitution of the boards in further details below.

It is worth noting that the society had no sources of income except for the member fees. The statutes did not allow companies to be members at a higher fee, and there is no evidence that the society approached the growing number of chemical industrial companies with the purpose of finding subsidies for the activities of the society (in the pre-Second World War period). At the same time, the admission rules were rather strict as only the executive committee could admit new members upon the recommendation of a member. It is, however, uncertain if the rule was used restrictively, *i.e.* to prevent some groups from joining. But examination of the lists of members reveals that all new members had documented training in chemistry and that secondary school teachers and amateurs were not to be found in the society, for one reason or another. Teachers and amateurs clearly belonged to other social networks. In 1893, the first group of students of chemistry, since the establishment of the society, was admitted entrance.[18] According to the 1900 statutes, the purpose of the society was to "further the collegiate connections between Danish chemists", with no definition of "chemist". Use of the term "collegiate", however, indicates that membership was intended for fellowmen, namely for those who earned a living by applying their chemical knowledge. The new statutes did, on the other hand, soften the admission rule. Now, anyone "occupied with chemical work" (a phrase not closer defined) could apply for

---

[16] Nielsen, forthcoming.
[17] Biilmann (1929), 142.
[18] Winther (1962), 105. In the list of members of 1879 as reprinted in Christensen (1902), several polytechnical students are included.

membership by informing the committee of the wish. He (or she) was subsequently admitted unless one of the members put forward "proper objections" (again without further details).[19]

According to the earliest existing balance sheets, the income of the society was spent on renting a suitable room for the meetings and on administrative expenses such as stamps.[20] This modest spending created a small surplus, although the society has never become truly wealthy. The first application of the surplus was made in 1883 when Odin T. Christensen produced a catalogue of chemical periodicals and lexicons in the Copenhagen area to make it easier for the members to locate relevant printed sources. The catalogue was printed at the expense of the society, but the members had to pay to get a copy. This publication was in fact the only publication of the society before the 1920s.[21] The financial situation could in no way support the regular printing of a periodical and thus the Danish chemists had to make do with foreign chemical journals.

The Danish Chemical Society did not publish its own periodical until the 1920s. Even when the publication began in 1926, it had to rely economically on connections to the chemical industry. In fact, the journal (*Kemisk Maanedsblad*, later *Dansk Kemi*) could not have been published without collaboration with the journal *Nordisk Handelsblad for kemisk Industri* (*Nordic Trade Paper for Chemical Industry*) that appeared six years earlier. The joint venture resulted in a periodical that included both relevant information for the members of the chemical society, such as résumés of lectures presented to the society, and news of relevance to the industry, such as descriptions of advanced laboratory and industrial apparatus and price lists of chemicals.

In contrast to many other national chemical societies of Europe, the Danish Chemical Society never obtained a permanent residence of its own. For this reason, the society has never established a book collection or library, and the state of affairs of its archival material has also suffered from a lack of permanent location.

4.4 Regular Routines and Events

The regular meetings were the pivoting point of the society. The ambition was to arrange one meeting every month in the academic terms, *i.e.* from October to April. The connection to the teaching term gives a clear indication of the close, although informal and individualized, relationship between the society and the university-level teaching institutions. This bond was further strengthened and rendered visible when from 1914 the meetings were held in the chemical auditorium of the university.[22] The use of lecture rooms might also have influenced the atmosphere at the meetings in such a way as to disfavour the

---

[19] Archive of the Danish Chemical Society (ADCS), box DCS.01,1. Located at the Steno Institute at the University of Aarhus.

[20] Archive of the Danish Chemical Society (ADCS), box DCS.01,1. Located at the Steno Institute at the University of Aarhus.

[21] Nielsen (2000), **1**, 180 ff.

[22] Nielsen (2000), **1**, 95.

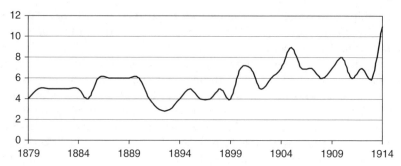

Figure 4.2   Number of meetings per year in the Danish Chemical Society, 1879–1914.[26]

active involvement of the younger, less-experienced members because the meetings resembled the ordinary teaching at the universities.

During the first decades, it was often difficult to acquire speakers, and often the members of the board took the task upon themselves. In 1885, Odin T. Christensen was hired to give surveys of newly published chemical literature at meetings where no speakers could be found. In the period 1901–1908, presented papers were rewarded with a fee in order to increase the number of presentations.[23] But if the number of papers was very unsatisfying, this did not apply to the quality of the presentations. The lectures that *were* given often contained results from the research frontier, nationally or internationally. In the period 1879 to 1914 (included), an average of 5.7 meetings were arranged annually, but the number varied from 3 per year in 1892 and 1893 to 11 in 1914, and in general the number rose after the turn of the century (see Figure 4.2).[24] The speakers chose subjects from their own research, mainly in inorganic and analytical chemistry. The focus on pure chemistry was by no means accidental. This is demonstrated by the solution of a situation that arose in 1882, when a polytechnic graduate asked permission to present a paper on the production and storage of super-phosphates. The executive committee decided, however, that such a technological paper should not be presented in the chemical society but in a technological society (and such did exist in Copenhagen). The board furthermore declared that delivered papers should focus on pure chemistry and technical details should, if included, only play a minor role.[25] The decision was later affirmed with respect to the limited role of analytical chemistry as a suitable topic for discussion. If one takes into consideration the general lack of papers, the firm decision is remarkable. The society – or at least its executive committee or other influential members – clearly wanted to define the society as a learned society with emphasis on pure science. And this self-image was consistently applied, also in relation to the election of lecture topics.

[23] Nielsen (2000), **1**, 95.
[24] Nielsen (2000), **11**, 79–80.
[25] Christensen (1902), 5.
[26] Nielsen (2000), **2**, 79–80.

All gatherings ended with a dinner that was to be paid for independently. The social element was important in itself, and during the First World War the dinner tradition grew stronger. The social dimension was such a crucial element that it makes perfect sense to compare the chemical society with a bourgeoisie club. In both cases, there were specific rules of admission based on a certain educated or cultured background, and the forums provided opportunities to meet men of similar interests, preferably with a fork and knife in hand; a situation very similar to the one in the Swedish Chemical Society.

Despite having no female members, ladies were invited to participate in the summer excursions, arranged from the 1890s. All excursions focused on industrial plants and visits included one to a super-phosphate and sulfuric acid factory in 1888 and another to the Carlsberg Brewery and Laboratory in the winter of 1889–1890. The first woman was admitted in the society in 1898, one year after the first two females graduated from the Polytechnic College.[27] The first woman on the executive committee – Agnes Petersen, married Delbanco – was elected in 1914, and the same year she became the first woman to deliver a paper in the society. There were never many women in the society and they played only minor roles as speakers or members of the executive committee – quite similar to the subordinate roles the female scientists often played in the laboratory.

4.5   National and International Relations

Throughout the nineteenth century, there seems to have been no contact between the chemical society and similar scientific or technological societies domestically or internationally. In the first part of the new century, this situation slowly changed, and international links were created before national ones. During the year 1911, the isolation of the society was importantly diminished in two ways. First, the spirit of internationalism reached Denmark and the society joined the newly formed *International Association of Chemical Societies*. In 1912, that association had thirteen members, including eleven European countries, the United States and Japan.[28] One of the three Danish representatives, Einar Biilmann, came to play a crucial role in the adoption of a democratic management system for the international association; a system that gave all countries – large or small – the same number of votes in the association.[29] The association initiated many projects concerning the establishment of international standards of constants, notation and nomenclature. These initiatives were economically supported by the Belgian industrialist Ernest Solvay, but the work of the committees was terminated with the outbreak of the First World War. The Danish Chemical Society had a hard time after the war,

---

[27] Nielsen (2000), **1**, 174.

[28] In spite of the association's name, the members were not national societies but countries as such. Thus, one country had but one representation in the association no matter the number of chemical societies in the country.

[29] Fennell (1994), 5.

Figure 4.3 Middle pages of banquet menu from the silver jubilee of the Danish Chemical Society in 1904, displaying several of the chemical institutions in Denmark and some of the most prominent chemists at the time. The topmost portrait is that of Sophus Mads Jørgensen, the first president of the society.

positioning itself in the hostile climate between the chemical societies of the countries that had been at war. The society made a big effort to stay neutral (as Denmark had been during the war) and at the same time settle the differences between the former enemies, but this story is outside the scope of this chapter.[30]

The second initiative, taken in 1911, was the establishment of regular contacts to a Swedish chemical society situated in Lund, Scania. This *Kemisk Mineralogiska Föreningen* (Chemical Mineralogical Society) had existed since 1868, and it thus predates both the Danish and the Swedish chemical societies.[31] But it was not until 1911 that Biilmann and the Swede Bror Holmberg started a tradition of regular joint meetings of both professional and social character. The meeting incorporated the Scanian chemists into the Danish chemical community, and the chemists started to visit each other single-handedly too. These Lund–Copenhagen meetings can be seen as indicators of a growing interest in establishing both domestic and international links to sister organizations. This tendency became stronger after the First World War, when it became a habit of the Danish Chemical Society to invite prominent chemists from abroad to lecture in Denmark, and often the invitations were co-arranged with other Danish scientific societies. A pre-runner of this tradition was a visit in 1910 by the German physical chemist Walther Nernst, induced by the chemical society, *Fysisk Forening* (the Danish Physical Society) and a third organization named *Danmarks naturvidenskabelige Samfund* (Denmark's Association for the Natural Sciences).

Nationally, the Danish Chemical Society had no permanent relations to parallel scientific societies, and the society led a very inward-looking existence. Even the Silver Jubilee in 1904 seems merely to have included the members (Figure 4.3).[32] Perhaps the only exception to this secluded practise arose in 1906 when the society became involved in the issue of establishing a consistent Danish chemical nomenclature throughout the school system (primary, secondary and university levels). A committee, consisting primarily of university-level researchers, prepared a suggestion that became accepted by both professor Emil Petersen, who wrote textbooks in chemistry for the university freshmen, and other teachers of chemistry at the university and Polytechnic College *and* by the Danish pharmacopeia committee. The suggestion, however, suffered a severe blow when the Ministry of Education refused to recommend it to be used in the primary schools. The ministry argued that it would be a considerable restraint of the teachers' latitude of teaching to impose one specific nomenclature. The bureaucrats of the Ministry seem to have been ignorant of the vital importance of achieving a universal, or at least nationally consistent, language of chemistry; and though the society became very active in the development of uniform chemical nomenclature on an international scale, it has never been successful in doing the same at home.[33]

---

[30] Nielsen (2000), **1**, chapter 5; Nielsen (1998).
[31] See chapter 14 in this book.
[32] The archival sources do not contain any list of participants, only a menu from the jubilee banquet.
[33] Nielsen (2000), **1**, 185–188.

The society did, however – despite little effort – manage to obtain a certain reputation as a sound and competent society that public institutions could consult. All the same, the society only took on such obligations when approached directly in a specific case. As early as 1907, the society was involved in a matter concerning import duties on goods to be used in Danish chemical industries, but more such cases appeared after the First World War, although they were never great in number. Because the society was never the initiator, it appears that consulting activities were never central to the image of the society. Moreover, only a minority of the members were actively involved in the cases. The acceptance on the part of the society to become involved in consulting businesses may nevertheless have raised the reputation of the society in the eyes of the Danish chemical industry. In these cases, the society – in spite of its declared refutation of applied science – took active part in the solution of some of the problems of the industry.[34]

4.6   Profile, Demarcation and Power Balancing

Above, I mentioned that of the three groups of students (university graduates, polytechnic graduates and pharmacists) the large majority were pharmacists, while the number of university-educated chemists was very small. A parallel pattern applies to the membership group of the Danish Chemical Society. The two largest groups of members were people with an educational background in pharmacy or chemical engineering. As Table 4.1 shows, these two groups taken together contributed more than 70% of the membership. The pharmacists contributed one third of the members and the proportion of polytechnic graduates was slightly larger, varying from 40 to 50% in the period. These large proportions seem extraordinary when one realizes that the Danish Chemical Society cannot have been the first choice of affiliation for the two groups. *Dansk Apotekerforening* (The Danish Pharmaceutical Association) was established in 1844. The pharmaceutical assistants had their own society from 1873. Even with the foundation of *Dansk Ingeniørforning* (the Association of Danish Engineers) in 1892, the proportion of engineers in the Danish Chemical Society did not decrease.[35]

It is, however, worth noticing that although the chemical engineers and the pharmacists constituted the large majority of the members of the Danish Chemical Society and the university-trained chemists were but a minority (no more than 15% and dropping), the society was from its outset devoted to pure, not applied, chemistry. As mentioned above, the executive committee in 1882 pointed out that the Danish Chemical Society was a forum for pure chemistry, where technological themes were being marginalized. The same pattern emerges when one studies the composition of the board of the society. As displayed in

---

[34] Nielscn (2000), **1**, 198.

[35] According to the rules of the Association of Danish Engineers, members had to have a formal training in engineering from either the Polytechnic College or Den kgl. militære Højskole (the Royal Military High School); Harnow (1998), 203.

Table 4.1 Distribution of the membership group of the Danish Chemical Society, 1879–1914 (selected years).[36]

Year	Pharmacists	Polytechnic graduates	University graduates	Medical Graduates	Misc./ Unknown	Total number
1879	32% (19)	41% (24)	7% (4)	5% (3)	15% (9)	59
1901	37% (42)	39% (44)	12% (14)	0	12% (13)	113
1915[a]	31% (47)	52% (79)	10% (15)	1% (1)	7% (10)	152

[a] No membership lists are available between 1901 and 1915.

Table 4.2 Distribution of executive committee members of the Danish Chemical Society, 1879–1914.[38]

	Pharmacists	Polytechnic graduates	University graduates	Total number
Total number	25% (5)	45% (9)	30% (6)	20
Total years in the committee	17% (21)	43% (54)	40% (50)	125
Number of presidents	0	1	5	6
Total years as president	0	3	22	25

Table 4.2, the university-educated chemists held 30% of the chairs in the executive committee in the period. And if one counts the number of man-years of the board, the proportion is even bigger, namely 40%. Five of six presidents in the period were university graduates, one was a polytechnic graduate while, on the other hand, the large group of pharmacists never had one of them elected president.[37] All presidents were professors of chemistry, and this indicates that the select few who studied chemistry at the university aimed for the top. In fact, the main part of the chemical graduates in the nineteenth century later obtained doctorates and professorships. Therefore, that small group contained a disproportionately high number of professors. On the other hand, the groups of pharmacists and polytechnic graduates also included professors who could have been elected president.

I would like to suggest two possible reasons for the dominant position of the very small group of university-trained chemists within the society. The first is that the Danish Chemical Society wanted to gain prestige by what appears to be a deliberate choice of profile. The strong emphasis on basic science in the

---

[36] Based on list of members in the archive of the Danish Chemical Society. The absolute number is given in parentheses. It has not been possible to find information on the educational background for a number of the members. These are thus listed under Misc./unknown, grouped with a very few that had more than one degree or a degree in agriculture.

[37] The six presidents were Sophus Mads Jørgensen (1879–1902 and again 1903–1906), as mentioned, Odin Christensen (1902–1903), Søren Peter Lauritz Sørensen (1906–1907), the polytechnic graduate Julius Petersen (1907–1910), Einar Biilmann (1910–1913), and Christian Winther (1913–1916).

[38] Nielsen (2000), **2**, 83.

education of the pharmacists and engineers left them with either a dislike of such useless knowledge, in relation to their future employment possibilities, or with a sense of being scientists themselves. It is my belief that it was this second group who joined the Danish Chemical Society. In spite of their varying employments in applied and technological chemistry they wanted to sustain the link to academia after graduation.

My second suggestion is that the members could have their interest in applied and technical chemistry covered by other societies and therefore wanted and expected to be kept informed of the developments of pure chemistry by being a member of the Danish Chemical Society. With this approach and as most of the polytechnic graduates and pharmacists were consumers rather than producers of pure chemical knowledge, it should not be surprising to find the university graduates in the lead.[39]

Applied and analytical chemistry were explicitly outside the scope of the society, but the members did not mind because there were other forums in Copenhagen that dealt with these aspects. The Association of Danish Engineers and the Danish Chemical Society, for instance, did not compete on members, but had from the start different strategies and aims. While the society was focusing on the cognitive development of pure chemistry, the engineering association much more aggressively sought to further the public recognition of non-military engineering and the social standing of the polytechnic graduates. The chemical society, on the other hand, chose to live a quiet life and it did very little to manifest itself in the minds of the general public, as mentioned above. Comparing with Lundgren's chapter (Chapter 14), one sees here a close resemblance to the Swedish Chemical Society.

4.7   Conclusion

If a definition of who the Danish chemists were in the nineteenth and early twentieth centuries is to be based on the distribution of the membership groups of the Danish Chemical Society, pharmacist or chemical engineer is a very accurate answer. But if one focuses instead on the lecture programme of the society, one gets the image of a group of people with an exclusive interest in the cognitive aspects of pure chemistry. There was, however, no built-in conflict between the two perspectives, as I have demonstrated.

Throughout the period, the "Danish" of the "Danish Chemical Society" meant "Copenhagen". Although the society was and is a national society, it was not until the establishment of a second university in Denmark in 1928 that the Danish chemical community really expanded outside the vicinity of the capital. Even today, chemists working at the provincial universities feel neglected by the Danish Chemical Society, which still holds almost all of its meetings in Copenhagen. In fact, the Danish name of the society, *Kemisk*

---

[39] See also Nielsen (2007).

Forening, does not indicate any intentions of covering the whole country. The national ambitions only exist in the official English name.

As long as the Danish chemical community essentially worked in Copenhagen, the Danish Chemical Society provided an informal forum for fellow-minded men and, later, women. The meetings gave professional stimulation, and an opportunity to meet old acquaintances and present colleagues. From the 1910s onwards, the society further delivered an international dimension when meetings across The Sound with the west Swedish chemists started and later with a (although somewhat irregular) programme of international speakers.

Discussing, finally, the early history of the society within Thomas Gieryn's terminology, it is clear that the society did not have an outspoken or even clear policy of demarcation in the establishment of boundaries between chemistry and non-chemistry and between chemistry and physics, pharmacy, technology, *etc.*[40] Concerning the boundary between chemistry and non-chemistry, the admission rules were very open and potentially very restrictive at the same time. One needed approval by other members to be admitted, but no groups were explicitly excluded from applying. In practice, one finds no secondary school teachers in the society, and I therefore assume that there was a tacit knowledge in the Danish chemical community of when it was suitable to seek admission. One might notice that other groups of scientists, in particular physicians and physicists, were welcomed, although they only joined the society in small numbers.

Concluding, I would call the hidden agenda of the Danish Chemical Society inclusive, but with some modifications that were known as tacit knowledge to the involved people. I would like to suggest that this is a typical strategy of a small-scaled chemical community, either in a small country or in a country with little chemical industry. It is a way of securing the survival of the society, because more formalized, written strategies of monopolization might have excluded such great numbers of potential members that the remaining group would be too small to survive. I do realize, though, that different, more restrictive strategies were chosen in other small-scaled chemical communities, such as Norway and Finland.[41]

References

Biilmann, Einar (1929), Kemisk Forening og Kemien i Danmark i de sidste 50 Aar, *Kemisk Maanedsblad* **10**, 141–147.

Christensen, Odin T. (1902), *Meddelelser fra Kemisk Forening 1900–1901*, Kemisk Forening, Copenhagen.

Enkvist, Terje (1972), *The History of Chemistry in Finland 1828–1918*, Societas Scientiarum Fennica, Helsinki.

Fennell, Roger (1994), *History of IUPAC 1919–1987*, Blackwell Science, Oxford.

---

[40] For further introduction see the preface of this volume. See also Nielsen (2007).
[41] On the history of the Finnish Chemical Society see Enkvist (1972) and Pyykkö (1991).

Hansen, Svend Aage (1970), *Early Industrialisation in Denmark*, Københavns Universitets Fond til Tilvejebringelse af Læremidler, Copenhagen.

Harnow, Henrik (1993), The Danish Engineer in Transition – the Reformation of Danish Engineering Education c. 1890–1933, in Dan Ch. Christensen, ed., *European Historiography of Technology*, Odense University Press, 164–174.

Harnow, Henrik (1998), *Den danske ingeniørs historie 1850–1920. Danske ingeniørers uddannelse, professionalisering og betydning for den danske moderniseringsproces*, Copenhagen.

Kofod, Helmer, ed. (1967), *Danmarks farmaceutiske Højskole 1892–1967*, Danmarks Farmaceutiske Højskole, Copenhagen.

Kragh, Helge (1998), Out of the Shadow of Medicine: Themes in the Development of Chemistry in Denmark and Norway, in D. Knight and H. Kragh, eds., *The Making of the Chemist. The Social History of Chemistry in Europe, 1789–1914*, Cambridge University Press, 235–263.

Nielsen, Anita Kildebæk (1998), Kemi, kontakter og krig: Internationale kemiske relationer – især efter 1. Verdenskrig, *Dansk Kemi* **79**(12), 18–23.

Nielsen, Anita Kildebæk (2000), *The Chemists. Danish Chemical Communities and Networks, 1900–1940*, Ph.D. thesis, Aarhus University.

Nielsen, Anita Kildebæk (2007), Periodicals and Professionalization. Fashioning a Danish Chemical Community in the Nineteenth Century, *Centaurus* **49**, 199–226.

Petersen, Hans Jørgen Styhr (1996), The Emergence of the Danish Chemical Industry. The Rôle Played by Chemists, in E. Homburg, A. S. Travis and H. G. Schröter, eds., *The Chemical Industry in Europe, 1850–1914. Industrial Growth, Pollution, and Professionalization*, Kluwer Academic Publishers, Dortrecht, 29–43.

Pyykkö, Pekka (1991), Finska Kemisksamfundet 1891–1991, *Kemia-Kemi* **18**, 1001–1004.

Wagner, Michael F. (1993), Danish Polytechnical Education between Handicraft and Science, in Dan Ch. Christensen, ed., *European Historiography of Technology*, Odense University Press, 146–163.

Winther, Chr. (1962), En kemikers Erindringer, *Dansk Kemi* **43**, 102–107.

CHAPTER 5

FRANCE: The Chemical Society of France in Its Formative Years, 1857–1914: Disciplinary Identity and the Struggle for Unity

ULRIKE FELL AND ALAN ROCKE

The words "*Société chimique*" were first used in June 1857, when two young Italians and one young Frenchman, all studying for the *licence* degree, joined together in Paris for mutual assistance, and applied this name to their new club. Within 18 months, this tiny, private self-study club of mostly foreign chemistry students in Paris was transformed into France's national chemical society, parallel to the Chemical Society of London (founded 17 years earlier). And within a few years, the *Société chimique* formed a principal means of advancing the interests of chemists in France.[1] This chapter examines the mixed successes achieved during the first 60 years of this important European scientific society.[2]

5.1 From Self-study Club to Scientific Society

The principal initiator and first president of what was at first more of a club rather than a scientific society was Giacomo Arnaudon, from Torino, at that time serving as préparateur to Michel Eugène Chevreul at the Gobelins

---

[1] *Bulletin de la Société Chimique* (1858–1862); Paquot (1950); Jacques (1953); Gautier (1957); Jacques and Bykov (1959); Fox (1980), 269–72; Carneiro (1992), 88–90, 230–39; Fell (1998); Fell (2000).
[2] The authors are grateful to Dr. Marika Blondel-Mégrelis, Anita Kildebæk Nielsen and Soňa Štrbáňová for helpful comments on this chapter.

Creating Networks in Chemistry: The Founding and Early History of Chemical Societies in Europe
Edited by Anita Kildebæk Nielsen and Soňa Štrbáňová
© The Royal Society of Chemistry 2008

dyeworks;[3] the other two people involved were E. Collinet, then working in Jean-Baptiste Dumas's laboratory at the Sorbonne, and Giuseppe Ubaldini, from Faenza, a student of Antoine Jérôme Balard and Marcellin Berthelot at the Collège de France. They agreed to meet every Tuesday evening at a café in the Cour du Commerce in the Quartier Latin. At their fifth weekly meeting, on the last day of June 1857, they voted to call their new club the *Société chimique*.[4] Seven weeks after the founding, a printed handbill listed a dozen members in the club, a set of statutes and the necessary official authorization from the imperial government of Napoléon III.[5] This was a social as well as a scientific club; at their meetings, "many a scientific discussion dissolved in cigarette smoke and beer foam."[6]

Three among the dozen members happened to be older and more experienced than the obscure students who formed most of the group: a Norwegian agricultural chemist named Anton Rosing, a Russian chemistry teacher named Leon Nikolaevich Shishkov and an Austrian organic chemist named Adolf Lieben. Rosing and Shishkov were taking study tours in Paris, both working in Dumas's laboratory, while Lieben was doing a postdoctoral project in Adolphe Wurtz's laboratory. "Together with [Rosing and Shishkov]," Lieben later reminisced, "I joined a club, shortly after its founding, that consisted of young French chemists, mostly students, who met regularly together, seeking to further their education by presenting reports on the progress of chemistry, and engaging in theoretical discussions. The leadership of the society soon fell to us [three], who were older and relatively more advanced ..."[7]

It was Lieben who established the first connection between this group and Wurtz's laboratory, a connection that proved critical for the future of the new club. Wurtz had established his private (officially unauthorized) teaching/ research laboratory in 1854, soon after his appointment as professor of chemistry in the *Faculté de médecine*. By the summer of 1857 he probably had around a dozen workers in his lab – especially foreigners, including (in addition to Lieben) the Irishman Maxwell Simpson and the Scot Archibald Couper. Before the end of that year these men were joined by the Russian Aleksandr Mikhailovich Butlerov and the Italian Agostino Frapolli. Most of these foreigners were postdoctoral workers on educational travel tours.

During the winter of 1857–1858 and the ensuing spring, the four just-named members of Wurtz's group joined Lieben in the *Société chimique*, as did Wurtz himself on 29th May.[8] The society thereafter took on the appearance of being

---

[3] Jacques and Bykov (1959), 1205.

[4] Gautier (1957), 7; Paquot (1950), 59 (citing MS procès-verbaux currently held by the Société Française de Chimie).

[5] Paquot (1950), 1–3 (reproduction of printed handbill with statutes and names of members as of 18th August 1857, presumably required as a condition of government authorization). To become a member, one had to be nominated by an existing member, and approved by a majority of current members by secret ballot. Dues were two francs per month.

[6] Lindet (1898), i.

[7] Lieben (1906), 3; Zeisel (1916).

[8] Frapolli on 1st December 1857, Butlerov on 22nd December, Couper and Simpson on 5th January 1858 and Wurtz on 29th May. Paquot (1950), 59–62; Jacques (1953); Jacques and Bykov (1959).

an extension of Wurtz's group. But even before this turn of affairs was consummated, the three older foreigners in the *Société chimique* had already shaken things up.[9] At the end of December, Rosing was elected president, and he would remain in that office until the end of May 1858, shortly before his departure from Paris. Then, on 10th March, Rosing proposed to broaden both the goals of the society and its membership.[10] At the next meeting a week later, two mature Parisian chemists with modest reputations, Aimé Girard and Alfred Riche, were presented and elected members of the society. The society decided to sponsor a small chemical library, and the theoretical discussions increased in intensity. That these events were all part of a concerted plan was confirmed many years later in Lieben's reminiscences, from which we quoted above.[11] It is hard to see all this as anything other than how Lieben described it: a parliamentary takeover of a French student club by a foreign clique, in order to create a permanent French scientific society, followed almost immediately by the deliberate hand-off of the society by the foreigners to carefully selected Frenchmen.

On 29th May 1858, Wurtz himself was elected a member, and the meeting place was changed to Wurtz's own workplace. From this point on, Wurtz was deeply involved in the affairs of the society. His student and subsequent president of the society, Armand Gautier, later recalled, "His vivacity, his cheerfulness, his natural cordiality quickly won him friends of all who approached him. In this fashion Wurtz became the soul and the true founder of the *Société chimique*."[12]

Coincidentally, Rosing, Shishkov, Lieben and Butlerov all left Paris virtually simultaneously, in May and June 1858. Couper lasted only a couple more months in Paris. The turnover was remarkable; whereas before this time one reads nearly all foreign names in the society's doings, after this time the names were nearly all French. On 24th July, the Wurtz student and society member Philippe de Clermont wrote to Butlerov about the society's news, then summarized by saying that "everything has changed, and it's only we Parisians now who are nailed to the spot and continue to keep alive the memory of those who honored our city with their presence."[13]

The transition is also marked by the now steady influx of Parisian chemists of reputation. Wurtz, Charles Friedel, Marcellin Berthelot, Charles Barreswil, Jean Barral and François Cloëz, among others, all became members in May and June; in all, the first ten months of 1858 saw an increase in membership from 14 to 63. Girard was elected president for the last seven months of 1858. Simultaneously, Wurtz consolidated plans for the publication by the society of a *Répertoire de chimie pure* and a parallel *Répertoire de chimie appliquée*, the former edited by Wurtz and the latter by Charles Barreswil. These publications were placed under the explicit auspices of the society, but without drawing any

---

[9] The following events are all described in the MS procès-verbaux, in Paquot (1950), 60–61.
[10] *Ibid.*, 61–62.
[11] Lieben (1906), 3–4.
[12] Gautier (1957), 17.
[13] Cited in Jacques and Bykov (1959), 1207.

financial support from it. A *Bulletin de la Société chimique* was also planned, but initially it was simply intended as printed minutes.[14]

The initial idea for the twin *Répertoires* had apparently come from Rosing and Lieben, and one of their last tasks in Paris was negotiating with publishers.[15] Since the main purpose was to publish interpretive abstracts of foreign papers in French (although re-publications of French papers were included as well), Wurtz looked abroad for collaborators – of whom he had many possibilities, even if he were to have chosen only from among former members of the society. Wurtz's initial foreign correspondents included Rosing, Lieben, Shishkov, Frapolli, August Kekulé and Alexander Williamson. The first monthly issues of the twin *Répertoires* were published in October 1858. Wurtz's *avant-propos* stated that his new journal "will offer the reader a compendium, succinct and as complete as possible, of discoveries that every day enrich the science, and which have been published elsewhere. Critical comments will not be excluded here; need I add that they will always be measured and benevolent?"[16]

These two journals were successful; Wurtz reported to Butlerov in July 1859 that he had 500 subscribers – a substantial number for this type of periodical.[17] But apparently there was internal dissension in the society – perhaps sparked by Wurtz's "measured and benevolent" critiques, or perhaps by the tension between atomist and anti-atomist camps – and both versions of the *Répertoire* ceased publication by 1864. The *Bulletin de la Société chimique* began publication almost as an afterthought in the fall of 1858 (Figure 5.1), then absorbed the functions of the two *Répertoires*, and continued as the principal journal of the society. A society tradition was also created for longer invited lectures, to be subsequently printed under the series imprint *Leçons de chimie professées en [1860, etc.] à la Société chimique de Paris*. These special lectures began in January 1860, and continued in this form for ten years. On 10th November 1858, society meetings were changed from weekly to biweekly. This was also about the time that the *Bulletin* began its transformation from a mere record of minutes to a true society "transactions" – an organ for the original publication of scientific papers. Wurtz was editor behind the scenes, and there can be no doubt that he was responsible for these moves.

Meanwhile, chemists of reputation continued to flock to the society; before the end of December 1858, Louis Pasteur, Henri Sainte-Claire Deville, Friedrich Beilstein, Auguste Cahours and Louis Troost all became members, and the total membership reached 101. At the last meeting of 1858, Girard announced that some members of the society wished to expand its goals and activities considerably; others, such as Arnaudon, preferred to maintain the limited functions consistent with the origins of the group. Put to a vote, 36 of the 52 members present favoured allowing the society to "expand the circle of its

---

[14] Paquot (1950), 4, 62; *Bulletin de la Société chimique* (1858–62), 1. Gautier (1957), 7, states that the membership was 24 at the end of 1857.

[15] In a 12th July 1858 letter to Butlerov, Rosing stated that it was his idea; he added that he "had made Girard my successor". Jacques and Bykov (1959), 1208; Storer (1868); Lieben (1906).

[16] Wurtz (1858–1859).

[17] Bykov and Jacques (1960), 117.

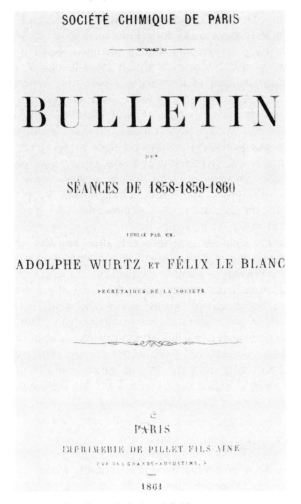

SOCIÉTÉ CHIMIQUE DE PARIS

BULLETIN

DES

SÉANCES DE 1858-1859-1860

PUBLIÉ PAR MM.

ADOLPHE WURTZ ET FÉLIX LE BLANC

SECRÉTAIRES DE LA SOCIÉTÉ

PARIS

IMPRIMERIE DE PILLET FILS AINÉ

RUE DES GRANDS-AUGUSTINS, 5

1861

Figure 5.1   Title page of *Bulletin de la Société Chimique*.

activities." But the new group needed a strong, prominent president to really make a splash. Girard then revealed that Jean-Baptiste Dumas had agreed in advance to accept the presidency of the society if the previous motion succeeded, and all present approved this second motion by acclamation. Pasteur and Cahours were named vice-presidents, and Wurtz took the crucial role of secretary, in charge of all publications. At a stroke, the quondam student self-study club had become an organization with pretensions to be the national society for all of French chemistry.[18]

New versions of the society's statutes and by-laws were approved in summer 1859. The name was henceforth *Société chimique **de Paris***, and the maximum membership was set (by the imperial government) at 500. A prospective new

---

[18] *Bulletin de la Société chimique* (1858–1862), 6–7.

member only needed to be proposed by two existing members; he paid 10 francs upon entry, and 36 francs annual dues. Foreigners were welcome, all sorts of chemistry were to be represented and no distinctions were made between members. As many as 20 honorary members could be named.[19]

A prospectus for the new society was printed in the autumn of 1858; it is undated and unsigned, but the text was almost certainly written by Wurtz. The author summarized the early months of the club's history, then noted that "some foreign chemists saw in the new society a means of establishing a bond between French and foreign chemistry; they brought their cooperation to bear for this purpose, and soon, thanks to their energy and activity, the society entered into a path of progress and growth from which it has not since strayed, and on which one must hope it will continue to tread. ... Little by little, the spirit of the society changed: originally simply a means of pursuing instruction, it rose by degrees to a more important rank. As it is constituted today, its goal is to unite into a scientific association chemists not only in France, but also abroad. ... [The society] invites participation from all chemists, from whatever part of the science they have come; it rejoices to admit to membership the least of us with as much alacrity as the greatest. It knows but one school, that of progress."[20]

This ringing endorsement of openness, democratic ideals, and unfettered fellowship in science was without doubt wholeheartedly meant. And indeed, membership was now open to all, at least in principle: an applicant only needed to be presented by two *sociétaires* and to pay the annual fee to be admitted. Still, Wurtz had a definite viewpoint that he hoped would be promoted by the activities of the society. In Section 5.4 we will further explore the role that this agenda played in the early history of the society.

5.2   The Members

Once the former study-group had become a veritable *société savante*, membership grew rapidly, to 238 in 1862 and to 285 in 1870. However, around the mid 1860s the society's growth levelled off, probably due both to adverse economic conditions and to demographic factors. A decade later membership began to grow again, slowly. The election of Marcellin Berthelot to presidency (in 1875 and again in 1882) and the wide-ranging dynamics of the academic reform movement, set in motion in the late 1870s after the republicans came to power in France,[21] may have had an influence in this new phase. The numbers suggest a critical impulse in the further development of the society: by the turn of the century the membership of the *Société chimique* had passed the 1000 mark, and by 1914 it was more than 1100.

But in comparative perspective, the growth of the *Société chimique* was not impressive. Its German and British sister societies, the *Deutsche Chemische Gesellschaft* and the Chemical Society of London, each had more than 3100

---

[19] *Ibid.*, 67–76 (meeting of 22nd July 1859).
[20] Jacques and Bykov (1959), 1206; Paquot (1950), 4.
[21] Paul (1985); Nye (1986).

members in 1910. Clearly, demography played a role: whereas in 1914 there were 68 million Germans and 46 million British, France's population was only 40 million.[22] Moreover, the *Société chimique* had to compete with numerous local societies that experienced a period of rapid growth after the French disaster of 1870–1871. The most serious competition derived from the so-called industrial societies, which usually offered a wide spectrum of activities, adapted to the specific needs of local chemists and industrialists. Modelled on the prestigious *Société industrielle de Mulhouse* – founded in 1826, but officially under German control after the annexation of Alsace-Lorraine – the *Société industrielle de Rouen* was established in 1872, in order to bring together chemists and manufacturers from the local textile industries.[23] During the 1870s, seven such industrial societies were created in France, together with another five pharmaceutical societies, six medical societies and seventeen associations of general scientific interest. Altogether, close to 150 new local learned societies came to life during this decade alone, followed by an additional 123 new institutions in the 1880s.[24]

Also slowing the growth of specialist societies such as the *Société chimique* was the enduring authority of elite traditional institutions such as the Parisian *Académie des sciences*, for membership in specialist societies did not come with the kind of prestige and recognition which the powerful academies were able to confer.[25] Another important institution that may possibly have been responsible for syphoning off potential clientele was the *Société de pharmacie de Paris*, founded in 1803.[26] As in other European countries, French chemistry was traditionally closely linked to the pharmaceutical and medical professions, and many chemists had originally been trained as pharmacists. Accordingly, they had plenty of opportunity to meet colleagues within well-established pharmaceutical institutions, and so an extra stimulus may have been needed to support the new *Société chimique*.

The *Société chimique* recruited its members from a wide array of occupational fields, although the academic sector was by far the strongest. In 1870, almost 40% of members were working in academia, while only 16% worked in industry. For 2.5% of the members the only occupational specification stated was "ingénieur"; probably most of them also worked in industry. Medicine and pharmacy accounted for around 13% of the members, and an additional 5% worked in other sectors. We were able to find no biographical information for nearly 26% of the members. In fact, the 1870 membership list did not include any occupational information; these data had to be obtained piecemeal from a wide range of biographical sources.[27] These kinds of sources tend to derive

---

[22] Nye (1986), 237.

[23] Lecouteux (1994).

[24] Fox (1980), 252.

[25] Crosland (1992), 1–10, 68–72.

[26] Bouvet (1936).

[27] The main sources were the *Index biographique français*, Poggendorff, *Dictionary of Scientific Biography* and *Dictionnaire de biographie française*. Many data were extracted from Letté (1993). Additional information was taken from specific sources, for instance publications of other scientific or professional associations or obituaries published in scientific journals. Valuable information on members of the *Société chimique* was also provided by Ernst Homburg.

from an academically controlled written culture, and they focus on academic elites. Although these tools are very useful in prosopographical research, they often fail to identify the more peripheral members of the community.

The membership list for 1914 provided more information than that for 1870, for the 1914 list includes data on the respective occupations for 644 of the 880 members resident in France. Two caveats should be made on the interpretive value of these data. First, there is inevitably some overlap between the different categories of occupations. For instance, a member listed simply as "pharmacien" would be classified in pharmacy and medicine even though he might actually be working in teaching or research. The same arbitrariness weakens the use of the term "ingénieur", as it fails to identify the actual sector of activity. A second caveat concerns the limited comparability of the two populations (1870 and 1914), due to the heterogeneity of the sources. Since information was obtained by means of the membership list in one case and secondary sources in the other, a straight numerical-quantitative comparison of the two populations would be misleading in some respects.

However, some qualitative observations may be made. In 1914, academia still accounted for the largest fraction of the membership, around 26%, although this sector was not as dominant as in earlier years. The institutional changes that French science had undergone in the preceding years – including massive expansion of the faculty system, decentralization and promotion of applied sciences – are reflected in the society's occupational structure. Thus, even though the pre-eminent Parisians' institutions were still important, more members came from the faculties and specialized colleges in the provinces. The rise of the chemical engineering profession is also significant. Whereas in 1870 engineers formed an insignificant part of the society's affiliates, by 1914 they totalled at least 8.5% of the members. The membership list gives no information about the actual work place of these so-called *ingénieurs*, but we can confidently assume that most of them were working in the industrial sector. This has to be taken into consideration when evaluating the proportion of members that were categorized as industrial chemists, manufacturers and businessman. This group formally amounts to 18% of the total membership in 1914, but one can safely assume that the true percentage is higher than that. About 15% of the members worked in the medical or pharmaceutical sector, and another 5.5% worked in other areas. The list does not give any information on the field of occupation for the remaining 27% of the members.

Future research should expand the scope of this first analysis to a comprehensive prosopography. By retrieving data on the education and the full careers of the members, as well as on their social backgrounds and political beliefs, much could be learned about the social constituency of the *Société chimique* and the general history of the French chemical profession. Ideally, the investigation should be extended also to those chemists who were not members of the society, since one cannot discern the true standing of the *Société chimique* without knowing the total number of chemists in France, as well as their educational backgrounds, their occupations and their geographical distribution.

5.3   Geographical Demarcations

Born out of purely local initiative, the *Société chimique* was soon able to attract many non-Parisians (including a large number of foreigners). The society's boundary-work consistently aimed at expanding its area of control beyond the metropolitan region and gaining authority over the entire French chemical landscape. Simultaneously, in order to reinforce its status as a national representative body, the society promoted its agenda on the international scene.

5.3.1   Becoming a National Society

Soon after its founding, chemists from the French provinces increasingly joined the society. The share of the so-called "*membres non-résidents*" (which included both provincial members and foreigners) grew from one third in 1861, to about half by 1870, to about 60% in 1914. In fact, the society's expansion in the 1880s and 1890s was in large part due to the influx of provincial chemists. However, the weakness of the non-resident contingent was evident when it came to decision-making and distribution of power. Until the early twentieth century the statutes cemented an openly centralized organizational structure, control of the society unquestionably remaining in the hands of the Parisian elite. Academic chemists from the capital traditionally dominated the executive council of the society. Around the turn of the century, of 30 council members no more than 4 came from the provinces, and not one of the 30 presidents of the society prior to its 50th anniversary in 1907 was resident outside of Paris.

This pattern began to change in 1905, when Maurice Hanriot (president of the society in 1899 and 1912) proposed to raise the proportion of non-residents in the council, only stipulating that their number should not surpass the Parisian contingent.[28] The same year, new statutes were approved, which officially raised the number of non-resident council members to 15.[29] Only in 1910 did the society abolish the policy of dividing its membership into two categories, *membres résidents* and *membres non-résidents*, a policy that had played a large role in maintaining the centralized structure of the society.

Decentralization was also fostered by the creation of local branches in the provinces. As early as 1895 the Alsatian chemist Albin Haller had proposed to open a local section in Nancy.[30] In 1898 the chemistry division of the *Société d'agriculture, sciences et industrie* in Lyons requested the creation of a branch of the *Société chimique*.[31] Other sections were created in Toulouse (1902), Lille (1902), Montpellier (1905) and Marseille (1905). In general, these branches were formed by already existing informal associations, study groups or disciplinary sections of unspecialized learned societies.

If boundary-work generally involves the crossing, deconstructing and reconstructing of boundaries between one discipline and another, the same holds true

---

[28] Procès-verbaux des séances du Conseil, 27.2.1905, A. SFC. See also Paquot (1950).
[29] Paquot (1950), 6–7.
[30] Procès-verbaux des séances du Conseil, 26.11.1895, A. SFC.
[31] Procès-verbaux des séances du Conseil, 21.11.1898, A. SFC.

for the geographical boundaries that existed within the French chemical profession. The goals of the governing *bureau* were clear-cut: to attract more members and thus increase its influence, both in geographical and in episte-mological terms. The opening of local sections represented an attempt of the *Société chimique* to dismantle original boundaries and reassemble them in a different form, that is, within a larger national body. When in 1906 the society, in the face of a tradition of provincial resistance, changed its name to *Société chimique de France*, it was given symbolic importance.[32] Officially, from now on the society stood for the entire French chemical community.[33]

However, in the provinces the headquarters' aspirations of taking control met with resistance, and extensive negotiations were needed to overcome the provincials' tendencies to assert their independence from the capital. In 1910 Haller pointed out that many members of local sections still had not joined the national Parisian organization, which was against the statutes. The strongest resistance was exerted in Lyon, where only 29 of 51 members had joined the national organization; moreover, the vice-president and the secretary were not among them. One year later the Lyons branch stated that its members could not be forced to join the *Société chimique de France*, given that the local section had existed before its merger with the national society. A continuing gulf separated the Parisian leaders from chemists in the provinces.

More broadly, the society's authority as a national organization had been since its very beginnings weakened by the French tradition of proliferation of branches and of dedicated provincial societies of all kinds – literally hundreds of which existed in the late nineteenth century. While during the Second Empire these societies were at the front line in the struggle for provincial autonomy, they would later develop into platforms for intellectual decentralization. Their activism was synchronized with the efforts of the regional faculties to break the monopoly of the traditional academic elite, notably the *grandes écoles*. Studies by Harry Paul, Terry Shinn, George Weisz, Robert Fox and others have revealed how the French faculty system underwent dramatic changes, especially during the 1880s and 1890s.[34] Provincial faculties were expanded and gained a significant degree of financial autonomy. Geographical decentralization was allied with a growing emphasis on technical and applied issues. Institutes for applied sciences were established in provincial cities like Grenoble, Toulouse, Lille and Lyons. Albin Haller, who later was to become one of the leading figures of the *Société chimique*, founded Nancy's *Institut chimique* in 1890 with the backing of local industries.[35] Clearly, these institutional changes in the French academic landscape were beneficial for the society, since the creation of new institutes resulted in a fast-growing number of trained chemists all over the country. On the other side, the parallel awakening of provincial autonomy defied the society's call for national unity.

---

[32] Fell (1998). A final name change, to *Société Française de Chimie*, took place in 1984. The current membership is about 4500.
[33] Gautier (1907).
[34] Paul (1985), Shinn (1979), Fox and Weisz (1980).
[35] Nye (1986), 41.

5.3.2 Becoming an International Player

We have seen that the origins of the *Société chimique* were tightly intercon-
nected with the foreign specialist community in Paris, and the next section of
this chapter will show how the society's endeavours were directed in important
ways toward international scientific movements. Indeed, as the international
arena increasingly turned into a showcase for a nation's triumphs, international
visibility became a key factor in the society's quest for disciplinary authority,
thus trying to secure its domestic role as a national symbol of chemistry. Again,
the concept of boundary-work describes best the ensemble of discursive and
non-discursive practices that served to demarcate the society's area of control.
By accentuating the frontier between national and international, between
outside and inside, the society attempted to secure its status as the national
representative both inside and outside the country. In fact, the *Société chimique*
played a key role in the movement to internationalize chemistry in the late
nineteenth and early twentieth centuries. Adolphe Wurtz was one of two
principals (Kekulé was the other) in the organization of the 1860 Karlsruhe
Congress, the first broadly international chemical congress. While this event
might still be characterized as an informal gathering of individual *savants*, the
1880s and 1890s witnessed the emergence of a more formalized congress
routine, primarily in the area of chemical nomenclature. Much of this activity
was led by the *Société chimique*.[36]

However, under the impetus of economic and technological competition, the
culture of international congresses became increasingly established in the area
of applied chemistry. When the society organized in 1900 a *Congrès international de chimie pure* in conjunction with the Paris Universal Exhibition, this
event attracted but little attention.[37] By contrast, the International Congress of
Applied Chemistry, held in Paris the same year and organized by the *Association des chimistes de sucrerie et de distillerie de France et des Colonies*, attracted
more than 1800 delegates.[38] Clearly, this younger, industrially oriented society
had been able to outdo the *Société chimique*, seriously challenging the latter's
claim of authority. A total of eight International Congresses of Applied
Chemistry were held all over the world between 1894 and the First World
War. Even though members of the *Société chimique* actively participated, the
society played only a marginal institutional role, thus revealing its weak status
in the industrial world. The *Société chimique* tried to regain terrain as an
international player when taking the initiative for the *International Association
of Chemical Societies* in 1911. When Albin Haller, then president of the society,
first proposed the creation of such an organization in 1910 to the members of
the *Conseil*, he most likely had in mind the gain of prestige that could be drawn
from such an endeavour.[39]

---

[36] Verkade (1985); Fell (1998).
[37] Witt (1907).
[38] Moissan and Dupont (1902).
[39] Procès-verbaux des séances du conseil, 25.10.1910, A. SFC.

5.4   Disciplinary Demarcation

At first glance, the goals of the French chemical society did not appear to differ significantly from the objectives of any other chemical society, that is, "the advancement and the pursuit of studies in general and applied chemistry".[40] In order to accomplish these aims (as we have seen), the society organized regular meetings, established a specialized library, organized public lectures and, last but not least, launched its own publications, one of which – the *Bulletin* – evolved from a modest newssheet into a respectable journal of (by 1914) some 2000 pages per year.

One might think that these endeavours were steered purely by scientific criteria, driven by the dignified pursuit of wisdom, and subjected to nothing but the essential norms of objectivity, falsifiability and truth. But there was more to it than that. What can be discerned behind the cover of disinterested scientific activity are strategic attempts of the leaders of the society to demarcate their field of authority and to actively influence the course of chemistry in France, while at times even pursuing a very personal agenda.

In this context the early role of Wurtz in organizing, with the help of the *Société chimique de Paris* and its publications, a deliberate campaign for the second chemical revolution is quite revealing for the society's difficult struggle for epistemic authority. A second example of the society's arduous attempt to expand its area of disciplinary authority was the boundary-work it embarked upon at the intersection between "pure" and "applied" chemistry.

5.4.1   A Campaign for the Second Chemical Revolution

In Wurtz's opinion, the chemical revolution associated with Antoine Lavoisier's name had been followed by a second revolution, one which had been little noticed but was just as momentous for the future of the science – and one which was just as French as the first. This revolution, in Wurtz's telling, had been pioneered by the iconoclastic French chemists Auguste Laurent and Charles Gerhardt. Partly due to their unconventional politics and lack of tact, partly due to legitimate scientific issues, Laurent and Gerhardt made little headway among elite Parisian chemists, and both had died before the age of 40 (Laurent in 1853, Gerhardt in 1856). As a young chemist trying to make his way through a difficult professional world, Wurtz had also been relatively indifferent to the chemical reforms being urged by these two outcasts – until shortly before Gerhardt's death, when Wurtz suddenly became convinced of the advantages of the new system. From that moment until his death in 1884, Wurtz devoted himself to a campaign dedicated to win his collegial community over to the new ideas.[41]

What chemists of the 1850s referred to as Gerhardt's system emphasized the unity between inorganic and organic chemistry, and attempted to make use of

---

[40] See for instance "Statuts" (1865).
[41] Rocke (2001).

all possible methods to determine real atomic weights and molecular formulas. This movement was opposed by those who saw the future of chemistry in a system that was free of hypotheses, and founded on empirically determined equivalent weights. For Gerhardt and the "atomist" camp, the molecule of water was H_2O and the atomic weight of oxygen was 16; for the "equivalentists", water was HO and oxygen was 8 – and so on. Gerhardt's ideas had been received coldly by his compatriots, but during the course of the 1850s his system gained considerable ground among progressive younger chemists abroad, especially in Great Britain and Germany. Although the reform had been born in France, the best route for French chemists to adopt the new chemistry after Gerhardt's death was to pay more attention to foreign research, which was becoming increasingly Gerhardtian. Strongly oriented as it was to the international scene, the *Société chimique* could thus serve as a centrally important tool in promoting the reformed chemistry.

We have already chronicled Wurtz's leadership of the latter stages of the "foreign takeover" and then hand-off of the *Société chimique*. This was obviously consonant with his campaign, both because of the foreign flavour of the society, but also more importantly because of the reformist predilections of its earliest leaders. Rosing and Shishkov were both favourably inclined to the new system of Gerhardt, even though their research director, Dumas, was not. Then Lieben formed the link to Wurtz's laboratory, providing an influx of new foreign *and* reformist society members. Finally, Wurtz entered the society himself, and brought its physical meeting place to his own laboratory.

A further step was Wurtz's taking the position of secretary and editor of the society's journals, and then using that vantage point to good effect in publishing papers supporting reform.[42] Establishing an academic journal meant deciding which articles and authors got published, giving the editors the control over what chemistry was supposed to be and what it was not supposed to be. Under the authority of Wurtz and those influenced by him, the society's publications served as a valuable instrument in the society's attempt to expand its epistemic authority.

Wurtz's foreign correspondents for his journals – men such as Rosing, Lieben, Shishkov, Frapolli, Kekulé and Alexander Williamson – were all in the reformist camp. When Wurtz wrote to his friend Williamson to ask if he would be a foreign editor for his *Répertoire*, he expressed his hope that his journal would gain a better hearing in France for the "new ideas".[43] More substantively, the *Bulletin* began, from 20th November 1858 on, to publish papers by members who used the system of atomic weights and molecular formulas that had been championed by Laurent and Gerhardt – the first such published appearance of these reformed weights in France since the deaths of the two iconoclasts. Ironically, the first published examples of the reformed weights and formulas by Wurtz himself were not seen until the early months of

---

[42] But Wurtz's campaign extended beyond the confines of the *Société*; the Karlsruhe Congress was another element of it. See Rocke (2001).

[43] Wurtz (1858); "J'espère que ce journal pourra contribuer à répandre les idées nouvelles."

1859. It is striking, and highly significant, that Wurtz waited to reveal himself as a Gerhardtian until the *Société chimique* gave him an unobstructed editorial platform on which to do so.[44]

However, Wurtz never tried to prevent his scientific rivals such as Berthelot and Deville from joining the society, and even becoming its president (Berthelot himself occupied the presidency for five annual terms). On the contrary, Wurtz welcomed them in. To have attempted to create a society packed with members of a single camp, excluding others, was clearly quite impossible, and would have been very much contrary to Wurtz's collegial and democratic instincts. Moreover, in order to succeed as a disciplinary society and to keep control over the ontological domain of chemistry, the *Société chimique* had to be able to absorb conflicting theoretical views. In that sense, the society's leaders were wise enough to leave open their publications even to obstinate defenders of equivalents.

But there is no question that in the *Société chimique* Wurtz had helped to create an important new organization with many allies in leadership positions, and an editorial platform from which to broadcast his views. No fewer than nine students of Wurtz became presidents of the *Société chimique* – Philippe de Clermont, Charles Friedel (three times), Auguste Scheurer-Kestner, Edmond Willm, Armand Gautier (three times), Edouard Grimaux (three times), Joseph LeBel, Maurice Hanriot (twice) and Alphonse Combes – and the society's editorial board was generally packed with members of his circle.[45]

Unfortunately, in the long run none of this seemed to help. To quote Gautier once more: "Without closing its doors to those in other camps, the brilliant students of Wurtz treated more coldly those who did not think and write as they; this is why, at the time of its early development and exuberant youth, the *Société chimique* did not suffer a malady of growth and decline." Rather than leave the organization, Gautier noted, many of the opponents of atomism in the society simply continued along their own path while maintaining good personal relations with the Wurtz camp, and the theoretical fragmentation of French chemistry thus continued unabated, until the turn of the century. This fragmentation was highly confusing to students, and no doubt damaged the overall success of French chemistry in these years.[46]

5.4.2 Between "Pure" and "Applied" Chemistry

Traditional historiography stresses the limited interactions between French industry and academia, and suggests that many industrialists came only late to understand the importance of such cooperation. In some respects the society's history confirms this assertion, but it also contradicts it in other ways. Although the *Société chimique* was dominated by the academic wing, the society's leaders did seek support from manufacturers and businessman, whose financial

---

[44] *Bulletin de la Société chimique* (1858–1862), 2–5.
[45] Carneiro (1992), 230–39.
[46] *Ibid.*; Rocke (2001), 235–67, 301–31.

backing was vital in helping the society to prosper. Regular contacts were established with the manufacturers' association *Chambre syndicale des produits chimiques* (today's *Union des industries chimiques*). On the occasion of joint celebrations, representatives of both the industrial and the academic side effusively declared their amiability and good will, advertising "the union between laboratory chemists and factory chemists".[47]

While seeking contact with the industrial world, the chemists were nonetheless careful to draw a demarcation line between *pure* and *applied*. The boundary-work here was subtle and complex. On the one hand, the chemists asserted that university-based science yields "basic" rather than "applied" knowledge; on the other, they emphasized that university-based science is essential for technological and national progress, thus they needed the (financial) support from the industrial world. The two assertions were not necessarily contradictory. The society was engaging here in a kind of strategic boundary-work to integrate the putatively "inferior" applied sector under the authority of the "superior" pure chemistry, in order to expand the latter's authority. The society's awards provide an impression of the symbolic value that was tacitly attributed to "pure" *versus* industrial and "applied" chemistry. For instance, in 1914 out of 13 honorary members only one – Ernest Solvay – was an industrialist. All of the others were university professors. And when in 1883 Charles Lauth was elected president of the society, he was the first to come from the manufacturing world. During the society's first half-century, only three industrialists were elected president – Joseph LeBel, Auguste Scheurer-Kestner and Lauth.[48]

Systematic lobbying by the society's leaders for support of industrialists began in the 1880s, when the *Société chimique* started an ambitious campaign aimed at increasing its influence and authority. Money was needed: the library had to be updated and the *Bulletin* improved, and as a true learned society the *Société* ought to have resources to grant research awards. The strategy aimed at sufficiently "scientizing" industrial chemistry in order to attract donor support. Boundary-work not only involved the rhetorical distinction between "pure" and "applied" (gain of prestige), but also an expansion of the boundaries of what was considered scientific enough to merit the attention of a venerable learned society. The society's demarcation strategies were thus not limited to constructing boundaries as means of exclusion, but also encompassed boundaries as means of communication, as interfaces rather than barriers. Here, the society relied on the initiative of mediators between the "scientific" and "non-scientific" worlds. In general these were chemists who had a foot in both camps, such as the Alsatian Charles Lauth, who started his career in industry, but never lost sight of the academic world and later became formally linked to it as the director of the Parisian Municipal School for Industrial Physics and Chemistry (*École municipale de physique et de chimie industrielles*). As a member

[47] Grimaux (1890).
[48] Gautier (1957).

of the society's council, Lauth continuously called for intensifying relations with industrial chemists, in order to increase the influence of the society.[49]

In 1880 the society's leaders set out to find sponsors among industrialists and businessmen. A circular letter was addressed to potential benefactors, appealing to the patriotic feelings of the recipients. France's national chemical society, the authors pointed out, was lagging behind its German and British sister organizations both in membership and in funding; thus, the *Société* was not able to contribute to the progress of French science and industry to the same extent, or to accomplish "what each of the rival societies are able to achieve in their countries." The letter, signed by the society's president Charles Friedel, by its honorary president Jean-Baptiste Dumas and by Wurtz (among others), called attention to the danger "of seeing, as a result of more vigorous endeavours abroad, what has been something like our national patrimony having been taken away from us and turned against us in the peaceful battles of civilization".[50] The initiative clearly carried the imprint of Wurtz, who throughout his career had always proven to be particularly alert when it came to international competition in the industrial applications of chemistry. And it is not surprising that the society initiated this campaign under the leadership of Friedel, one of Wurtz's former students.

A new membership category was created for sponsors, referred to as *membres donateurs*, whose names were published annually at the top of the membership list. A second subscription was organized in 1894 under the presidency of the industrialist Scheurer-Kestner (likewise a former Wurtz student), when the society almost faced bankruptcy. During the 1880s and 1890s the *Société* was able to collect a total of around 180 000 francs.[51]

In order to expand its influence and consequently the boundaries of what was to be considered as part of formal chemical science, the society started systematically to promote what one could call the scientization of industrial chemistry. In 1888, possibly as a preparative measure for the upcoming Paris international exposition, the *Société chimique* set up a specialized section for industrial chemistry, and organized separate sessions for chemists interested in industrial topics. Another attempt to extend the society's influence into industrial chemistry was made in the 1890s, when the council formed plans for a *grande école* for industrial chemistry (discussed below, in Section 5.5). While this initiative had some positive effects, the introduction of the industrial section did not meet with real success. The specialized sessions attracted disappointingly few participants,[52] and industrial chemists repeatedly expressed their discontent about their interests not being adequately represented by the *Société chimique*. In 1893 the council met to consider possible measures to discourage industrial members from leaving the society. It was agreed to

---

[49] Lauth, procès-verbaux des séances du Conseil, 17.1.1885. A. SFC.

[50] Circular by the *Société chimique*, not dated, *ca.* 1880, when Charles Friedel was president, signed by Dumas, Friedel, Berthelot, Cloëz, Deville, Gautier, Girard, Pasteur, Wurtz and a few others. A. SFC.

[51] Fox (1980), 272.

[52] Procès-verbaux des séances du Conseil, 6.2.1891, A. SFC.

expand the industrial section of the *Bulletin* and to give more weight to industrial members within the *Bureau* and the *Conseil*.[53] But the participation of industrial chemists continued to decline, and in 1896 the council decided to discontinue the specialized meetings.[54]

As to the reasons for this setback, one has to keep in mind that boundaries – such as that between "pure" and "applied" – are dynamic, and subject to continual negotiation. The boundaries had to be defended against competing interests outside and inside the terrain of control. Within the academic wing of the *Société chimique* there was a considerable degree of resistance to the industrial endeavours of the society. Thus, Léon Lindet, who was otherwise known as being sympathetic to questions of applied science, declared that in industrial chemistry there was barely anything interesting worth publishing, and that relevant information could as well be exchanged by personal contact.[55] Grégoire Wyrouboff, professor at the *Collège de France*, averred that trying to realize "the union between science and industry" within a single society was not beneficial for either of the two factions. He even suggested that it would be advantageous for the society to be financially independent of industrial funding.[56]

Within the industrial wing, dissatisfaction grew out of the inability of the society to truly represent its interests; the *Société chimique* continued to represent an academic approach to chemistry, as it was still dominated by the Parisian academic elite. Furthermore, the confines of industrial chemistry were heavily contested by other organizations, such as local industrial societies or the venerable *Société d'encouragement pour l'industrie nationale*, while the specific professional interests of industrial chemists were to some degree represented by the alumni associations of their respective engineering schools or colleges. This multitude of competing entities in the field of industrial chemistry, rather than the alleged indifference to industrial matters, may also have delayed in France the creation of an autonomous national society for industrial chemistry, comparable to the organizations that emerged in Great Britain and Germany in the second half of the nineteenth century. An equivalent *Société de chimie industrielle* eventually emerged, but only in 1917.[57]

5.5   Professional Demarcation

Authority over educational standards and control of entry to professional positions, that is, autonomy with respect to the state, are commonly seen as characteristic features of a "profession." However, this traditional concept of professionalization, largely cultivated by Anglo-American macro-sociological and historical research, cannot be easily applied to the uniqueness of the French

[53] Procès-verbaux des séances du Conseil, 1.2.1893, A. SFC.
[54] Procès-verbaux des séances du Conseil, 26.11.1896, A. SFC.
[55] Procès-verbaux des séances du Conseil, 1.2.1893, A. SFC.
[56] Procès-verbaux des séances du Conseil, 1.2.1893, A. SFC.
[57] Fell (2000).

educational system. In France, the evolution of professional structures was by no means conditioned by gaining autonomy from governmental authorities. On the contrary, the French state played an important role in the professionalization of scientific disciplines. Professional groups needed legitimization through the state, embodied by the *Ministère de l'Instruction publique* or the *Ministère du Commerce*.[58] Furthermore, French chemical education was characterized – and still is today – by a multitude of different qualifying institutions regulating access to professions, with no unified standards for professional accreditation; the large range of *universités*, *grandes écoles* and research institutions have led to an equally large range of professional traditions and identities, educational profiles and occupational patterns.

Thus, when one speaks of the French chemistry profession one has to take into account all of its variety. So the *Société chimique* rarely acted as a genuine professional organization. The society was ineffective (indeed, it never really systematically *sought* to be effective) in representing and promoting the profession with the French government, or even more broadly in providing an effective group identity for the French chemical profession.

Nevertheless, the society occasionally did make timid attempts to defend the professional interests of its members. For example, in 1891 the council discussed a project presented by a representative of the *Chambre syndicale des produits chimique* to found a new *École de chimie pratique et industrielle*. The aim was to provide chemists with special knowledge related to industrial and technical matters.[59] A commission charged with the further elaboration of the plan was formed, and in the same year a booklet appeared (most probably written by members of the commission), publicly requesting the creation of a *grande école de chimie*.[60] Although the proposal had no immediate repercussions, it seems to have had some influence in promoting the introduction of a fourth curriculum year within the already existing *École municipale de physique et de chimie industrielles*, as well as the establishment of a laboratory for applied chemistry at the Paris Faculty of Sciences.

While the council did discuss educational measures on the local scale, the society's professional lobbying was insignificant on the national level. The society did not take measures to harmonize the miscellany of educational standards, or to establish uniform licensing examinations for chemists. In fact, not all of the members of the council considered educational matters to be any part of the society's responsibilities. Wyrouboff, a chemist and philosopher of Russian descent, suggested that since the *Société chimique* was a "société savante" and not a "société pédagodique", "it should deal with questions of high science only".[61]

This learned society did little to raise the social situation of its professionals. In pre-war France, professional interests of chemists and chemical engineers were generally dealt with through the alumni organizations of the different colleges,

---

[58] Geison (1984).
[59] Procès-verbaux des séances du Conseil, 20.6.1891, A. SFC.
[60] Anonymous (1891).
[61] Procès-verbaux des séances du Conseil, 7.12.1891, A. SFC.

grandes écoles or universities. The internal sense of solidarity of these groups, their *esprit de corps*, was too potent for the *Société chimique* to be able to compete as a national professional lobby group. Unlike in Germany or Great Britain, where professional groups such as the *Verein Deutscher Chemiker* or the *Institute of Chemistry* acquired an important membership and powerful positions, there was no need or space for a French equivalent before the First World War.

5.6 Conclusion

As we have seen, in spite of the *Société chimique*'s efforts to become a representative body for French chemistry, the society was by no means recognized as such, as it was unwilling or unable to represent the social and professional interests of its members. The drawing and policing of boundaries between who could and could not access the chemical profession was not in the hands of the venerable learned society. The *Société* never played within the French chemical community a role comparable to its German or British sister organizations. The society's most fundamental function was to provide disciplinary channels of specialized communication, *via* the publication of the *Bulletin* and the organization of lectures and congresses. What the society *was* able to give the emerging chemical community was an organizational structure, encouragement of interactions between practitioners and sharing of expertise, and embodiment of a certain commonality of values.

As a disciplinary society the *Société chimique* had to do boundary-work: demarcate its field of expertise, erecting rhetorical boundaries between what chemistry was supposed to be and what it was not supposed to be.[62] The society's disciplinary boundary-work included both monopolizing and expansion of authority. Especially under Wurtz's charismatic leadership, the society tried to take epistemic control in order to advocate certain scientific ideas, as the example of his arduous campaign for atomism shows. The society furthermore pursued both disciplinary and geographical expansion, while trying to suppress competing authority claims made by industrial chemists and the provincial sections, among others.

To be sure, the strategic attempts of the *Société chimique* to demarcate its field of authority cannot be explained solely by features that set science apart from other cultural practices – such as Popper's falsifiability, Kuhn's paradigmatic consensus or Merton's social norms of science. What seemed to constitute "chemistry" as represented by the society was not so much embodied by obvious "essential" characteristics of science such as methods, rules, tools or language; rather, what defined "chemistry" *versus* "non-chemistry" was an array of contingent circumstances and strategic behaviour, driven by social forces, personal interests and local contexts, both internal and external to the scientific enterprise. However, it would be misleading to think of the society's chemists as always following a calculated strategy of demarcation. Instead, in

---

[62] Gieryn (1983); Gieryn (1999), 4–5.

practice, their boundary-work was often done unreflectively, as a result of everyday routines and negotiations.

If the success of the chemical discipline depended on its ability to diversify and differentiate, the success of the *Société chimique* strongly depended on its ability to represent all various occupational and scientific areas in a unifying "ideology" of chemistry. The society had to be able to articulate differences and to assimilate them under a shield of unifying rhetoric. In fact, the society drew its authority from the elision of antagonisms such as "pure" *versus* "applied", "national" *versus* "international", "Paris" and "province", or "atomists" *versus* "anti-atomists". Through assimilating these opposing poles within a hierarchical structure of values and authorities, the society was able to construct, under the lead of its academic wing, a kind of collective identity among French chemists. However, the society's authority was not strong enough effectively to counter the theoretical and professional fragmentation of French chemistry.

References

A. SFC=Archives de la Société française de chimie, Paris.

Anonymous (1891), De la nécessité de la création d'une grande école de chimie pratique et industrielle sous le patronage de la société chimique de Paris, Paris 1891, cited in J.-C. Guédon, Conceptual and institutional obstacles to the emergence of unit operations, in W. F. Furter, ed., *History of Chemical Engineering*, ACS, Washington, 1980, 45–75.

Bouvet, M. (1936), *Histoire de la pharmacie en France des origines à nos jours*, Occitania, Paris.

Bulletin de la Société Chimique (1858–62).

Bykov, G. V. and Jacques, J. (1960), Deux pionniers de la chimie moderne, Adolphe Wurtz et Alexandre M. Boutlerov, d'après une correspondence inédite, *Revue d'histoire des sciences* **13**, 115–134.

Carneiro, Ana (1992), *The Research School of Chemistry of Adolphe Wurtz*, Ph.D. dissertation, University of Kent, Canterbury.

Crosland, Maurice (1992), *Science under Control: The French Academy of Sciences, 1795–1914*, Cambridge University Press.

Fell, Ulrike (1998), The Chemistry Profession in France: The Société Chimique de Paris/de France, 1870–1914, in D. Knight and H. Kragh, ed., *The Making of the Chemist: The Social History of Chemistry in Europe, 1789–1914*, Cambridge University Press, 15–38.

Fell, Ulrike (2000), *Disziplin, Profession, und Nation: Die Ideologie der Chemie in Frankreich vom Zweiten Kaiserreich bis in die Zwischenkriegszeit*, Leipziger Universitätsverlag.

Fox, Robert and Weisz, George, ed. (1980), *The Organization of Science and Technology in France, 1808–1914*, Cambridge University Press.

Fox, Robert (1980), The Savant Confronts his Peers: Scientific Societies in France, 1815–1914, in Fox and Weisz, ed., *The Organization of Science and Technology in France, 1808–1914*, Cambridge University Press, 241–282.

Gautier, Armand (1907), Extraits des procès-verbaux des séances, 5 January, *Bulletin de la Société Chimique*, ser. 4, **1**, 103.

Gautier, Armand (1957), Conférence faite à la demande du conseil de la Société, 17 Mai 1907, in *Centenaire de la Société Chimique de France (1857–1957)*, Masson, Paris, 1–89.

Geison, G. R., ed. (1984), *Professions and the French State, 1700–1900*, University of Pennsylvania Press, Philadelphia.

Gieryn, T. (1983), Boundary-work and the demarcation of science from non-science: Strains and interests in professional ideologies of scientists, *American Sociological Review* **48**, 781–795.

Gieryn, T. (1999), *Cultural Boundaries of Science: Credibility on the Line*, University of Chicago Press.

Grimaux, E. (1890), *Bulletin mensuel de la Chambre syndicale des produits chimiques*, 104.

Jacques, J. (1953), Butlerov, Couper, et la Société Chimique de Paris, *Bulletin de la Société Chimique* **1953**, 528–530.

Jacques, J. and Bykov, G. V. (1959), Nouveaux matériaux concernant l'histoire de la Société Chimique de Paris, *Bulletin de la Société Chimique* **1959**, 1205–1210.

Lecouteux, P. (1994), *La Société industrielle de Rouen (1872–1939). Une sociabilité spécifique?*, Thèse de doctorat, Université Paris IV, Sorbonne.

Letté, M. (1993), *Les annales de chimie et de physique, 1864–1873. Essai d'analyse quantitative d'un journal au service de la science officielle* (Mémoire rédigé pour la presentation du D.E.A. en histoire des Sciences à l'EHESS sous la direction de Jean Dhombres, 2 vol.), Paris.

Lieben, A. (1906), Erinnerungen an meine Jugend- und Wanderjahre, in *Festschrift Adolf Lieben zum fünfzigjährigen Doktorjubiläum und zum siebzigsten Geburtstage von Freunden, Verehrern und Schülern gewidmet*, Winter, Leipzig, 1–20.

Lindet, L. (1898), Notice sur la vie et les travaux de Aimé Girard, *Bulletin de la Société Chimique*, series 3, **19**, i–xxvi.

Moissan, H. and Dupont, F. (1902), *Exposition Universelle Internationale de 1900, IVe Congrès International de Chimie Appliquée, Paris, 1900*, Paris.

Nye, Mary Jo (1986), *Science in the Provinces*, University of California Press, Berkeley.

Paquot, Charles, ed. (1950), *Mémorial de la Société Chimique de France, 1857–1949: Histoire et développement de la Société Chimique depuis sa foundation*, Société Chimique, Paris.

Paul, Harry (1985), *From Knowledge to Power: The Rise of the Science Empire in France, 1860–1939*, Cambridge University Press.

Picard, A. (1891), *Exposition universelle internationale de 1889 à Paris. Classe 45, Produits chimiques et pharmaceutiques*, Ministère du Commerce, de l'industrie et des colonies, Paris.

Procès-verbaux des séances du Conseil, 27.2.1905, A. SFC.

Rocke, A. (2001), *Nationalizing Science: Adolphe Wurtz and the Battle for French Chemistry*, MIT Press, Cambridge.

Shinn, Terry (1979), The French Science Faculty System, 1808–1914, *Historical Studies in the Physical Sciences* **10**, 271–332.

Société industrielle de Mulhouse (1876), *Bulletin special, publié à l'occasion du 50me anniversaire de la fondation de la Société*, Paris.

Statuts de la Société chimique de Paris, par décret du 27 novembre 1864 (1865), *Bulletin de la Société Chimique*, ser. 2, **3**, 3–5.

S[torer], F[rank] (1868), obituary of Anton Rosing in *American Journal of Science and Arts* **96**, 148–149.

Verkade, P. (1985), *A History of the Nomenclature of Organic Chemistry*, Reidel, Boston.

Witt, O. N. (1907), Ueber die Grenzen der angewandten Chemie und die Aufgaben unserer Congresse, in E. Paternò and Villavecchia, V. ed., *Atti del VI Congresso internationale di chimica applicata, Roma*, Rome, **1**, 92–97.

Wurtz, Adolphe (1858), letter to Alexander Williamson, 19 June, Harris Collection, Bloomsbury Science Library, University College London.

Wurtz, Adolphe (1858–59), Avant-propos, *Répertoire de chimie pure*, **1**, 5.

Zeisel, S. (1916), Adolf Lieben, *Berichte der Deutschen Chemischen Gesellschaft* **49**, 835–892.

CHAPTER 6

GERMANY: Discipline – Industry – Profession. German Chemical Organizations, 1867–1914

JEFFREY ALLAN JOHNSON

6.1 Introduction

In 1914 there were two principal German organizations for chemists, each with several thousand members: the *Deutsche chemische Gesellschaft zu Berlin* (DCG,[1] German Chemical Society of Berlin, founded 1867) and the *Deutsche Gesellschaft fur angewandte Chemie* (DGAC, Society for Applied Chemistry, founded 1887 and reorganized in 1896 as the *Verein deutscher Chemiker*, VDC, Association of German Chemists). Why had a second large organization emerged, particularly given the DCG's dramatic growth over its first two decades? How, moreover, did the DCG and VDC come to represent, respectively, the disciplinary and industrial-professional faces of chemistry in Germany, what were the resulting differences in their membership and what was their relationship both to each other and to other chemical organizations? The following analysis is intended to answer these questions.

   The analysis is organized in three main parts: first, an examination of the establishment of the DCG, its membership, its relationship to the German chemical industry and the society's goals and functions (especially in regard to publications). The next part considers the DGAC/VDC, which emerged as a major competing group mainly seeking to recruit the rapidly growing numbers of German industrial chemists by focusing on questions of applied chemistry, technology and what could be called professionalization, despite the

---

[1] "Zu Berlin" was dropped after 1876.

Creating Networks in Chemistry: The Founding and Early History of Chemical Societies in Europe
Edited by Anita Kildebæk Nielsen and Soňa Štrbáňová
© The Royal Society of Chemistry 2008

problematic nature of this term.[2] Contrasting views on disciplinary or professional identity in each society resulted in significantly different memberships in each, the DCG heavily academic and international, the VDC industrial and national, albeit with considerable overlaps among them. In connection with this point, the chapter also briefly considers some other organizations that included chemists as members in 1914 (see Table 6.1). These competing groups are of interest mainly in regard to their impact on the strategies of the VDC or DCG. The final section examines each group's relationship to the international structure of chemistry, again revealing significant differences.

6.2 Development of the *Deutsche chemische Gesellschaft*, 1867–1914

6.2.1 Foundation and Membership

Presiding over the inaugural meeting of the DCG on 11th November 1867, August Wilhelm Hofmann announced a two-fold purpose for the new society. First, it would serve as a "neutral ground" on which chemists could meet to debate the latest developments in their discipline, with many stimulating and beneficial effects on their own work and on the "progress of science in general". Hofmann spoke from experience as former head of the Royal College of Chemistry in London and past president of the Chemical Society of London, which the DCG was using as a model. He had only recently become professor in Berlin and director of the chemical laboratory of the Berlin Academy of Sciences as well as of an adjacent new university teaching institute. But Hofmann also stressed that in this "especially favourable" time of "wonderful" industrial growth, "theory and practice" were closer than ever before; thus the DCG's second goal was to "give the representatives of theoretical and applied chemistry an opportunity for a mutual exchange of ideas, and in so doing to re-seal the alliance between science and industry".[3] Here he echoed phrases he had written five years earlier in London, as one of the scientific leaders in the new field of synthetic coal-tar dye chemistry, in which he and his students had come to exemplify the scientific-industrial "alliance" since the introduction of mauve in 1856.[4] Hofmann's expectations for the DCG's mutual benefits to both academic and industrial chemists proved to be true, certainly in the early decades.

Hofmann himself had not, however, organized the constituent meeting. Claiming unfamiliarity with the local situation as a new arrival in Berlin, he had left that task to a committee consisting of mostly younger chemists, active in both academe and industry. The chair was the organic chemist Adolf Baeyer,[5] then instructor at the *Gewerbeakademie* (Commercial Academy),

[2] See Jarausch (1990), 9–14; McClelland (1991), 6, 26; Burchardt (1980); Johnson (1990).
[3] *Ber. DCG* (1868), **1**, 3. On the DCG in general, with some discussion of the other organizations, see Ruske (1967); for Hofmann (from 1888, von Hofmann) – Meinel and Scholz (1992).
[4] Johnson (1992), 169.
[5] Later enobled as Ritter von Baeyer; succeeded Justus Liebig as professor of chemistry at the University of Munich, 1873; first complete synthesis of indigo, an important natural dye compound, 1878; Nobel Prize, 1905.

Table 6.1   Principal German chemical organizations, 1867–1914.

German name	English name	Abbreviation (if used)	Year founded	Year renamed (if before 1914)
Main Organizations				
Deutsche chemische Gesellschaft	German Chemical Society	DCG	1867	
Verein zur Wahrung der Interessen der chemischen Industrie Deutschlands	Association for Promoting the Interests of the Chemical Industry of Germany	VzW	1877	
Deutsche Gesellschaft für angewandte Chemie (became VDC)	German Society for Applied Chemistry	DGAC	1887	1896
Verein deutscher Chemiker	Association of German Chemists	VDC	1896	
Deutsche Elektrochemische Gesellschaft (became DBG)	German Electrochemical Society		1894	1902
Deutsche Bunsen-Gesellschaft für angewandte physikalische Chemie	German Bunsen Society for Applied Physical Chemistry	DBG	1902	
Ausschuss zur Wahrung der gemeinsamen Interessen des Chemikerstandes	Committee to Promote the Common Interests of the Chemical Profession	AWGIC	1905	
Other organizations mentioned in the chapter				
Versammlung (later, Gesellschaft) deutscher Naturforscher und Ärzte	Assembly (later Society) of German Natural Scientists and Doctors		1822	
Verein analytischer Chemiker (became DGAC)	Association of Analytic Chemists		1877	1887
Freie Vereinigung bayerischer Vertreter der angewandten Chemie (became the F. V. Deutscher Nahrungsmittelchemiker)	Free Association of Bavarian Representatives of Applied Chemistry		1883	1901
Verband selbständiger öffentlicher Chemiker	Association of Independent Public Chemists		1896	
Verband der Laboratoriumsvorstände an Deutschen Hochschulen	Association of Laboratory Directors at German Universities and Colleges		1897	
Hofmannhaus-Gesellschaft	Hofmann House Society	HHG	1898	Dissolved 1908
Verein weiblicher Chemiker	Association of Female Chemists		1900	
Bund der technisch-industriellen Beamten	League of Technical and Industrial Officials	BUTIB	1904	
Freie Vereinigung Deutscher Nahrungsmittelchemiker	Free Association of German Foodstuffs Chemists	FVDN	1901	
Verein deutscher Kalichemiker	Association of German Potash Chemists		1913	

one of the original components of the later *Technische Hochschule* (TH, College of Technology). The DCG's initial statutes were drawn up by Hofmann's former assistant Carl A. Martius (later von Martius), already an active entrepreneur in the Berlin dye industry, and Martius's friend C. Hermann Wichelhaus, *Privatdozent* (unsalaried lecturer) in chemistry in Berlin, both of whom were key members of Baeyer's organizing committee. To these men, Hofmann was a highly productive organic chemist and the key leader whose prestige and abilities could ensure that the DCG did not simply remain a Berlin chemical society, but one that would grow to national significance at a time when Prussia was emerging as the leader of a unifying, and rapidly industrializing, German nation. Not least among Hofmann's qualifications was his close relationship to the family of the Prussian crown prince, whose wife he had tutored in London. In 1867 Berlin had become the capital of the North German Federation; by 1871 it would be capital of the newly unified German Empire. Following unification, German national institutions, including scientific and technological agencies and organizations, only gradually emerged alongside or in place of their Prussian equivalents in Berlin. Concurrently several chemical companies appeared in the 1860s and 1870s, in Berlin and along the Rhine and Main rivers to the west, which formed the core of the new German synthetic dye industry. Hence the DCG and its leaders had the opportunity to play an influential role in shaping both a nationally organized discipline of chemistry and a scientifically oriented chemical industry.

Reflecting Hofmann's proclaimed goal at the initial meeting, businessmen and technical or industrial chemists (not engineers) made up a large share of the initial membership of the DCG – about half of the first 103 members.[6] Hofmann was elected first president by acclamation and served as president or vice-president every year until his death in 1892. His friends, associates and former students in industry and science played a central role in the early development of the DCG. Key leadership posts went to several industrial chemists including Martius (secretary until 1872, and six terms as vice-president) and Ernst Schering, founder of what became the Schering chemical corporation in Berlin (treasurer until 1880). Wichelhaus served as first editor of the DCG's journal, *Berichte der Deutschen Chemischen Gesellschaft* (Reports of the DCG, Figure 6.1), followed in 1883 by Hofmann's senior assistant, Ferdinand Tiemann.[7]

The initial statutes called for the DCG to promote the "entire field of chemistry", and in the first decade this was clearly understood to include both scientific and industrial concerns. In regard to the former, the DCG quickly grew to include not merely Berliners or even Germans, but a truly international membership. By the end of 1872, 30% of the members were located outside the borders of Germany; eight years later, the proportion of international members had reached about 40% (Table 6.2).[8] At the same time, Hofmann made it his

---

[6] DCG-Verzeichnis (1868).
[7] Lepsius (1918), 178–179.
[8] DCG-Verzeichnis (1873), DCG-Verzeichnis (1881).

Figure 6.1   Title page of the first issue of *Berichte*, 1868, which became one of the foremost European chemical journals.

goal to incorporate as many German academic chemists as possible, and he reported with pride in 1872 that the DCG was having its greatest success among the younger generation. Whereas, out of 63 German academics specializing in "pure chemistry", only about 70% of the full professors were members, 82% of the associate professors had joined, and 93% of the unsalaried lecturers.[9] In order to establish themselves as the German national chemical society, the leaders of the DCG changed the membership policy in 1876. Whereas the original statutes had reflected the DCG's origins by distinguishing between members in Berlin and those outside (with a small category of "participants",

---

[9] *Ber. DCG* (1872), **5**, 1116.

Table 6.2   Academically trained German chemists and membership of DCG and VDC, 1868–1914.

End of Year:	(1) Acad.-trained German chemists (rounded)	(2) DCG members	(3) % of (2) not in Germany (estimate)	(4) DCG members in Germany (est.)	(5) (4) as % of (1)	(6) VDC members (>90% in Germany)	(7) (6) as % of (1)
1872	1000 (est.)	827	30%	579	58%		
1880	1600 (est.)	2265	40%	1359	85%		
1895	3000	3208	44%	1800	60%		
1900	4300 (est.)	3410	42%	1978	46%	1120	37%
1907	5800	3554	39%	2168	37%	2096	49%
1913	7000 (est.)	3393	37%	2138	31%	3403	59%
						5261	75%

Sources: (Col. 1, 1895, 1907): *Statistik des Deutschen Reichs, Neue Folge* (1897), **103**, 366; (1910), **203**, 255; Col. 1, other dates: estimates by author (all figures in Col. 1 should be seen as minimal estimates); Col. 2: Lepsius (1918), 186–187; Col. 3–4: DCG-Verzeichnis (1873, 1881); 1895: Duisberg (1896), 97; 1900–1914: estimates by author based on sampling of new members listed in *Ber. DCG*; Col. 6: *Z. angew. Chem.*, Aufsatzteil (1914), **27**, 586.

intended for students), the new statutes distinguished only between regular and associate members. One became an associate member by being nominated by two regular members, if there were no objections by other regular members. After a year, the associates who kept up their membership automatically became regular members. Aside from this, and a small number of honorary members, the DCG's membership recognized no distinctions of rank or professional qualification. In principle, anyone interested in chemistry could join, and any member could present papers and (if accepted) publish in the *Berichte*.[10] Print was the DCG's preferred medium of communication with its members outside Berlin, because it held all its meetings in Berlin and had no local organizations elsewhere; as a result, it lacked a regional presence in Germany. The annual *Versammlung deutscher Naturforscher und Ärzte* (Assembly[11] of German Natural Scientists and Doctors), in contrast, met in a different city each year; its chemical section, often heavily attended, became the forum where most German chemists tended to interact personally (particularly helpful for those who could rarely travel to Berlin to attend the DCG's sessions). Nevertheless, by 1880 the DCG had probably enrolled over 80% of the academically trained chemists in Germany (Table 6.2).

From 1868, the DCG normally met twice a month for research presentations, which were then published in the subsequent issue of its *Berichte*. Hence, originally this journal was indeed "reports" from Berlin, but it very quickly expanded (Table 6.3), its papers reflecting the DCG's growing national and international membership. The *Berichte* became a vehicle not only for the spread of "modern" views in structural synthetic organic chemistry, but above all for recording the enormous volume of experimental work resulting from the new approaches, which made possible the rapid determination of thousands of new structural formulas for simpler organic compounds. The *Berichte* competed successfully with older journals like the *Journal für praktische Chemie*, edited by the anti-structuralist Hermann Kolbe in Leipzig, who, despite being named an honorary member of the DCG, stubbornly opposed the new society (for reasons including extreme anti-Semitism, which Hofmann and his colleagues in the DCG generally did not share).[12] Along with the original papers, the *Berichte* began publishing abstracts in 1880. A series of annual "summary lectures" covering major areas of research began in 1890; initially most speakers were leading organic chemists, with an occasional presentation on other fields, including industrial chemistry.[13]

In its early years the DCG led a somewhat nomadic existence within Berlin, as it had no permanent offices or meeting hall. Hofmann, with a legendary ability to persuade his associates to volunteer their time and efforts, kept the DCG's costs to a minimum. After 1884 the senior assistants' offices in his institute served as the DCG's headquarters.[14] Hofmann, who enjoyed

---

[10] DCG statutes (1876), 1328–1329.
[11] The Assembly, which first met in 1822, later changed into *Gesellschaft* – Society.
[12] Rocke (1993), 354–362.
[13] Lepsius (1918), 184–186.
[14] Pinner (1900), IV–VI.

Table 6.3 DCG journals and finances, 1868–1913.

End of year:	Members DCG	DCG income (rounded, Marks)	DCG expenses (rounded, Marks)	DCG capital (rounded, Marks)	Ber. scientific papers	Ber. pages for papers (excluding abstracts)	Ber./ Zentralblatt abstracts	Ber./ Zentralblatt pages for abstracts
1868	257	3900	2200	3200	97	282	–	–
1870	617	8600	6300	5700	277	992	–	–
1880	2265	40900	37600	18800	563	2473	644 (B)	n/a
1890	3440	94700	76400	136100	784	3851	1257 (B)	1202
1895	3208	97700	77900	206800	636	3317	1516 (B)	1634
1900	3410	150400	149600	529800	636	3826	6344 (Z)	2612
1905	3624	186800	171100	762600	732	4220	8282 (Z)	3596
1910	3391	190000	195000	959900	559	3643	10587 (Z)	4112
1913	3393	161000	152300	1284600	523	4052	11219 (Z)	4376

Sources: Lepsius (1918), 186–187, Ber. DCG (1892), **25**, 3675, (1897), **30**, 3184; (1907), **40**, 5029; (1914), **47**, 597–598.

socializing, frequently presided over festive meetings, culminating in the "*Benzolfest*" of 1890 honouring August Kekulé, one of the founders of structural chemistry, on the 25th anniversary of his theory of the benzene ring.[15] There was much to celebrate; the DCG had grown steadily over its first two decades and was now by far the largest academic chemical society in the world, with more than 40% of its members outside Germany (Table 6.2).

6.2.2 The *Deutsche chemische Gesellschaft* and the Chemical Industry in the Hofmann Era

German industrial chemists and members of the DCG played a major role in organizing the German delegation to the Vienna exposition of 1873, which was prominently featured in the *Berichte* and highlighted the rapidly growing synthetic dye industry. Much of this growth was fuelled by easy credit in the boom years following unification, and the absence of a national patent law that allowed German entrepreneurs to copy products and processes patented in Britain or France. In 1877, however, the DCG's board submitted to the Reichstag a petition to amend the government's draft patent law; the change permitted chemical patents only for production processes, not for the products themselves.[16] This fostered the development of in-house industrial research laboratories in Germany and the rise of an innovative, science-based chemical industry in Germany.[17] The *Berichte* helped by publishing patent reports as well as scientific papers.

Although chemists' work on the patent bill might have led naturally to the creation of a division of industrial chemistry in the DCG, as Hofmann's friend Martius proposed, Hofmann advised against it. Why? It seems clear that Hofmann did not want the DCG to represent business interests, nor did many businessmen see it as the appropriate organization for this. As one later recalled, the DCG did not attract "a major proportion of industrial elements" during the 1870s, "because scientific research was then still essentially in the hands of professors and professional scholars ... despite the close relations between science and technology."[18] Moreover, despite his own rhetoric about academic–industrial cooperation, Hofmann and most of his academic colleagues primarily sought to enhance the DCG's international scientific reputation; hence academic chemists, and organic chemists in particular, dominated the DCG.

Thus the *Verein zur Wahrung der Interessen der chemischen Industrie Deutschlands* (VzW, Association for Promoting the Interests of the Chemical Industry of Germany) was established in November 1877 as an independent economic interest group and trade association whose major goal was to influence the government on trade policy, patent law and other areas, including issues of

---

[15] Lepsius (1918), 41–45; on Kekulé (later Kekulé von Stradonitz), see Anschütz (1929).
[16] Johnson (1992), 175–176.
[17] *cf.* Homburg (1992).
[18] Verein zur Wahrung ... (1902), 406.

chemical education for industrial chemists.[19] There was some overlap in the leaderships of the VzW and the DCG, because leading Berlin businessmen in the VzW, including Hofmann's friends Martius, Schering and Julius F. Holtz (a director of the Schering firm) continued to play central roles in the leadership of the DCG. In fact Holtz succeeded Schering as treasurer of the DCG and served until 1910, during which time he was also the chairman of the VzW. He helped the DCG to accumulate a significant capital during its first two decades of growth by investing its surplus dues (Table 6.3).[20] Leaders like Holtz ensured that the German chemical industry retained its interest in and support for the scientific side of chemistry, as reflected in the *"Anilinfest"* of 1890. This honoured Hofmann on the 25th anniversary of his coming to Berlin and celebrated his pioneering work on aniline dye compounds, which had helped lay the foundations of the coal-tar dye industry.[21] Preparations for an equally festive jubilee for the DCG in 1892 were interrupted, however, by Hofmann's death in May.

6.2.3 The *Deutsche chemische Gesellschaft* under Fischer, 1892–1914: Institutional Growth and New Publications

As Emil Fischer came to Berlin after Hofmann's death, with his professorship and its many related academic duties came also the leadership of the DCG. Less sociable than Hofmann, Fischer took the presidency less often, but still became the dominant personality in the organization.[22] With his organizing talent, the DCG gained a permanent headquarters and greatly expanded its publication activities by 1914, without significantly increasing its membership.

The DCG was confronting repeated deficits (1885, 1886, 1892), resulting from the expense of publishing larger numbers of scientific papers in the *Berichte*, plus more comprehensive abstracts of a growing literature, exacerbated by a decline in dues (membership had peaked in 1887 at 3614). To avoid raising dues, in 1896 Fischer persuaded the DCG to relieve the strain on the *Berichte* by publishing fewer original papers and by purchasing the *Chemisches Zentralblatt* (Chemical Central Page) as its weekly abstracts journal, available to members at an additional cost. The DCG also agreed to continue Friedrich Beilstein's *Handbuch der organischen Chemie* (Handbook of Organic Chemistry), drawing upon the *Zentralblatt*'s abstracts. The organic chemist Paul Jacobson assumed a new position as general secretary, with primary responsibility for managing publications and editing the *Berichte*.

With the expanded publication projects of the DCG, the old offices in the institute became increasingly inadequate, but separate offices would be expensive. Here Holtz's connections played a crucial role, as he coordinated a

---

[19] Ungewitter (1927), 5; Ruske (1967).
[20] On Holtz see Kraemer (1911).
[21] Lepsius (1918), 45–46.
[22] Emil Fischer was awarded the Nobel Prize in 1902 for his pioneering work on the structure of the sugars. See Hoesch (1921), Fischer (1987); there is still no recent full-length biography.

campaign for industrial contributions and loans through an independent funding organization, the *Hofmannhaus-Gesellschaft* (HHG, Hofmann House Society), which built the first headquarters building for the DCG, the *Hofmannhaus*, which opened in 1900. The DCG finally had a permanent meeting place and a home for its growing library and publication activities. The VzW initially also occupied a floor in the new building, as did the *Berufsgenossenschaft der chemischen Industrie* (employee insurance company of the chemical industry, also chaired by Holtz). The DCG purchased the building in instalments until 1908, when the HHG was dissolved and the other two organizations moved into their own headquarters next door.[23]

In 1907 the DCG assumed responsibility for editing *Justus Liebig's Annalen der Chemie* (Annals of Chemistry), which took the place of a proposed second journal Fischer had not wanted to establish. As the *Berichte*'s deficits recurred, in 1910 the DCG again reorganized, reducing costs, raising some fees and seeking industrial contributions to subsidize additional reference publications supplementing the slow-appearing *Beilstein* volumes.[24]

The DCG now organized a consortium with the editors of most of the principal chemical journals. They agreed that the *Berichte* would publish shorter papers of general and immediate interest, while longer articles and series of articles would go to an appropriate specialized journal. The consortium thereby established certain journals as the principal voices for their subdisciplines within the general disciplinary framework represented by the *Berichte* and *Chemisches Zentralblatt*, and issued a joint plea to authors to cooperate with their efforts to coordinate and concentrate the literature. But the group was incomplete; Wilhelm Ostwald's *Zeitschrift für physikalische Chemie* (Journal for Physical Chemistry) briefly entered the consortium, then dropped out again. The DCG was dominated by organic chemists, and physical chemists evidently doubted the *Berichte*'s ability to judge their contributions. As Ostwald later recalled, his critics often dismissed him as "no chemist, because I never produced a new substance". This reflected demarcation issues at the boundaries of chemistry and physics.[25]

With a much larger institutional presence by 1914, the DCG continued to foster academic chemistry as a discipline with a much broader range of publications than in 1895, but it was attracting a shrinking proportion of the total number of academically-trained chemists in Germany (as well as a decreasing share of foreign members [Table 6.2]). It was now competing for members with a younger organization, which used a quite different approach to attract a rapidly increasing group of chemists: middle-class, academically trained, salaried employees (plant managers and researchers) in the big firms of the chemical industry.

---

[23] Pinner (1900); Lepsius (1918), 103, 183; Ruske (1967), 102–108.

[24] Lepsius (1918), 124–136, 187; *Ber. DCG* (1892), **25**, 3674–3676; Wallach (1910); DCG statutes (1911); Ruske (1967), 122–126.

[25] *Ber. DCG* (1910), **43**, 2790, 2815; Ostwald (1926–1927), **2**, 48; *cf.* Gieryn (1983). Ostwald, one of the leading physical chemists of the nineteenth century, was professor at Leipzig University (1887–1906) and Nobel Laureate for chemistry (1909).

6.3   Deutsche Gesellschaft für angewandte Chemie and Verein Deutscher Chemiker

6.3.1   A German Society for Applied Chemistry, 1887–1895

For chemistry students in German universities after 1880, prospects for an academic career (for which there were only a couple of hundred positions in the universities and *technische Hochschulen* [THs, colleges of technology]), or even secondary school teaching (limited because the dominant group of humanistic secondary schools did not usually teach chemistry, apart from general physical science) quickly receded in favour of industry. From 1880 to 1890 the proportion of chemistry students in Berlin looking toward business careers rose from about a third to perhaps three-quarters.[26] Many sought jobs in the dye industry. Spurred on by the new patent law, by the declining prices of older dyes and by the opportunities for large-scale, systematic research in fields like the azo dyes, industrial innovation had accelerated so rapidly that German companies could no longer depend chiefly upon academic scientists or publications in the *Berichte* for their product ideas. Instead they developed in-house industrial research facilities or "inventing laboratories" supported by large patent departments, creating the first modern corporate research and development bureaucracy, staffed by chemists trained in universities and THs.[27] By 1895 74% of the approximately 3000 people identified as "chemists" by the German occupational census were salaried employees, and at least 9 companies, all dye producers, employed more than 20 chemists each, with a total of 250 at the 3 largest firms.[28]

The need for qualified chemists led the VzW to take up the problem of higher technological education in 1886. The THs had mostly instituted a *Diplom* examination (roughly comparable to a master's degree, which did not exist in the German context); this covered the areas of industrial concern, but their chemistry students tended to bypass the examination, which conferred no title or social status, going on to a university to earn a quick doctoral title. The THs' response was to seek academic parity with the universities, including the right to award the doctoral title. The VzW proposed instead a uniform German state licensing examination for all chemistry students, at the universities or at the THs.[29]

At this point a new organization appeared, which examined this issue from the perspective of the employees and teachers. The *Deutsche Gesellschaft für angewandte Chemie* (DGAC, German Society for Applied Chemistry) was founded in 1887 when the *Verein analytischer Chemiker* (Association of Analytic Chemists), a small group founded ten years earlier, decided to reorganize into an organization with broader goals and membership, including chemists in private and public testing laboratories, industrial employers and

---

[26] Johnson (1992), 178.
[27] Meyer-Thurow (1982); Homburg (1992).
[28] Johnson (1990), 125.
[29] Scholz (1990); Burchardt (1976), 776–777.

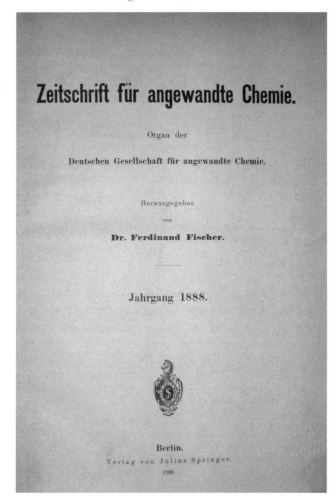

Zeitschrift für angewandte Chemie.

Organ der

Deutschen Gesellschaft für angewandte Chemie.

Herausgegeben

von

Dr. Ferdinand Fischer.

Jahrgang 1888.

Berlin.
Verlag von Julius Springer.
1888.

Figure 6.2   Title page of *Zeitschrift für angewandte Chemie.*

employees (mostly chemists, some engineers), and professors of technical and applied chemistry (most at the THs, a few at the universities).[30] They adopted a journal edited by Ferdinand Fischer, who taught technical chemistry at TH Hannover (later at Göttingen University), which was joined with another journal in 1888 as the *Zeitschrift für angewandte Chemie* (Journal for Applied Chemistry, Figure 6.2) published biweekly (later weekly).[31] The society assumed full ownership in 1904, as Berthold Rassow (professor of chemical technology at Leipzig University) took over the editorship.[32] Fischer was

---

[30] Lacking a complete list of members, the author has not attempted quantitative estimates.

[31] Ruske (1967), **26**, 39–44; Lepsius (1918), 179.

[32] *Z. angew. Chem.* (1898), **11**, 803–804; Rassow (1912), 17. Predecessor journals were the *Repertorium der analytischen Chemie* (Reference Journal of Analytical Chemistry, 1881–1887) and Fischer's *Zeitschrift für die chemische Industrie* (Journal for the Chemical Industry, 1887).

among those who also advocated the establishment of *Bezirksvereine* (district associations), which, in contrast to the DCG, gave the new organization a regional presence and attracted more younger industrial chemists who could not easily travel to Berlin. In 1889–1914 the district associations increased from 4 to 23, with three-quarters of the regular members (as well as 345 non-voting associate members with incomplete academic training, a category introduced in 1900). Again unlike the DCG, the DGAC held annual conventions in a different German city each year. While the main focus of the journal, and the presentations at the district associations and annual conventions, was on technical questions, the DGAC also addressed chemical education, and by 1890 it had formally requested an official licensing examination. To the younger employees, who quickly became the majority of its members, the DGAC was a means to improve their status and minimize competition by excluding chemists without academic credentials.[33] In pursuit of this goal and further growth, the DGAC reorganized in 1896.

6.3.2   A Professional Association for German Chemists: From *Deutsche Gesellschaft für angewandte Chemie* to *Verein Deutscher Chemiker*, 1896–1914

"The situation of the employed chemists had not improved in the course of time ... The chemistry profession was not yet recognized."[34] Thus Berthold Rassow, the official historian of the *Verein Deutscher Chemiker* (VDC, Association of German Chemists), writing in 1912, described the situation a quarter-century earlier at the founding of the DGAC. The 1912 jubilee history maintained that the VDC had dealt with these issues through vigorous "activity in the interest of the chemical profession",[35] *i.e.* professionalization. This strategy had expanded the organization in two senses, adding both members and a broader range of institutional goals.

As the DGAC grew (Table 6.4), an energetic leader emerged, exemplifying the interaction of corporate and chemists' interests: Carl Duisberg, a young university-educated chemist of modest origins who rose from director of research at Bayer in 1887 to director of the firm by 1900. In 1896, as leader of the *Rheinischer Bezirksverein* (Rhenish District Association) of the DGAC, he played a key role in reorganizing the DGAC into the VDC, whose goal extended beyond "applied chemistry" to becoming the main professional organization for all German chemists.[36] Whereas the 1888 statutes had not defined qualifications for membership (except for support by an existing member), the 1896 statutes limited individual membership to "chemists and other academically trained persons"; but they also permitted "governmental agencies, firms, and associations with similar goals" to become corporate

---

[33] Johnson (1990), 125–126; *Z. angew. Chem.* (1900), **13**, 871–872; (1914), **27**, 587.
[34] Rassow (1912), 4; Johnson (1990), 123, 138 n1.
[35] Rassow (1912), title of chapter V, 72.
[36] Johnson (1990), 126; on Duisberg see Flechtner (1981).

Table 6.4 *Deutsche Gesellschaft für angewandte Chemie/Verein Deutscher Chemiker*, journal and finances, 1890–1913.

End of year:	Members DGAC/ VDC	Income from dues (rounded, Marks)	Cost of journal (rounded, Marks)	Z. angew. Chem., pages	VDC capital (rounded, Marks)
1890	568	n/a	n/a	762	n/a
1895	1129	20700	11600	753	17100
1900	2406	51300	29200	1338	26900
1905	3282	69900	39400	2088	50200
1910	4437	82300	54300	2575	60400
1913	5261	101800	63600	2538	232900

Sources: Rassow (1912), **36**, 64; *Z. angew. Chem.*, Aufsatzteil (1914), **27**, 593–594.
Note: capital for 1913 reflects contributions to the Jubilee Fund of 1912.

members. The district associations gained formal recognition and influence, electing delegates to the VDC's *Vorstandsrat* (Council), which in turn elected the *Vorstand* (board), with a chairman, vice-chairman, two secretaries and a treasurer.[37] After this, the main power clearly rested with the factory owners and senior employees of large corporations, who could usually dominate the district associations, composed increasingly of lower-ranking, dependent employees (whose share in the VDC's membership rose from some two-thirds in 1896 to nearly 90% by 1909).[38] Academic chemists continued to play a role in the leadership, with one or two professors on every board, but always in the minority after 1896; only prominent industrial chemists and directors became chairmen after 1890, such as the pioneering dye chemist Heinrich Caro and later Duisberg himself (1907–1912).[39]

The *Berzirksverein Frankfurt* (Frankfurt District Association), located in one of the biggest centres of the German chemical industry, was somewhat unusual in tending to elect academic chemists rather than corporate directors to its leadership. From its founding in 1893, the Frankfurt group took the lead in pressing the leaders of the DGAC to expand their efforts to promote the "professional interests of the chemists", including issues like working conditions and patent law.[40] To Duisberg, these were calls for chemists to raise their "professional consciousness" and obtain "a social status similar to that of the representatives of other academically educated occupational groups like government officials, physicians, pharmacists, attorneys, engineers, *etc.*"[41] Not social but educational reform, however, was to be the way to this status. He joined Ferdinand Fischer in mobilizing the VDC's growing strength behind the proposals for a state examination and more technical chemistry courses in the

---

[37] VDC statutes (1888), (1896).
[38] *Z. angew. Chem.* (1896), **9**, 113; **22** (1909), 2541.
[39] Travis and Reinhardt (2000), 332–336; Rassow (1912), 57–62.
[40] Wentzki (1912), 219.
[41] *Z. angew. Chem.* (1896), **9**, 155–156.

universities, whereby Duisberg intended not only to promote the professional goal of raising academic standards for chemists, but also to shape their training "according to the interests of industry".[42]

Fearing that a state examination would devalue the doctorate, leading to the loss of research assistants and revenue from dissertation fees, in 1897, Adolf Baeyer, Emil Fischer and Wilhelm Ostwald organized the *Verband der Laboratoriumsvorstände an Deutschen Hochschulen* (Association of Laboratory Directors at German Universities and Colleges). This group administered its own, unofficial *Verbandsexamen* (association examination) as an intermediate qualifying test for all potential degree candidates, including those at the THs. The new examination helped to reform introductory training for chemistry students, nearly two-thirds of whom went on to complete doctorates before 1914. The VDC's proposal for a state examination, meanwhile, became a dead issue in 1899 when the Prussian THs, soon followed by those in other states, won the right to award doctoral degrees and to confer the title *Diplom-Ingenieur* (certified engineer) for passing the *Diplom* examination. The one group of chemists to have an official licensing examination, in view of their connection to health regulation, were the foodstuffs chemists (and pharmaceutical chemists).[43]

Stymied in their licensing campaign, Duisberg and other leaders of the VDC now sought to enhance the secondary education of chemists as a means toward professional recognition. Until 1907 chemistry was the only major discipline in the Prussian universities (aside from agriculture and dentistry) in which completion of a nine-class classical or modern secondary-school course with an *Abitur* (leaving examination) was not required to obtain a doctorate.[44] With the economic boom of the late 1890s, increased demand for industrial chemists was apparently having a democratizing effect on chemistry students, so that by around 1900 barely half of the candidates for the *Verbandsexamen* had taken the *Abitur*, whereas two-thirds of all industrial chemists reported having one in the mid 1890s.[45] Thus in 1900 the statutes of the VDC were revised again, explicitly restricting membership to "academically-trained chemists". This also served to exclude those women chemists who from 1900 began taking short laboratory courses at the Agricultural College of Berlin, leading to low-paying analytical jobs in the sugar industry.[46] In 1902 and 1907 the VDC warned students without an *Abitur* against trying to get jobs in industry; the second warning acquired greater force, as the *Abitur* became mandatory for all chemistry doctoral students in Prussia (except a few certified pharmacists studying who had written "outstanding dissertations").[47]

The VDC's opposition to women chemists did not essentially change before 1914, even after the Prussian universities were formally opened to women in

---

[42] Duisberg (1896), 107–108.
[43] Johnson (1985), 247–257, 262–269.
[44] Johnson (1985), 266.
[45] Duisberg (1896), 106; Titze (1983), 65.
[46] *Z. angew. Chem.* (1900), **13**, 871; Johnson (1998), 6.
[47] *Z. angew. Chem.* (1902), **15**, 990–993; (1907), **20**, 1477; Rassow (1912), 76; Johnson (1985), 266–267.

1909 (elsewhere this had come earlier, and a handful of women had already earned doctorates in chemistry). The VDC thus did not recognize the *Verein weiblicher Chemiker* (Association of Female Chemists, founded 1900, 49 members in 1908), but it may have admitted its first female member, the daughter of a former chairman, in 1910 (the DCG did so in 1877).[48]

In order to maintain the VDC as a central organization for all (male) chemists regardless of specialization, a revision of the statutes in 1907 introduced "specialty groups" with the same rights to representation on the VDC's council as the district associations.[49] These groups, often led by academic chemists or the representatives of the principal firms in particular branches of industry, were to deter the formation of competing associations of specialists. By now there were at least three main competitors (*cf*. Table 6.1): the *Freie Vereinigung Deutscher Nahrungsmittelchemiker* (Free Association of German Foodstuffs Chemists), previously the *Freie Vereinigung bayerischer Vertreter der angewandten Chemie* (Free Association of Bavarian Representatives of Applied Chemistry, 1883–1901); the *Verband selbständiger öffentlicher Chemiker* (Association of Independent Public Chemists, founded 1896); and the *Deutsche Bunsen-Gesellschaft für angewandte physikalische Chemie* (DBG, German Bunsen Society for Applied Physical Chemistry), previously the *Deutsche Elektrochemische Gesellschaft* (German Electrochemical Society, 1894–1902).[50] The DBG had about 700 members, the others probably fewer. The foodstuffs and public chemists (who operated municipal or independent private testing laboratories) had, however, gained official recognition with the passing of national regulatory legislation (from 1879) and a national licensing examination (1894); a journal of foodstuffs chemistry began to appear in 1886.[51] The latter fields received little attention from the DCG, whose academic organic chemists often disdained "smear" chemists.[52] Many German physical chemists also resented the dominance of classical structural organic chemistry and the dye industry in the DCG and VDC, but the DBG's independent approach also correctly reflected expectations for broader industrial applications than the electrochemistry that had been their original focus.[53] By 1914 the VDC had 12 specialty groups, but competing organizations continued to form, such as the *Verein deutscher Kalichemiker* (Association of German Potash Chemists), despite a specialty group in their field.[54]

---

[48] Johnson (1998), 6–7; *cf. Ber. DCG* (1877), **10**, 246 (listing Dr.phil. Lydia Sesemann in the physical chemistry laboratory at Leipzig).

[49] VDC statutes (1907), 399.

[50] Rassow (1912), 19, 24–25; Reinhardt (2003), 11; Jaenicke (1994).

[51] *Vierteljahresschrift über die Fortschritte der Chemie der Nahrungs- und Genußmittel* (Quarterly on the Progress of Foodstuffs Chemistry, 1886–1898), became the *Zeitschrift für Untersuchung der Nahrungs- und Genußmittel* (Journal for Testing of Foodstuffs). The competing journal of the public chemists (*Zeitschrift für öffentliche Chemie*, Journal of Public Chemistry) began in 1895.

[52] *cf*. Johnson (1985), 259–260.

[53] Jaenicke (1994), 46–55. The DBG's journal began as *Zeitschrift für Elektrotechnik und Electrochemie* (Journal of Electrotechnology and Electrochemistry, 1894), then *Zeitschrift für Elektrochemie* (1895–1903), and finally *Zeitschrift für Elektrochemie und angewandte physikalische Chemie* (Journal for Electrochemistry and Applied Physical Chemistry).

[54] *Z. angew. Chem.* (1914), **27**, 588, 591–593.

6.3.3 Professionalization on a Broader Basis: Interaction with the *Deutsche chemische Gesellschaft* and Other Groups

The concerns of the *Verein Deutscher Chemiker* with professionalization had meanwhile begun to broaden amid growing concerns about the deteriorating economic opportunities for chemists. One source for these concerns can be seen in the Imperial occupational census returns of 1907, showing that although the total number of people calling themselves chemists had nearly doubled (to 5800) since 1895, 86% were now dependent employees (independent positions increased by only 50). Between 1895 and 1912 the number of what the VDC called "big firms" in the chemical industry (employing more than 20 chemists) rose from 9 to 20, and their share of industrial chemists rose from about one-third to more than half of all chemists in the chemical industry narrowly defined; hence fewer and fewer of these chemists might become directors of such firms. Duisberg linked the growing economic worries among younger chemists to concerns about patent law and the rights of employee-inventors raised in the convention of the VDC in 1906.[55] Accordingly, in 1905–1906 he helped to initiate an *Ausschuss zur Wahrung der gemeinsamen Interessen des Chemikerstandes* (AWGIC), Committee to Promote the Common Interests of the Chemical Profession), in which the VDC cooperated with the foodstuffs and public chemists' organizations, as well as the DCG, in trying to raise chemists' "professional consciousness" and improve relations with the state and Imperial bureaucracies. This would require "precisely distinguishing between academically educated chemists, vocationally trained *Chemotechniker* [chemical technicians], and empirically schooled *Chemikanten* [skilled laboratory and production workers] and laboratory helpers". The technicians presented particular problems for the VDC, because they worked closely with works chemists to operate the increasingly complex production processes of big plants, and despite their lesser level of training their salaries often approached those of younger industrial chemists. Hence, the AWGIC surveyed salaries among industrial chemists (through the VDC) as well as academic assistants (through the DCG), and later addressed official fee schedules for consulting, expert testimony and public laboratory analyses (leading, by 1914, to new state and federal laws benefiting chemists).[56] By 1906 the assistants at Prussian universities had collectively requested increases in their salaries,[57] and the radical *Bund der technisch-industriellen Beamten* (BUTIB, League of Technical and Industrial Officials, founded in 1904) was aggressively seeking to recruit industrial chemists as well as technicians. This alarmed the VDC's leaders.[58]

The BUTIB hoped to unite technical employees of all types in a broad-based movement to demand political and economic reforms. Their propaganda criticized the relatively low salaries as well as restrictive employment contracts,

---

[55] Johnson (1990), 128–130.

[56] *Z. angew. Chem.* (1907), **20**, 513–514, 1512–1513; (1914), **27**, 600–601, 613–614; Johnson (1990), 132.

[57] Burchardt (1980), 338.

[58] Johnson (1990), 129–131.

which it condemned as "intellectual serfdom".[59] In response, leaders of the VDC worked out a broader program to address the problems faced by chemists in modern industry. The VDC had set up a job placement register in 1900 and a *Hilfskasse* (mutual assistance fund) in 1903. Duisberg and his colleagues now established an agency for legal advice (not representation) to individual chemists on issues like contracts, patent rights and the "competition clause", which prohibited employees who left a firm from taking a job with a competitor for several years (the *Karenz* – waiting-period), under penalty of forfeiting several years of salary. The VDC then established a "social committee", made up of employers and screened employees, which developed legislative petitions for salaried employees' insurance, patent reform and the regulation of the *Karenz*. In 1911 the Reichstag passed an insurance law, while bills in the other two areas were still pending when war began in 1914. The VDC's social committee also prepared a model, non-mandatory hiring contract for use by the leading firms in the chemical industry, and incorporating provisions to which the VzW had already agreed as a basis for heading off more far-reaching legislation or collective bargaining. It specified that the *Karenz* must be paid (at an undefined rate), and that the maximum standard fine for breach of contract would be three times the employee's last salary.[60]

On becoming chairman in 1907, Duisberg denounced the BUTIB's "trade-union" methods and called instead for the professional unity of all "chemists with a higher education ... against all outsiders. There is no need to let economic reasons separate us into two enemy camps, employers and employees".[61] His rhetoric and his new policies were apparently effective. The first two years of the BUTIB's existence (1904–1906) were the slowest years of growth for the VDC since 1898; but although it then lost nearly 8% of its members in 1907, more than twice as many new members also joined.[62] Under that year's reorganization, the VDC gained a general secretary (Rassow) with a paid editorial and business staff. The VDC then grew more rapidly, passing the DCG in membership. In 1912 its new Jubilee Fund, based on contributions, gave it greater financial security and less dependence upon dues to support its publications and other activities (Table 6.4). By stressing educational requirements and social status, the VDC excluded from its ranks those who could not afford to complete a secondary education (but had previously found it possible to study chemistry), and exaggerated the social gap between chemists and "outsiders", including chemical technicians. But it also became the largest chemical organization in Europe, including perhaps 75% of roughly 7000 German chemists by 1914 (Table 6.2). Such were the fruits of professionalization.

---

[59] Johnson (1990), 131.
[60] Rassow (1912), 15–16, 18, 38–51; Johnson (1990), 133–134.
[61] *Z. angew. Chem.* (1907), **20**, 1497–98.
[62] *Z. angew. Chem., Aufsatzteil* (1914), **27**, 586.

6.4   The German Chemical Societies and the International Structure of Chemistry to 1914

The DCG under Hofmann had a pronounced international outlook, as reflected not only in its general membership but even more in its honorary members, of whom 19 out of 28 were non-Germans; this trend continued after Hofmann's death with 20 honorary members outside Germany (including some Germans working abroad) of the 25 total designated by 1914, despite some initial resistance to honouring French chemists that was overcome by the insistence of Emil Fischer.[63] In contrast, the VDC took a much more restrictive approach, honouring only one non-German. Although the VDC did have foreign district associations in Belgium, New York and (from 1913) Switzerland, it was less welcoming to Eastern Europeans; from 1904 to 1914 it repeatedly advocated restricting the admission of foreign students, especially Russians, to German universities and THs.[64]

More significant was the DCG's cooperation with other national chemical societies in promoting international cooperation. DCG members attended an 1892 international conference in Geneva called by the International Commission for Nomenclature established in 1889 in Paris, but little more was done in this area until the creation of the *International Association of Chemical Societies* (IACS) in Paris in 1911. Along with the French and British societies, the DCG was a co-founder, delegating Wilhelm Ostwald (then a very strong proponent of international scientific cooperation) to hold the rotating presidency until April 1912. By then 15 societies from 13 nations had joined (including the DBG, but not the VDC). The IACS sought to facilitate international agreements on questions of common concern such as inorganic and organic nomenclature, by coordinating the work of member organizations on these matters, or through international commissions. The 1914 war disrupted these promising efforts.[65]

The DCG had previously established an atomic weights commission in 1898, which then invited other national chemical societies to form an international atomic weights commission in 1900. At the time, the major issue in contention was which element to use as the integral value for atomic weights, $H = 1$ or $O = 16$, as increasingly precise determinations had shown a puzzling, non-integral ratio between the two (this was before the modern understanding of atomic structure and isotopes). As a physical chemist, Ostwald strongly supported the oxygen standard, while the more qualitatively minded professors of organic chemistry and the dye chemists in the VDC tended to support the $H = 1$ standard. Despite the opposition, the former prevailed within both the DCG and the international commission, which later joined the IACS and issued periodic updates on atomic weight determinations.[66]

---

[63] Ruske (1967), 94–96; Lepsius (1918), 180.

[64] Rassow (1912), 56; *Z. angew. Chem.* (1914), **27**, 587, 602–604.

[65] Lepsius (1918), 136–137. A report on the first year of the IACS, with statutes and lists of member societies, is in IACS (1912).

[66] Landolt, H., Ostwald, W. and Seubert, K. (1901); Clarke *et al.* (1914); see also Görs (1999).

Another part of the emerging structure of international science was the biennial (later triennial) International Congress for Applied Chemistry, which had begun in the 1890s to consider questions related to agricultural and analytical chemistry. As the congress broadened in scope, both the DCG and VDC (along with other organizations such as the VzW) sent delegations to meetings in Paris (1900), London (1909) and New York/Washington (1912), and they jointly hosted the 1903 congress in Berlin (which was held in the Reichstag building).[67] The war disrupted this trend as well.

6.5 Discipline and Profession in the German Chemical Societies before 1914: Conclusions

What can one conclude from the foregoing about the roles of the principal German chemical organizations in shaping the discipline and profession of chemistry before 1914? Rudolf Stichweh has pointed out that the "fundamental social act" of a scientific discipline is publication, primarily for colleagues; hence the promotion of disciplinary communication would be a primary goal of a scientific society dominated by academics, such as the DCG. Continued growth of an academic discipline such as chemistry, however, requires "secondary professionalization", whereby a broader profession, many of whose members might not work in an academic context, emerges from the discipline. Although the central goal of a professional association is "controlling the border between the profession and its social environment", above all in regard to "questions of training the rising generation", a secondary professional group (like the VDC) typically has great difficulty standardizing its professional qualifications in a competitive marketplace dominated by organized employers whose aims are not primarily scientific.[68] Stichweh's model helps to clarify the relationship between the DCG and VDC, which developed in contrasting but also complementary patterns, both mediated by the German chemical industry. The DCG promoted chemistry primarily as an academic discipline, controlled by academic chemists, albeit in "alliance" with industry. The latter served as an important source of financial subsidies for the DCG's growing organization and publication projects, while becoming the main source of demand for academically trained chemists. The VDC promoted "applied" chemistry, partly as an academic discipline, but increasingly as an industrial science, to be developed and controlled by industrial chemists and the research-intensive firms they represented as owners or senior employees. Although the DCG soon outgrew its founding circle of members in Berlin, it kept its activities there and thus could attract its members chiefly by the quality of its publications, first the *Berichte* and later the *Zentralblatt* and others. It succeeded in attracting a large membership, mainly academic chemists, from Germany and around the world; but its statutes never specified academic qualifications.

The VDC, in contrast, adopted a regional organization and held its annual meetings throughout Germany in order to attract large numbers of German

---

[67] Rassow (1912), 89–92.
[68] Stichweh (1994), 314–317, 327–329.

industrial chemists, to whom it offered a program of promoting professional status. As a leader of the VDC stressed in 1910, "One can only speak of a profession when all members are equal to one another in education and social position. The income level and economic position, whether dependent or independent, make no difference in this respect." Professionalization thus had the not-inconsiderable advantage (for the corporate directors in the VDC) of dividing the younger industrial chemists from the more radical "circles of middle and lower officials".[69] Hence the profession of chemistry was to be defined by uniform educational qualifications, both in secondary and higher education, as in the established professions such as law and medicine. Like the goal of official recognition through a licensing examination, this was not achieved by 1914, but the VDC's efforts attracted enough members to make it the largest chemical association in Europe.

The greater uniformity and high quality of scientific education in the German system, particularly after the reforms at the turn of the century, and the general identification of specific educational attainments with a specific social status, had considerably simplified the problem of defining who was to be considered a "professional". Because German industry increasingly demanded academically trained chemists, it was relatively easy to equate academic status with occupational qualifications, thereby drawing lines between insiders who could be admitted to a professional organization like the VDC and outsiders with lesser training (including those in the BUTIB) who could not. The VDC thus avoided the hierarchical system of grades of membership, which developed in some other national contexts, and instead adapted to the title-oriented stratification that already existed in German society. Moreover, by creating two different national chemical organizations rather than the single united association that the DCG had seemed destined to become in 1867 (Hofmann's "academic-industrial alliance"), German chemists were also adapting to a social context that gave special status to professors. This made it seem logical, almost necessary, to separate the functions of disciplinary representation (DCG), dominated by academic chemists, and professional representation (VDC), controlled by industrial chemists. This division continued until a single *Gesellschaft Deutscher Chemiker* (Society of German Chemists) emerged after the Second World War. Interdisciplinary differences proved even harder to overcome, as the DBG continued as a separate organization for physical chemists.[70]

Abbreviations

(For translations of names, and dates of founding, see Table 6.1)

AWGIC	*Ausschuss zur Wahrung der gemeinsamen Interessen des Chemikerstandes*
Ber.	*Berichte* (see Reference List)

---

[69] *Z. angew. Chem.* (1910), **23**, 658–659.
[70] Ruske (1967), 210–214; Jaenicke (1994), 138–143.

BUTIB	Bund der technisch-industriellen Beamten
DBG	Deutsche Bunsen-Gesellschaft für angewandte physikalische Chemie
DCG	Deutsche chemische Gesellschaft
DGAC	Deutsche Gesellschaft fur angewandte Chemie
IACS	International Association of Chemical Societies
TH, THs	Technische Hochschule, Technische Hochschulen
VDC	Verein deutscher Chemiker
VzW	Verein zur Wahrung der Interessen der chemischen Industrie Deutschlands
Z. angew. Chem.	Zeitschrift für angewandte Chemie (see References)

Acknowledgments

I would like to thank the Chemical Heritage Foundation for a travel grant supporting my participation in the workshop in which I presented the first version of this chapter. I also wish especially to thank the editors of the book, Soňa Štrbáňová and Anita Kildebæk Nielsen, as well as other contributors and colleagues (in particular Ernst Homburg and Carsten Reinhardt), for their helpful comments and suggestions, which have significantly improved the chapter.

References

Abelshauser, W. *et al.* (2004), *German Industry and Global Enterprise. BASF: The History of a Company*, Cambridge University Press, New York.

Anschütz, R. (1929), *August Kekulé*, 2 vols., Verlag Chemie, Berlin.

Ber. DCG (1868–1914), **1–47**=*Berichte der Deutschen Chemischen Gesellschaft zu Berlin* ("*zu Berlin*" not used after 1876).

Bund der technischen Angestellten und Beamten (1929), *25 Jahre Technikergewerkschaft – 10 Jahre Butab*, Industriebeamten-Verlag, Berlin.

Burchardt, L. (1976), Wissenschaft und Wirtschaftswachstum: Industrielle Einflussnahmen auf die Wissenschaftspolitik im Wilhelminischen Deutschland, in *Soziale Bewegung und politische Verfassung: Beiträge zur Geschichte der modernen Welt*, Hg. U. Engelhardt, V. Sellin and H. Stuke, Klett-Cotta, Stuttgart, 775–783.

Burchardt, L. (1980), Professionalisierung oder Berufskonstruktion? Das Beispiel des Chemikers im wilhelminischen Deutschland, *Geschichte und Gesellschaft* **6**, 326–348.

Clarke, J. W. *et al.* (1914), Jährlicher Bericht des Internationalen Komitees für Atomgewichte für 1915, *Z. angew. Chem.* **27**, 653–654.

DCG Statutes (1876), Statuten der Deutschen Chemischen Gesellschaft, *Ber. DCG* **9**, 1327–1332.

DCG Statutes (1911), Statuten-Entwurf, *Ber. DCG* **44**, 156–168.

DCG-Verzeichnis (1868), Mitglieder-Verzeichniss der deutschen chemischen Gesellschaft zu Berlin, *Ber. DCG* **1**, 13–15.

DCG-Verzeichnis (1873), Verzeichniss der Mitglieder der Deutschen Chemischen Gesellschaft am 1. Januar 1873, *Ber. DCG* **6**, i–xx.

DCG-Verzeichnis (1881), Verzeichniss der Mitglieder der Deutschen Chemischen Gesellschaft am 1. Januar 1881, *Ber. DCG* **14**, i–lxiv.

DGAC (1893), Zur Geschichte der Deutschen Gesellschaft für angewandte Chemie, *Z. angew. Chem.* **6**, 555.

Duisberg, C. (1896), Ueber die Ausbildung der technischen Chemiker und das zu erstrebende Staatsexamen für dieselben, *Z. angew. Chem.* **9**, 97–111.

Fischer, E. (1987), *Aus meinem Leben*, Springer, Berlin (reprint, originally published 1922).

Flechtner, H.-J. (1981), *Carl Duisberg: eine Biographie*, Econ, Düsseldorf.

Gieryn, T. (1983), Boundary-work and the demarcation of science from non-science: strains and interests in professional ideologies of scientists, *American Sociological Review* **48**, 781–795.

Görs, B. (1999), *Chemischer Atomismus: Anwendung, Veränderung, Alternativen im deutschsprachigen Raum in der zweiten Hälfte des 19. Jahrhunderts*, ERS, Berlin.

Hoesch, K. (1921), *Emil Fischer, sein Leben und sein Werk*, Verlag Chemie, Berlin.

Homburg, E. (1992), The Emergence of Research Laboratories in the Dyestuffs Industry, 1870–1900, *British Journal for the History of Science* **25**, 91–111.

IACS (1912), Auszug aus den Protokollen der zweiten ... [April 1912] Tagung der Internationalen Assoziation der Chemischen Gesellschaften, *Ber. DCG* **45**, 1455–1463.

Jaenicke, W. (1994), *100 Jahre Bunsen-Gesellschaft, 1894–1994*, Steinkopff, Darmstadt.

Jarausch, K. H. (1990), The German Professions in History and Theory, in Jarausch and Cocks (1990), 9–24.

Jarausch, K. H. and Cocks, G., ed. (1990), *The German Professions 1800–1950*, Oxford University Press, New York.

Johnson, J. A. (1985), Academic Self-Regulation and the Chemical Profession in Imperial Germany, *Minerva* **23**, 241–271.

Johnson, J. A. (1990), Academic, Proletarian, ... Professional? Shaping Professionalization for German Industrial Chemists, 1887–1920, in Jarausch and Cocks (1990), 123–142.

Johnson, J. A. (1992), Hofmann's Role in Reshaping the Academic–Industrial Alliance in German Chemistry, in Meinel and Scholz (1992), 167–182.

Johnson, J. A. (1998), German Women in Chemistry, 1895–1925, *NTM: International Journal of History and Ethics of Natural Sciences, Technology and Medicine* **NS 6**, 1–21.

Kraemer, G. (1911), Julius Holtz, *Ber. DCG* **44**, 3395–3398.

Landolt, H., Ostwald, W. and Seubert, K. (1901), Dritter Bericht der Commission für die Festsetzung der Atomgewichte, *Ber. DCG* **34**, 4353–4384.

Lepsius, B. (1918), Festschrift zur Feier des 50jährigen Bestehens der Deutschen Chemischen Gesellschaft und des 100. Geburtstages ihres Begründers August Wilhelm von Hofmann, *Ber. DCG* **51**, Sonderheft.

McClelland, C. E. (1991), *The German Experience of Professionalization: Modern Learned Professions and their Organizations from the Early Nineteenth Century to the Hitler Era*, Cambridge University Press, Cambridge, UK, New York.

Meinel, C. and Scholz, H., ed. (1992), *Die Allianz von Wissenschaft und Industrie: August Wilhelm Hofmann (1818–1892)*, VCH Verlagsgesellschaft, Weinheim/Bergstr.

Meyer-Thurow, G. (1982), The Industrialization of Invention: A Case Study from the German Chemical Industry, *Isis* **73**, 363–381.

Ostwald, W. (1926–27), *Lebenslinien: Eine Selbstbiographie*, 3 vols., Klasing, Berlin; K. Hansel, ed. (2003), 1 vol., Verlag der Sächsischen Akademie der Wissenschaften, Leipzig.

Pinner, A. (1900), Bericht über die am 20. October 1900 erfolgte Einweihung des Hofmann-Hauses, *Ber. DCG* **33**, Sonderheft.

Rassow, B. (1912), *Geschichte des Vereins Deutscher Chemiker in den ersten fünfundzwanzig Jahren seines Bestehens*, O. Spamer, Leipzig.

Reinhardt, C. (2003), *'Öffentliche Chemie': Experten zwischen Staat und Öffentlichkeit in Deutschland, 1850–1980*, unpubl. ms., University of Regensburg.

Rocke, A. J. (1993), *The Quiet Revolution: Hermann Kolbe and the Science of Organic Chemistry*, University of California Press, Berkeley.

Ruske, W. (1967), *100 Jahre Deutsche Chemische Gesellschaft*, Verlag Chemie, Weinheim/Bergstr.

Scholz, H. (1990), *Zu einigen Wechselbeziehungen zwischen chemischer Wissenschaft, chemischer Industrie und staatlicher Administration, sowie deren Auswirkungen auf die Entwicklung der wissenschaftlichen Chemie in Deutschland …* (Dissertation B, Humboldt University, Berlin).

Stichweh, R. (1994), Professionen und Disziplinen: Formen der Differenzierung zweier Systeme beruflichen Handelns in modernen Gesellschaften, in his *Wissenschaft, Universität, Professionen. Soziologische Analysen*, Suhrkamp, Frankfurt/M., 278–336.

Titze, H. (1983), Enrollment Expansion and Academic Overcrowding in Germany, in K. H. Jarausch, ed., *The Transformation of Higher Learning 1860–1930: Expansion, Diversification, Social Opening, and Professionalization in England, Germany, Russia, and the United States*, University of Chicago Press, Chicago/Stuttgart, 57–88.

Travis, A. and Reinhardt, C. (2000), *Heinrich Caro and the Creation of Modern Chemical Industry*, Kluwer, Dordrecht, NL.

Ungewitter, C. (1927), Ausgewählte Kapitel aus der chemisch-industriellen Wirtschaftspolitik, 1877–1927, *Hg. Verein zur Wahrung der Interessen der chemischen Industrie Deutschlands*, Verlag Chemie, Berlin.

VDC statutes (1888), *Z. angew. Chem.* **1**, 335–336.

VDC statutes (1896), *Z. angew. Chem.* **9**, 391–394.

VDC statutes (1907), *Z. angew. Chem.* **20**, 392–399.

Verein zur Wahrung der Interessen der chemischen Industrie Deutschlands (1902), Fünfundzwanzig Jahre unseres Vereinslebens, *Die chemische Industrie* **25**, 405–417.

Wallach, O. (1910), Bericht der ... (Elferkommission) ... vom 28. Dezember 1910, *Ber. DCG* **43**, 3367–76.

Wentzki, O. (1912), Bezirksverein Frankfurt von 1893–1911, in Rassow (1912), 218–226.

Z. angew. Chem. (1888–1914), **1–27**=*Zeitschrift für angewandte Chemie.*

CHAPTER 7

GREAT BRITAIN: Chemical Societies and the Demarcation of the British Chemical Community, 1870–1914[1]

ROBIN MACKIE

7.1 Introduction

For much of the nineteenth century, Britain was the largest and most dynamic industrial economy in the world. The size of the British economy and of the imperial British state created many employment opportunities for chemists. It is thus not perhaps surprising that Britain was one of the first countries to develop both a professional chemical community and chemical societies. Moreover, self-governing societies, be they learned, campaigning, professional or interest-based, played a prominent role within British, and particularly middle-class, society.[2] By the First World War, British chemists operated in a dense institutional network: in 1912, *Official Chemical Appointments* (OCA), the directory of chemical posts published by the British professional association of chemists, the *Institute of Chemistry*, listed 24 "Societies and Institutions directly interested in the advancement of chemical science and technology".[3]

This complex social and professional world means that it makes little sense to discuss the demarcation of the British chemical community in terms of a single

---

[1] I would like to thank the editors of and contributors to this book and my colleagues at the Open University, Gerrylynn Roberts and Anna Simmons, for their helpful comments on this chapter.
[2] For the role of voluntary organizations in nineteenth-century in Britain, see Morris (1990). For professional associations, see Millerson (1964).
[3] *Official Chemical Appointments* (1912), 196.

Creating Networks in Chemistry: The Founding and Early History of Chemical Societies in Europe
Edited by Anita Kildebæk Nielsen and Soňa Štrbáňová
© The Royal Society of Chemistry 2008

organization. Instead, this chapter looks at the three key chemical societies, the *Chemical Society* (CS), founded in 1841, the *Institute of Chemistry* (IC), founded in 1877, and the *Society of Chemical Industry* (SCI), founded in 1881. Of these three organizations, only one, the Institute of Chemistry, defined itself as a professional body. Yet the role of each was shaped by those of the other two so that it was the constellation of these three organizations, rather than any single body, which gave form to the British chemical community in the years between 1870 and 1914. Indeed this organizational structure was to survive until the 1970s when, after discussions which had at times also involved the SCI, the CS and the IC merged with a number of smaller organizations to form the Royal Society of Chemistry. This chapter starts with the societies themselves, considering, in turn, their foundation and goals, their membership, and their activities.[4] To better understand their role, it then looks at what we know about chemists who did not join them. Finally, it asks how these societies contributed to the process of professionalization and demarcation in chemistry, which forms the core of this book.

7.2   The Foundation and Goals of the British Chemical Societies

7.2.1   The First Society: The Chemical Society

The decision to form the *Chemical Society* (CS) was taken at a meeting in London in February 1841. The goal was to bring together those interested in "the advancement of chemistry and those branches of science immediately connected with it".[5] Just months later, in April 1841, the *Pharmaceutical Society of Great Britain* was launched by pharmaceutical chemists as a response to proposed legislation which would have enshrined the medical regulation of pharmacy. Although too close a relationship between the two events should not be assumed, the contrasting directions of the two organizations reveal different goals. From its beginning, the role of the *Pharmaceutical Society* was to represent and defend an interest. To the small group of London-based chemists who launched the CS, on the other hand, it was the science that was central. Its founders included consultants, such as Robert Warington, the first secretary, and academics, such as Thomas Graham, Professor at University College, London (UCL), who became the first president. Their aim was to create a "scholarly forum" where men from different fields could meet and discuss chemistry.[6] From these beginnings, the CS went on to create an important library and a highly regarded learned journal, the *Journal of the Chemical Society (JCS)*.

---

[4] The chapter draws on the Open University project "Studies of the British Chemical Community, 1881–1972: the Principal Institutions", which has received substantial funding from the Leverhulme Trust. A major tool of the project is a biographical database which uses a wide range of sources to construct the educational and occupational routes followed by representative samples of chemists who joined the British chemical societies or were graduates of British universities. For details of the project and the sources used, see Mackie and Roberts (2001b).
[5] Quoted in Russell, Coley and Roberts (1977), 69.
[6] Bud and Roberts (1984), 49.

7.2.2   A Professional Body: The Institute of Chemistry

Since the CS was open to all those with an "interest" in chemistry, membership was by election and did not require any formal qualification. During the 1860s and 1870s, this became a matter of controversy with a minority in the CS arguing that a more formal threshold was required. The debate was provoked by changes in legislation which created a need for certified chemical expertise and by fears that status and fees were being undermined by poorly trained men using the title of Fellow of the CS to seek business. The debate was, therefore, about professionalization, and many of the most vocal critics of the CS policy were chemical practitioners who earned their living from government and consultancy work. In turn, academics were concerned that any certification should recognize academic study as well as practical experience. By the mid 1870s, demands for clearer demarcation had crystallized into a proposal for a professional association of chemists with entry based on evidence of both academic and practical expertise in chemistry. In calling for such an association, advocates drew on examples from other professions where entry was controlled and, in particular, on law and medicine.[7]

The debate culminated in the formation of the *Institute of Chemistry* (IC) in 1877. Although this organization was entirely independent of the CS, many of its founding members, such as its first president, Edward Frankland, Professor at the School of Mines, were prominent within it. Indeed, although the idea foundered for legal reasons, an early suggestion had been to create a category of "Practising Fellow" within the CS. The leaders of the two organizations saw them as performing different roles and as appealing to overlapping but different audiences: whereas the CS was a scientific and learned society, the IC was to be a professional organization built around accreditation and defence of professional status. It was hoped that these different roles, and the strong overlap between their councils, would allow the two organizations to complement each other and not compete. In calling the new organization an "Institute", the example of self-governing associations, which sought to regulate their professions, such as the *Institution of Civil Engineers* (1818) and the *Institution of Mechanical Engineers* (1847), was followed. The 1870s and 1880s saw the creation of many such professional bodies, including the *Institution of Electrical Engineers* (1871) and the *Institute of Chartered Accountants* (1880).

7.2.3   Chemists in Industry: The Society of Chemical Industry

A similar dynamic can be seen behind the formation of the *Society of Chemical Industry* (SCI) in 1881. If, in contrast to the CS and the IC, the first impetus came from chemists outside London, here too a new organization was created to cater for the needs of a distinct group of chemists, in this case chemists in industry. Once again, suggestions that this be managed within existing organizations were set aside in favour of the creation of a new one, albeit with

---

[7] This and the following paragraph rely heavily on Russell, Coley and Roberts (1977).

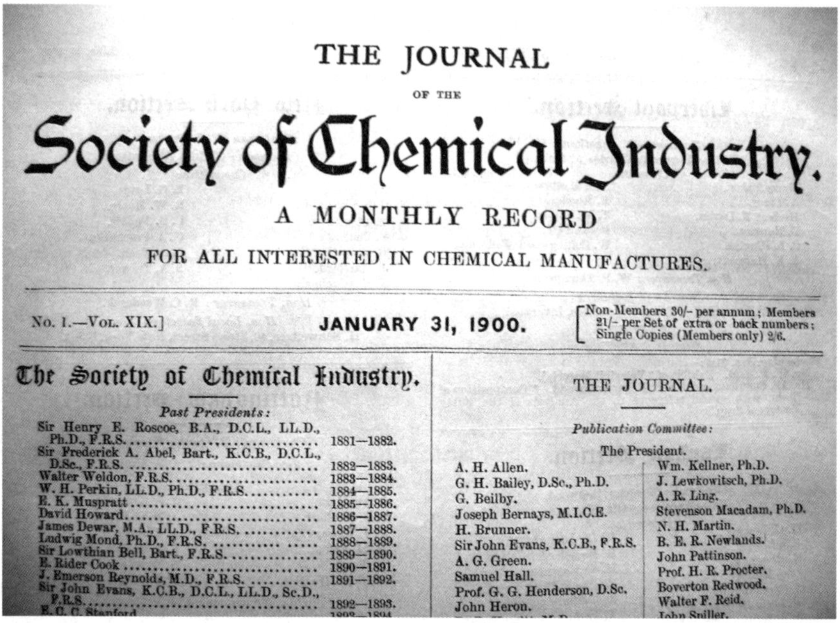

Figure 7.1 shows the heading of The Journal of the Society of Chemical Industry, 1900, containing:

THE JOURNAL

OF THE

Society of Chemical Industry.

A MONTHLY RECORD

FOR ALL INTERESTED IN CHEMICAL MANUFACTURES.

No. 1.—Vol. XIX.]   **JANUARY 31, 1900.**   Non-Members 30/- per annum; Members 21/- per Set of extra or back numbers; Single Copies (Members only) 2/6.

The Society of Chemical Industry.

Past Presidents:
Sir Henry E. Roscoe, B.A., D.C.L., LL.D., Ph.D., F.R.S. ... 1881—1882.
Sir Frederick A. Abel, Bart., K.C.B., D.C.L., D.Sc., F.R.S. ... 1882—1883.
Walter Weldon, F.R.S. ... 1883—1884.
W. H. Perkin, LL.D., Ph.D., F.R.S. ... 1884—1885.
E. K. Muspratt ... 1885—1886.
David Howard ... 1886—1887.
James Dewar, M.A., LL.D., F.R.S. ... 1887—1888.
Ludwig Mond, Ph.D., F.R.S. ... 1888—1889.
Sir Lowthian Bell, Bart., F.R.S. ... 1889—1890.
E. Rider Cook ... 1890—1891.
J. Emerson Reynolds, M.D., F.R.S. ... 1891—1892.
Sir John Evans, K.C.B., D.C.L., LL.D., Sc.D., F.R.S. ... 1892—1893.
E. C. C. Stanford ... 1893—1894

THE JOURNAL.

Publication Committee:
The President.
A. H. Allen.   Wm. Kellner, Ph.D.
G. H. Bailey, D.Sc., Ph.D.   J. Lewkowitsch, Ph.D.
G. Beilby.   A. R. Linz.
Joseph Bernays, M.I.C.E.   Stevenson Macadam, Ph.D.
H. Brunner.   N. H. Martin.
Sir John Evans, K.C.B., F.R.S. B. E. R. Newlands.
A. G. Green.   John Pattinson.
Samuel Hall.   Prof. H. R. Procter.
Prof. G. G. Henderson, D.Sc.   Boverton Redwood.
John Heron.   Walter F. Reid.
                                John Spiller.

Figure 7.1 Heading of *The Journal of the Society of Chemical Industry,* 1900.

strong involvement of CS and IC members. As with the IC, the leaders of the existing chemical societies ensured that the new one came to be defined as adding to the chemical community rather than dividing it.

By the 1870s, associations of chemists working in industry existed in the more industrial regions of England, in particular in Tyneside and Lancashire. The origins of the SCI can be traced back to events in South Lancashire in 1880–1881. The new society, however, was launched in London and the meeting was presided over by the Lancashire-based Henry Roscoe, who was Professor at Owens College in Manchester, a former president of the CS and a chemist of international reputation. The meeting was attended by many leading members of the CS and the IC, and was assured by Frederick Abel, then president of the IC (and later a president of the SCI), that the "object" of the new society, which was defined as "the advancement of manufacturing chemistry", was "entirely distinct" from that of the Institute.[8]

The SCI did indeed take a different path from both the Chemical Society and the Institute of Chemistry. In contrast to the IC, membership criteria were set broadly to include all "persons who by their attainments may be considered by the council to be eligible".[9] Whereas the CS and its journal focused on pure chemistry, the *Journal of the Society of Chemical Industry* (*JSCI*, Figure 7.1) developed into a major publication in applied chemistry. Above all, perhaps, the SCI differed from its two predecessors in being much less centred on

---

[8] "The Society" (1931), 11.
[9] *Ibid.*, 11.

London. From its beginning local sections were established in the industrial regions of the UK and later in Australia, Canada and the USA. Most of the day-to-day activities of the SCI were centred on these sections, and the SCI's Annual General Meeting (AGM) took on a different purpose. It was decided as early as 1883 that it should focus on "social intercourse, combined with the inspection of works of general scientific or special technical interest".[10] Throughout the period until the First World War, the rhetoric of the SCI emphasized its role as a social and networking organization – a "bond of union" between chemists in different industries and countries, between employers and men with technical expertise, and between science and industry.[11]

7.3 The Membership of the Societies

7.3.1 Numbers in the Three Societies

Table 7.1 gives figures for the membership of the three organizations at their launch and at five-yearly intervals from 1870 to the outbreak of the First World War. As can be seen, by 1885, the SCI was the largest of the three, and by 1890 its membership was larger than the other two organizations combined. If it grew the most rapidly, however, it was the only one not to grow constantly throughout the period, and by 1914 it had lost 400 members from its peak in 1908. Turnover was always higher than in the other two organizations. At the other extreme was the IC: in this period by far the smallest of the three, and admitting well under 50 new members in most years. Those who joined, however, mostly stayed. As the professional body, the IC had the toughest entry criteria; its modest numbers show that few chemists in this period chose to or could acquire professional certification. This situation was to change dramatically after the First World War. During the War, the *Institute* altered its entrance qualification by recognizing chemistry degrees as equivalent to its exam (thereby effectively creating two entry routes – a university route and an exam one). Thereafter its membership soared, passing that of the CS and the SCI in the 1920s and that of the two combined by 1960.

   As will be discussed below, not all members of the three societies were resident in the United Kingdom. Focusing first, however, on those who did live in the UK, the total membership of the three societies was calculated by comparing the membership lists for 1910/1912 and discounting the overlap between the societies. Just over 6% of the members of the total membership of the three societies were members of all three, and a further 20% were members of any two. Discounting this overlap gives a total for the three societies and resident in Great Britain and Ireland in 1910/1912 of just over 4600. The overlap in their councils was much greater, with over 80% of members of the IC council also in the SCI and over 90% in the CS. Even in the SCI, where council membership was most separate, well over half the council members

---

[10] Abel (1883), 303.
[11] Roscoe (1882), 1. For the early history of the SCI, see also Mackie and Roberts (2001a).

Table 7.1   Membership of the principal British chemical societies to 1914.

	Chemical Society	*Institute of Chemistry*	*Society of Chemical Industry*
At foundation	153 (1843)	392 (1878)	297 (1881)
1870	554	–	–
1875	801	–	–
1880	1034	424	–
1885	1360	436	2090
1890	1698	780	2595
1895	1999	874	2895
1900	2292	980	3459
1905	2785	1136	4326
1910	3073	1295	4299
1914	3205	1454	4142

Source: Membership data from the Annual Reports of the three societies.

were members of another chemical society. The leadership of all three chemical societies was drawn disproportionately from that small segment of their populations that was a member of all three.

The overwhelming majority of members were men. It was only in 1920 that the CS granted equal membership to women.[12] The IC admitted women from 1892 on and the SCI appears never to have excluded them, although all the founding members were male. However, on the eve of the First World War, only a handful of women (under ten in the IC) had joined either organization.

7.3.2   Home and Abroad

Many members of all three societies did not live in the United Kingdom. Table 7.2 gives the numbers and percentages living in Great Britain and Ireland and abroad on the eve of the First World War at a time when the overseas element in the societies was probably at its largest. As can be seen, significant proportions of all three societies, and particularly of the SCI, were based overseas. As Table 7.3 shows, overseas members were widely dispersed. Most of the overseas members of the CS and the IC lived in the British Empire in 1910/ 1912, whilst the overseas membership of the SCI was dominated by its large US cohort.

It would be misleading, however, to suggest that the membership of the societies can easily be divided into a "British" and a "foreign" membership. Clearly, some non-UK members will have joined despite having little connection to the UK and will have been chiefly interested in the journals. The SCI also had a large American section, which hosted the AGM in 1904 and 1912. However, most overseas members of the three societies appear to have had strong links with the United Kingdom. Many of those working in the British Empire, for

---

[12] Moore (1947), 96–7.

Table 7.2   Membership of the three societies, resident in the UK and abroad in 1910/1912.

	Total membership	Resident in UK	Resident in UK (%)	Resident abroad	Resident abroad (%)
CS 1910	3020	2463	81.6	557	18.4
IC 1912	1384	1208	86.3	176	12.7
SCI 1912	4198	2395	57.1	1803	42.9

Note: There is a slight discrepancy between the membership numbers for the CS given in the Annual Reports (used in Table 7.1) and the membership lists (used here). This appears to be due to the time of year that the figures were compiled.
Source: The membership lists of the CS (1910), the IC (1912) and the SCI (1912).

Table 7.3   Overseas membership of the three societies in 1910/1912: percentage by location.

	CS 1910	IC 1912	SCI 1912
Dominions	34.5	44.1	15.3
India	19.6	14.1	3.2
Rest of Empire	13.5	15.3	0.7
All Empire	**67.6**	**73.4**	**19.2**
USA	17.6	10.2	64.8
Europe	8.8	7.3	12.8
Rest of world	6.1	9.0	3.2
Non-Empire	**32.4**	**26.6**	**80.8**

Note: Percentages are based on full figures for the IC and samples for the CS and the SCI. Dominions here means Australia, Canada, Newfoundland, New Zealand and South Africa.
Source: As Table 7.2.

instance, had obtained qualifications at British universities either before leaving to work overseas or as citizens of the Dominions who had come to Britain to finish their studies. Furthermore, many members of all three societies had careers which took them to more than one country. Around 20% of all members of the three societies with British origins spent part of their career abroad, most of them working in two or more overseas locations. This geographical mobility took them not only to the various territories of the British Empire but also to areas of British influence such as Latin America and the Far East. The United States was also a major destination. Some chemists only went overseas for short periods, most often at an early stage of their career; others spent many years abroad, perhaps emigrating to one of the Dominions, or working for decades in the colonies before retiring to the UK. On the eve of the First World War, therefore, the membership of the three British chemical societies represented a chemical community, centred indeed on Britain, but of imperial if not global dimensions.[13]

---

[13] Mackie, Roberts and Simmons (2004).

7.4   Education and Occupations

Data on the education and occupations of chemists are based on representative samples of admissions to the three societies between specified dates and are, therefore, for consecutive cohorts of chemists and for their entire lives. Those who joined in either period will have continued to be active well into the twentieth century. 1886 was taken as the end-date for the first cohort since it marked the end of a period of grace in which the IC admitted members on the basis of long practice and reputation alone; 1917 saw significant changes in the entry route into the IC. The first cohort, therefore, corresponds roughly to the "founding" generation, at least in the IC and the SCI, whilst the second one includes individuals who joined in the period of steady growth up to the First World War.

7.4.1   Education and Qualifications

As previously noted, entry to the IC was by exam and membership was seen as a qualification. IC members were, therefore, by definition, qualified chemists. For admission to the CS and the SCI no level of qualification was specified. Many members of both societies were, of course, qualified either because they were also members of the IC, or because they held university degrees. Table 7.4 shows the percentage of members with university degrees, IC membership or neither.

The British higher education system was a complex one in which universities enjoyed great autonomy and differed enormously in character.[14] However, if the curriculum differed between universities, the overwhelming majority of degrees held by members of the chemical societies were in chemistry. Chemical engineering was barely taught at university level before 1914,[15] and pharmacy students followed a different qualification and career trajectory. The only British higher degree available before the First World War was the D.Sc., which was not necessarily based on research, and many of those with higher degrees had studied for a Ph.D. at a foreign (usually German) university. The final column in Table 7.4 reveals that a significant number in the CS and especially the SCI did not hold formal qualifications. This did not, however, mean that they had not studied chemistry at higher or further education establishments. In this period, few employers, particularly in industry, required degrees and many students left university before obtaining a qualification.[16] Britain also possessed a highly developed system of part-time and evening schools which taught large numbers of students.[17]

If university study and qualifications were not essential, they were becoming more common. A comparison of the two cohorts shows a substantial rise in the

---

[14] Sanderson (1972); Anderson (1992).
[15] The first postgraduate course in chemical engineering was offered by Battersea Polytechnic in 1909 and the first undergraduate by Glasgow University in 1923. However, numbers studying chemical engineering remained very small until the 1930s. Donnelly (1988), 573, 580.
[16] For the continued significance of study without a degree, see Donnelly (1991), 9.
[17] Pollard (1989), 115–213.

Table 7.4   Percentage of members of the three chemical societies with qualifications (1875–1917).

	n	With a university degree	With a DSc or PhD	With IC membership	With no university or IC qualification
CS (1875–1886)	1926	27	18	30	54
CS (1887–1917)	4427	42	16	26	46
IC (1877–1886)	499	36	22	100	–
IC (1887–1917)	1717	62	19	100	–
SCI (1881–1886)	2511	14	9	20	76
SCI (1887–1917)	8914	26	15	17	67

Note: Figures are based on systematic samples with a random start taken of all who joined the named institutions between the specified dates. n is here the size of the cohort from which the sample was drawn.
Source: Mackie and Roberts, 2001b.

overall percentages with university degrees and a decline in the percentages with no qualification. The trend among those with higher degrees is less clear, perhaps because the "founding" cohort included a disproportionate number of eminent chemists. In the case of the CS and the SCI there was a drop in the percentage with IC membership. If the British chemical community, as defined by the membership of the three societies, was a more highly qualified one in 1917 than in 1877, this was more the result of the rising percentage of university graduates than of recruitment into the chemists' professional association.

As one might expect, the level of qualification was considerably higher among the council members of the three societies. Fewer than 5% of CS council members and 30% of SCI council members in the 1887–1917 cohort had no qualification. Both IC and university qualifications were common, with a majority of CS council members holding a Ph.D., and most of the members of all the councils, membership of the IC.

7.4.2   Occupations

Table 7.5 gives figures on the different sectors in which chemists worked. The figures add up to well over 100% since they are for whole careers: many chemists worked in more than one sector over their lives, either holding simultaneous posts – for instance as a public analyst and a consultant – or, more commonly, consecutively. The first five columns show the principal sectors in which chemists worked. The final column covers a wide range of posts, including pharmacists, an important component of the early membership of chemical societies in some other European countries but a negligible group in the UK.

Table 7.5 serves not only to demonstrate the wide range of sectors in which chemists worked, but also challenges any expectation that the three organizations served distinct groups defined by workplace. The CS, with its emphasis on pure science, might have been expected to appeal particularly to those working

Table 7.5   Percentage of members of the three chemical societies working in different sectors (1875–1917).

	n	Academia	Schools	Consulting	Government	Industry	Other
CS (1875–1886)	1926	26	8	25	11	26	16
CS (1887–1917)	4427	35	19	25	18	49	14
IC (1877–1886)	499	47	10	52	33	47	14
IC (1887–1917)	1717	36	10	36	37	57	11
SCI (1881–1886)	2511	13	3	20	9	56	11
SCI (1887–1917)	8914	16	2	15	9	74	12

Note: "Academia" includes both universities and technical colleges; "consulting" means independent practice. Posts that fit none of the other categories (*e.g.* work as a medical doctor or for public utilities) are classified as "other". Percentages add up to well over 100, because many members held posts in more than one sector. See also Note to Table 7.4.
Source: As Table 7.4.

in education, yet it had members working in all the sectors defined. As one might expect, the IC was strong in consulting and government – sectors where its emphasis on certification might be expected to be most relevant – yet it had as many members who worked in academia and more in industry. Even the SCI, which defined itself in terms of industry, had significant numbers who worked in other sectors (and, indeed, up to a quarter of its members appear never to have worked in industry). Over time, there was some polarization between cohorts, with the CS showing rising percentages in academia and schools, and the IC a falling percentage in academia. Independent consultancy appears to have become less significant, particularly in the IC, although the figure for the first cohort may be the result of the rather narrow circles from which the founders were drawn. By contrast, government employment rose. Overall, the most striking change was the higher percentage working in industry in all three societies in the later cohort. Although these figures are per cohort, rather than at a specific moment in time, they confirm the trend towards industry highlighted by James Donnelly.[18]

In all three societies, there were important differences between the occupations of the councils and of the ordinary members. The careers of council members were rather more varied than those of ordinary members, with the overwhelming majority working in more than one of the sectors defined in Table 7.5. Furthermore, in each society the council had characteristics which set them apart from the wider membership. In the CS, academia was dominant, with almost 90% of council members spending at least part of their careers in education. In the IC, academics were also over-represented in the council compared to the broader membership, and so too were chemists working in government. In the SCI, industrial employment was more common than in the other councils, but here too academics and consultants were over-represented. If the councils can be seen as the elite of the chemical community, it was drawn disproportionately from chemists working in the universities, government or as consultants.

---

[18] Donnelly (1991), 8–14.

7.5   The Activities of the Chemical Societies

If the membership of the CS, the IC and the SCI had much in common, there were major differences between their activities. This is perhaps not surprising: the three societies were seeking to recruit members from the same pool of chemists, but had been set up for different purposes. Perhaps the clearest way to bring out these distinctions is to compare the activities of these three large and active organizations during one year. The year 1900 has been selected for this purpose.

7.5.1   The Chemical Society in 1900

The *Chemical Society* published two journals in 1900, both aimed primarily at members. Its first regular publication, the *Quarterly Journal* (1848–1862), had been replaced in 1862 by the *Journal of the Chemical Society* (*JCS*), which, since 1878, had consisted of two sections, *Transactions* and *Abstracts*. The second journal, *Proceedings of the Chemical Society* (*PCS*, Figure 7.2), was launched in 1885 with the aim of ensuring that members received promptly details of the papers presented at society meetings. *PCS* appeared shortly after meetings of the CS and contained information on the business conducted, including summaries of papers presented and short profiles of candidates for election. *JCS* appeared monthly and was largely devoted to scholarly papers. The

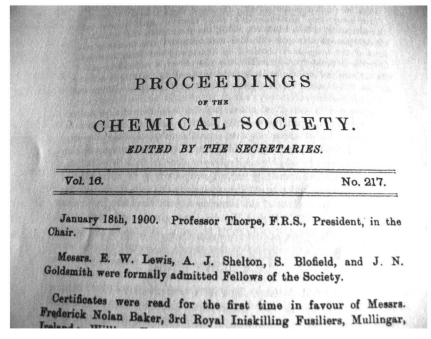

Figure 7.2   Heading of *Proceedings of the Chemical Society*, 1900.

number of papers processed was impressive: at the 1900 AGM, the president, Professor Thomas E. Thorpe, reported that in the previous year, 175 papers had been presented to the society, 83 of which had already appeared in the *Transactions* section together with 37 papers from previous years. Meanwhile *Abstracts* had reported on 3617 papers, mostly from foreign journals. Of these, 1477 were on organic chemistry, with 611 on analytical chemistry forming the second largest group.[19] Although no breakdown of the papers in *Transactions* is provided, the titles of articles suggest a similar distribution. In 1900, 111 authors published in *Transactions* and they included many eminent British chemists and several from abroad. Most were members of the CS, but at least 20, including two female contributors, were not.

Publication dominated the life of the organization: indeed the president reported that "the increasing mass of literature" was "a matter of much concern" and threatened "to assume unwieldy proportions".[20] In the financial year ending in 1900, of the total expenditure of £4994, £3389 was spent on the *JCS* and £172 on the *PCS*; the membership fees (which stood at £2 a year) accounted for £4088 of the annual income of £5371.[21] The *Chemical Society*'s other activities were centred on the fortnightly "scientific meetings", held from October to June at the society's Burlington House headquarters in central London. Several papers were presented at each meeting, whilst further papers were received but not read; meetings were often preceded or followed by informal social gatherings in nearby Piccadilly.[22] Burlington House also contained a substantial library. The two awards of the society – the Faraday Lecture and Medal (awarded since 1869 to eminent non-British chemists) and the Longstaff Medal (for distinguished research, first awarded 1881) – celebrated scholarly achievement. At the 1900 AGM, the latter was awarded to William Henry Perkin, Junior, Professor of Organic Chemistry at Owens College, Manchester.

The learned focus of the CS gave it the strongest international links of the three societies. From its inception, and starting with Justus von Liebig, some of Europe's leading chemists were elected as honorary Foreign Members; by 1900, there were 33 such members. The deaths of Foreign Members were commemorated by Memorial Lectures and these provided occasions to celebrate and reinforce links to chemists abroad. Thus, for instance, in 1900, Roscoe was invited to give a Memorial Lecture for his former professor, Robert Bunsen, a Foreign Member since the foundation of the society. The death of Lars Fredrik Nilson was commemorated in a lecture by Otto Pettersson, his colleague from the Free University of Stockholm.[23] Faraday Lectures, given by such distinguished chemists as Cannizzaro, Hofmann and Mendeleev, were a further opportunity to strengthen ties with the elite of Europe's chemists. Yet, in 1900, international ties appear to have operated primarily at a personal level. In that

---

[19] "Annual General Meeting" (1900), 557–558.
[20] *Ibid.*, 557–558.
[21] *Ibid.*, 586–588. For comparative purposes £1 in 1900 was worth approximately £65 in 2000.
[22] Moore (1947), 87–88.
[23] "Annual General Meeting" (1900), 556–557.

year, no mention of formal links with foreign chemical societies was made in any of the CS's journals and only with the formation of the *International Association of Chemical Societies* in 1911 do fraternal ties appear to have figured more prominently on their agenda.[24]

7.5.2   The Institute of Chemistry in 1900

The *Institute of Chemistry*, as it appears through the pages of the *Proceedings of the Institute of Chemistry* (*PIC*, published 1877 onwards, Figure 7.3) in 1900, was a very different organization. In his address to the AGM in March 1900, the outgoing president of the IC, Thomas Stevenson, Senior Analyst at the Home Office, reminded his audience that the purpose of the IC was to promote better education, to examine and "grant certificates of competency" and to "elevate the profession of Consulting and Analytical Chemistry".[25] He organized his annual report around these three themes – discussing the raising of educational standards, improved training in bacteriology, the results of the IC's exams, the increasing recognition of its diplomas and the representations made by the IC to the government in relation to the 1899 Sale of Food and Drugs Act. *PIC* focused narrowly on these concerns. In contrast to the *JCS*, there were only two issues a year and no scientific papers were published. Instead the journal reported on the preparation and examination of candidates for membership (exam papers were printed in full), changes in government regulations which might impact on the practice of chemistry and the administration of the organization.

With no tradition of scientific meetings, the only open meetings of the IC were the AGM and the annual dinner, a formal event to which leading figures from government, education and kindred learned and professional societies were invited. Between AGMs, the business of the IC was conducted by its council, a number of committees, of which the Nominations and Education Committee was by far the largest, and the full-time registrar and secretary. Most of the IC's annual expenditure of just over £1500 went on costs associated with the London office, which contained a library with works relevant to students of chemistry and laboratories for exams. Income came mostly from the membership (£957 in 1899; the annual subscription was a guinea – £1, 1 shilling) and examination fees (£310 in 1899). The cost of admission to the IC was high: 10 guineas for the intermediate and final exams for the Associateship, and a further 5 guineas to become a Fellow.[26] Membership of the IC was restricted to British and Empire citizens and the main overseas dimension of its work was to provide exams and professional services for candidates and members in the Empire. By 1912, the IC had fifteen Corresponding Secretaries in nine Dominions and colonies.[27]

---

[24] Moore (1947), 33–38, 229.

[25] Stevenson (1900), 20.

[26] "Balance Sheet for 1899"(1900), 16; *Official Chemical Appointments* (1912), 199.

[27] *Official Chemical Appointments* (1912), 200.

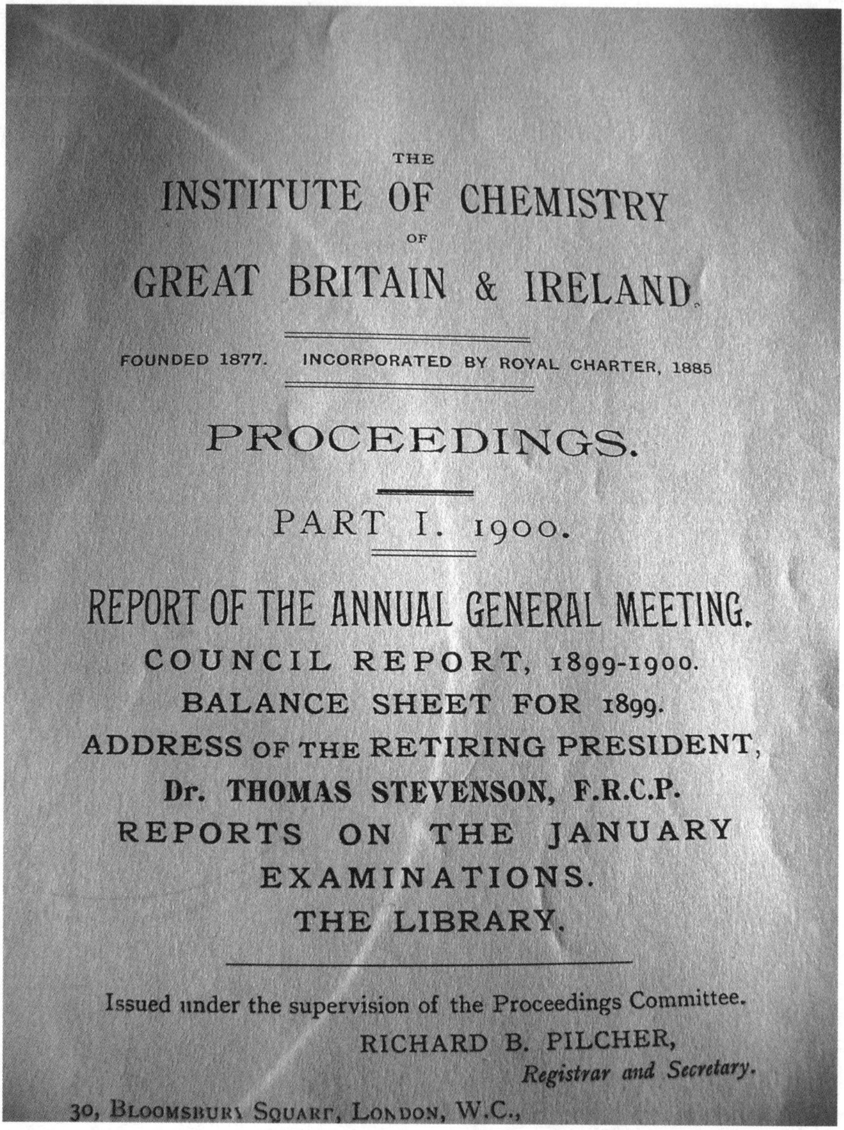
THE
INSTITUTE OF CHEMISTRY
OF
GREAT BRITAIN & IRELAND.

FOUNDED 1877. INCORPORATED BY ROYAL CHARTER, 1885

PROCEEDINGS.

PART I. 1900.

REPORT OF THE ANNUAL GENERAL MEETING.
COUNCIL REPORT, 1899-1900.
BALANCE SHEET FOR 1899.
ADDRESS OF THE RETIRING PRESIDENT,
Dr. THOMAS STEVENSON, F.R.C.P.
REPORTS ON THE JANUARY
EXAMINATIONS.
THE LIBRARY.

Issued under the supervision of the Proceedings Committee.
RICHARD B. PILCHER,
Registrar and Secretary.
30, BLOOMSBURY SQUARE, LONDON, W.C.,

Figure 7.3 Heading of *Proceedings of the Institute of Chemistry*, 1900, with the Council Report 1899–1900.

7.5.3 The Society of Chemical Industry in 1900

That the *Society of Chemical Industry* was less London dominated than the other two organizations is evident from the society's journal, the *Journal of the Society of Chemical Industry* (*JSCI*, launched 1882, monthly to 1902, then fortnightly). About half of each issue was taken up with reports from the local sections of the SCI, of which there were eight in 1900. If the largest of these was

London, with 865 members, the New York section was only slightly smaller, and Manchester had 450 members.[28] Most of the sections held at least one meeting a month at which papers were presented, mostly on technical subjects, and these papers were reported in the journal. Most speakers were members, and they included consultants and industrial chemists as well as academics. The other half of *JSCI* was devoted to abstracts, reports of new patents and trade reports. The abstracts, which were drawn from a wide range of journals, were on technical and applied chemistry, with sections ranging from chemical apparatus to brewing and photography.

The AGM was the only national meeting of the SCI and was held in a different location each year. It was a major event: in 1900, it was in London and lasted three days. On the morning of the first day, a business meeting was held where the highlight was an address by the outgoing president Charles F. Chandler, Professor at the Columbia School of Mines and the first US president of the SCI, on "Chemical and Technical Education in the United States". The SCI Medal was presented to Edward Schunck for his research on dyestuffs. The rest of the three days was given over to visits to a wide range of government and academic laboratories and factories in the London area. Nor was the social side neglected: as well as the annual dinner and a reception by the Lord Mayor, excursions were organized to the National Gallery of British Art and to Oxford. On the fourth day, a "party, numbering about 150, approximately one-third of which were ladies", set off for Paris to visit the International Exhibition and the Congress for Applied Chemistry. Events in Paris lasted a full week.[29] That the president of the SCI should be an American and that the AGM should include a trip to Paris and an international chemical conference is a reminder of how important international links were to the SCI – almost half of its members, after all, were not resident in the British Isles (see Table 7.2). Building international ties – particularly within the English-speaking world – was an important part of the rhetoric of the society.[30] Yet the emphasis was on international links within the SCI rather than external ones.

As for the CS and the IC, the membership provided the main source of the SCI's income. In the year ending in 1900, £3872 out of a total income of £5203 came from membership subscriptions, set at 25 shillings per year, with a further £568 coming from advertisements in *JSCI*. *JSCI* was also the main expense for the SCI, requiring £3225 out of a total expenditure of £5330, with most of the rest going on the London headquarters, headed by a salaried general secretary.[31]

7.6 Chemists Outside the Chemical Societies

This chapter has so far focused on the chemical societies and those chemists who chose to join them. Yet, membership of any of the societies was never a

---

[28] "Proceedings of the 19th AGM" (1900), 590.
[29] *Ibid.*, 636.
[30] Mackie and Roberts (2001a), 135–136.
[31] "Statement of Revenue" (1900), 495.

criterion for employment and the numbers who chose not to join any society were always significant. In Scotland, for instance, in 1912, 46% of holders of academic or government posts do not appear to have been members of the CS, the IC or the SCI. To be sure, many of these were in junior posts and may well have gone on to join one of the societies later in their careers. But, if this was the nature of progression, it suggests that in this period membership of one of the societies may not so much have been a necessary condition for, as a consequence of, a successful career. Certainly, there is little evidence that industrial employers expected professional membership or even university degrees in new appointments.

If admission to one of the three chemical societies was not an essential criterion for entry to the profession, as admission to the professional society was for doctors or lawyers, why then did some chemists choose to join and others not? It might be that the societies represented the elite of the profession, and those less highly qualified chemists or chemists in lower status jobs were excluded by formal or informal mechanisms. This might be seen as an issue of demarcation, as the societies used their entry requirements to keep out individuals whom they thought lacked the necessary chemical expertise or status. As we have seen, the IC examinations were intended to be used in this way and did in fact exclude chemists without the required training. There is little evidence, however, that the more informal CS and SCI admission requirements were so used.[32] In explaining why chemists joined societies, we need perhaps to focus less on the societies and more on the costs and benefits to individuals. As is evident from the preceding analysis, membership of the different societies differed in terms of both the effort and cost required to gain admission and the benefits that membership bestowed. Which society chemists chose to join and whether they chose to join any may have had most to do with how they saw their relationship to the chemical community. For some chemists access to scientific or technical information will have been important, for others accreditation or contacts may have played a greater role.

One way to explore this is to look at a group of chemists who had the necessary background to join but chose not to do so, that is university chemistry graduates who entered careers in chemistry but did not join any of the societies. The records kept by a number of British universities make it possible to do this, although the differing purposes for which the records were created and the way they have been compiled means that straightforward comparisons are not possible.

Table 7.6 gives estimates for the number of graduates joining the different chemical societies for a few major British universities. In the case of Edinburgh and Glasgow universities, the figures are for all graduates with chemical posts in 1912; for the English universities, they are for holders of first degrees and the society memberships are for any point in their lives. It is possible that differences in the percentages of graduates from the different universities

---

[32] For instance, all the candidates for membership to the CS in 1900 whose details were printed in *PCS* were elected, including some with few formal qualifications.

Table 7.6 Percentage of graduate "chemists" from different universities joining the three chemical societies.

	n	CS	IC	SCI	In none of three societies
Edinburgh University 1912	71	28	13	28	55
Glasgow University 1912	48	45	11	36	36
University College London, 1870–1918	266	39	45	27	43
Manchester University, 1904–1918	43	33	44	23	40
Royal College of Science, 1904–1918	26	27	38	35	46
Cambridge University, 1904–1918	12	67	50	42	8

Note: The figures for the two Scottish universities are based on registers of graduates compiled
annually by the universities for governance purposes and list details of all graduates up to the year
of publication. Data are for all individuals where the occupation (or, in a few cases, the workplace)
in the 1912 registers suggests that the graduate held a chemistry-related post. Membership of the
chemical societies was calculated by cross-reference with the membership lists of the chemical
societies for 1910/1912. The figures for the English universities are from published graduation lists
and are for chemistry graduates obtaining a B.Sc. or equivalent. In each case, names have been
checked against the membership lists of the societies and indicate membership of a society at any
point in their lives.
The figures for Edinburgh, Glasgow and UCL are for all graduates; those for the other universities
are based on samples.
Source: As Table 7.4.

joining one of the societies are due to the way figures were compiled, although
this does not explain the difference between Edinburgh and Glasgow, or the
considerable variation between the English universities. Other possible factors
include metropolitan bias (a possible explanation for the low IC percentages in
the Scottish universities), the sort of posts graduates of different universities
were likely to take up (which did vary enormously) or even the organizational
allegiances of academic staff. The figures do suggest, however, that significant
numbers of graduates chose not to join any of the chemical organizations.

In the case of Edinburgh and Glasgow universities, it was possible to make a
comparison between members and non-members. Although differences between graduates of the two universities were often more substantial, members
of the three chemical societies were more likely than non-members to have
higher degrees, more likely to be living outside Scotland and more likely to be
working in education. The percentages of members and non-members working
in government and industry were remarkably similar. One way to interpret such
figures might be to think in terms of need: chemists who worked in universities
or who left Scotland may have seen greater advantage in the certification of
their qualifications and a network of professional contacts than those who
found work closer to home. Although less research has so far been done on
comparing the careers of members and non-members among graduates of the
English universities, it appears that UCL graduates who joined one of the

societies were more likely to hold a higher degree than those who did not. Female graduates (14% of UCL graduates) were also far less likely to join one of the chemical societies.

7.7   Professional Demarcation

Classic accounts of the development of professions in Britain have emphasized self-regulation. Starting from the examples of medicine and law, such accounts have focused on qualifying institutions and their ability to control entry to the profession.[33] The founders of the IC were aware of this model when they launched their organization, and consciously followed it in certain respects. Yet by such standards their attempts to "professionalize" chemistry in the United Kingdom can only by judged a partial success. True, a qualifying institution was established and its exams became a standard for professional accreditation. But the IC did not win control of entry to professional posts, and many chemists, including leading ones, chose not to join it. Indeed, their membership as a proportion of working chemists in the UK may have been lower in 1914 than in 1886. Table 7.4 shows that the percentage of CS and SCI members with IC qualifications declined from the first to the second cohort, whilst Table 7.6 reveals that well over half of the graduates of the universities surveyed chose not to join the IC. For such graduates, the certification provided by their university qualifications appears to have been sufficient in their chosen careers. By the early twentieth century, with the number of graduates in chemistry rising more rapidly than the IC membership, there was a real risk that it would become increasingly irrelevant. This was a key factor in the decision of the IC to change its entry requirements in 1917 to facilitate entry for chemistry graduates.[34]

Models of professionalization which focus on the role of qualifying institutions are rightly criticized for relying too heavily on Anglo-Saxon examples.[35] Ernst Homburg argues that it is more useful to define professions in terms of occupations and see chemists as a "modern" occupation or profession, that is as one where skills can be applied in "very different positions on the basis of a more-or-less uniform professional training". Professionalization is the process by which posts requiring these professional skills come to be defined, first in education and then later in other sectors.[36]

If these definitions of profession and professionalization are used, what then was the contribution of the chemical societies? It is clear that new professional posts were being created in late-nineteenth-century Britain, but this was primarily the consequence of economic, social and political changes over which the societies could claim little control. However, if we shift the emphasis from demarcation in terms of limiting entry to demarcation in terms of defining roles

---

[33] Millerson (1964).
[34] Russell, Coley and Roberts (1977), 174–180.
[35] Gispen (1988); Lundgreen (1990).
[36] Homburg (1998), 42.

and extending boundaries, their role emerges. The threat of competition from unqualified practitioners was only intermittently a theme in internal debates, for instance, in the 1870s when the IC was founded and on the eve of the First World War, when ordinary chemists sought a more militant defence of their status.[37] A far more consistent theme was the utility of chemistry, the "universal" science. Chemical leaders were anxious to stress the value of their science across a wide range of fields, and in particular in industry where more chemists were finding employment. In talking up the value of the contribution that chemists could make, the leaders of the chemical societies can be seen as seeking to widen the boundaries of their profession's territory. When chemists joined the organizations, in addition to, or perhaps in preference to, other occupational groups they might have joined, from teachers to dye-makers, they can be seen as identifying with this process of defining a broad chemical profession. Chemical societies contributed to the demarcation of a successful chemical profession by persuading chemists and non-chemists that they represented an occupation with needed skills.[38]

In this process, the existence of several chemical societies served the British chemical community well in the pre-1914 period. The different organizations were able to pursue different but complementary goals. The CS retained its status as the leading scientific society. IC exams met the needs of the minority of chemists for whom professional accreditation was essential. And the SCI brought in many chemists who were unlikely to join either a purely learned body or one requiring qualifications they did not need. The risk, of course, was that multiple organizations would lead to the fragmentation of the community. That this was a concern of the leadership can be seen in the other major strand in their rhetoric: the unity of chemistry. This was discussed not only in discipline terms but also in occupational ones. A constant refrain was that any division between theoretical and practical science was a false one, that "in industrial science they simply applied the facts of science".[39] That fragmentation did not happen can indeed partly be explained by the efforts of the leadership, which, as we have seen, was far more integrated than the wider membership of the societies. The many leading chemists who held senior academic posts had a particular interest in preserving the unity of the community, since this ensured the widest opportunities for their students. Yet unity may also have been maintained because the membership of the societies was less segmented and more mobile than was perhaps understood. Although there were differences between the societies, all included members from a range of occupations. Moreover, significant numbers of chemists were active in several occupational sectors. To a degree which is perhaps surprising it appears that the rhetoric did indeed reflect reality: chemists did possess transferable skills which allowed them to work in education, government or industry and to switch between posts during their careers.

[37] Russell, Coley and Roberts (1977), 174–180; Donnelly (1996), 786–789.
[38] For an analysis of the development of professions in these terms, see Abbott (1988).
[39] Sir W. Roberts-Austen, quoted in "Opening of the John Cass Institute" (1902), 287.

7.8 Conclusion

As we have seen, the nineteenth century saw the foundation of three major chemical societies in the United Kingdom, all of which had developed by 1914 into broad-based and active organizations. That the organization of the chemical community was divided between three societies might have been a weakness, but in fact had many advantages. It is perhaps helpful to see the British chemical community as balancing between the rhetorics of utility and of unity, between desires to expand the boundaries of the professional sphere and fears of fragmentation. From this perspective, one can perhaps detect change over time. If, in the period before the First World War, the desire to expand was in the ascendant, later, in the more troubled environment of the inter-war years, more emphasis came to be placed on fears of fragmentation. One consequence of this was concerns about the price of division and the first real efforts to bring about mergers between the chemical societies.[40]

Abbreviations

AGM	Annual General Meeting
CS	*Chemical Society*
IC	*Institute of Chemistry*
JCS	*Journal of the Chemical Society*
JSCI	*Journal of the Society of Chemical Industry*
OCA	*Official Chemical Appointments*
PCS	*Proceedings of the Chemical Society*
PIC	*Proceedings of the Institute of Chemistry*
SCI	*Society of Chemical Industry*
UCL	University College, London

References

– (1900), Annual General Meeting, *Transactions of the Chemical Society* **77**, 555–590.
– (1900), Balance Sheet for 1899, *Proceedings of the Institute of Chemistry* Part 1, 16.
– (1900), Statement of Revenue and Expenditure, *Journal of the Society of Chemical Industry* **19**, 495.
– (1900), Proceedings of the Nineteenth Annual General Meeting, *Journal of the Society of Chemical Industry* **19**, 589–636.
– (1902), Opening of the Sir John Cass Technical Institute, Aldgate, *Journal of the Society of Chemical Industry* **21**, 287.
– (1912), *Official Chemical Appointments*, 4th edition, Institute of Chemistry, London.

---

[40] For changing views within the SCI, see Mackie and Roberts (2001a); for merger discussions between the wars, see Russell, Coley and Roberts (1977), 236–263.

– (1931), The Society of Chemical Industry, *Journal of the Society of Chemical Industry* Jubilee Number, 9–21.

Abbott, Andrew (1988), *The System of Professions. An Essay on the Division of Expert Labor*, Chicago University Press, Chicago.

Abel, Frederick A. (1883), President's address, *Journal of the Society of Chemical Industry* **3**, 302–316.

Anderson, Robert (1992), *Universities and Elites in Britain since 1800*, Cambridge University Press, Cambridge.

Bud, Robert and Roberts, Gerrylynn K. (1984), *Science versus Practice. Chemistry in Victorian Britain*, Manchester University Press, Manchester.

Donnelly, J. F. (1988), Chemical engineering in England, 1880–1922, *Annals of Science*, **45**, 555–590.

Donnelly, James (1991), Industrial recruitment of chemistry students from English universities: a revaluation of its early importance, *British Journal of the History of Science* **24**, 3–20.

Donnelly, James (1996), Defining the industrial chemist in the United Kingdom, 1850–1921, *Journal of Social History* **29**, 779–796.

Gispen, C. W. R. (1988), German engineers and American social theory: historical perspectives on professionalization, *Comparative Studies in Social History* **30**, 550–574.

Homburg, Ernst (1998), Two fractions, one profession: the chemical profession in German society, 1780–1870, in David Knight and Helge Kragh (eds.), *The Making of the Chemist. The Social History of Chemistry in Europe, 1789–1914*, Cambridge University Press, Cambridge, 37–76.

Lundgreen, Peter (1990), Engineering education in Europe and the USA, 1750–1930: the rise to dominance of school culture and the engineering professions, *Annals of Science* **47**, 33–75.

Mackie, Robin and Roberts, Gerrylynn K. (2001a), Un secteur à part? Les chimistes industriels et la *Society of Chemical Industry* dans le contexte de la communauté chimique britannique, in Ulrike Fell (ed.), *Chimie et Industrie en Europe. L'apport des sociétés savantes industrielles du XIXe siècle à nos jours*, Éditions des archives contemporaines, Paris, 127–147.

Mackie, Robin and Roberts, Gerrylynn K. (2001b and regularly updated), Biographical Database of the British Chemical Community, http://www.open.ac.uk/ou5/Arts/chemists/index.htm.

Mackie, Robin, Roberts, Gerrylynn K. and Simmons, Anna E. (2004), The circulation of expertise: British chemists abroad, 1890–1939, unpublished paper delivered at the Imperial Globalization Conference, Bristol, 10–11 September.

Millerson, Geoffrey (1964), *The Qualifying Associations: A Study in Professionalization*, Routledge & Kegan Paul, London.

Moore, Tom Sydney (1947), *The Chemical Society, 1841–1941, A Historical Review*, The Chemical Society, London.

Morris, R. J. (1990), *Class, Sect and Party. The Making of the British Middle Class, Leeds 1820–1850*, Manchester University Press, Manchester.

Pollard, Sidney (1989), *Britain's Prime and Britain's Decline, 1870–1914*, Edward Arnold, London.

Roscoe, Henry E. (1882), President's address, *Proceedings of the First General Meeting*, The Society of Chemical Industry, Manchester, 1–7.

Russell, Colin A., Coley, Noel G. and Roberts, Gerrylynn K. (1977), *Chemists by Profession. The Origins and Rise of the Royal Institute of Chemistry*, Open University Press, Milton Keynes.

Sanderson, Michael (1972), *The Universities and British Industry, 1850–1970*, Routledge & Kegan Paul, London.

Stevenson, Thomas (1900), Address of the Retiring President, *Proceedings of the Institute of Chemistry of Great Britain and Ireland*, Part I, 18–27.

CHAPTER 8

HUNGARY: Scientific Community of an Emancipating Nation: Chemical Societies in Hungary before 1914

ÉVA KATALIN VÁMOS

8.1 Introduction

The period indicated in the title of this chapter is, in Hungarian history, part of the epoch of the Dual Monarchy, or Dualism (1867–1918), in which Hungary was relatively independent and had good development chances. By the end of the nineteenth century its scientific institutions were firmly established in their forms still accepted today. In new university buildings, relatively well-equipped laboratories allowed research at European level. Exactly for that reason a number of the scientists that obtained university chairs at the turn of the century strived to address problems of their speciality of world interest from Budapest. They published the results of their research in journals of international reputation.

The scientific world of the epoch of Dualism focused, in Hungary, around a single centre, Budapest, and this remained preserved, so to speak, for nearly a century.

The school-founding Hungarian professors of chemistry of the turn of the nineteenth and twentieth centuries participated in study journeys abroad, wrote excellent textbooks and carried out research on an international level in the fields of physical chemistry, analytical and inorganic chemistry, organic chemistry, agricultural chemistry and industrial chemistry. The Hungarian pharmaceutical industry reached international fame. The important steps of the

Creating Networks in Chemistry: The Founding and Early History of Chemical Societies in Europe
Edited by Anita Kildebæk Nielsen and Soňa Štrbáňová
© The Royal Society of Chemistry 2008

progress of chemistry were, however, not directed from Budapest. The Nobel laureates of chemistry of Hungarian origin belonged to the next generations.

The present *Magyar Kémikusok Egyesülete* (Hungarian Chemical Society, MKE[1]) was founded in 1907. Before its coming into being – and still in parallel to it – another four societies and sections of societies, respectively, represented the science of chemistry, the chemical industry and the safeguarding of chemists' interests.[2] These were:

- the Class of Mathematics (later Mathematics and Natural Sciences) of the Hungarian Academy of Sciences (*A Magyar Tudományos Akadémia, MTA, Matematika és Természettudományok osztálya*, 1830);
- the Section of Chemistry of the Royal Hungarian Society for Natural Sciences (*Királyi Magyar Természettudományi Társulat* – KMTT – *Kémiai Szakosztálya*, 1841), later Section of Chemistry and Mineralogy (*Királyi Magyar Természettudományi Társulat Kémiai és Ásványtani Szakosztálya*, KAS, 1892);
- the Association of Chemical Industrialists (*Vegyészeti Gyárosok Országos Egyesülete*, VEGYOE,1904);
- the Section of Chemical Engineering of the Association of Hungarian Engineers and Architects (*Magyar Mérnök- és Építészegylet Vegyészmérnöki Szakosztálya*, – MÉ, 1906).

The work achieved in parallel by the above societies will be sketched in brief from their beginnings to the outbreak of the First World War. Special attention will be given hereby to the similarities/overlaps and differences in their work as well as to the aspects making it necessary to establish a fifth society of chemistry, MKE, which will be described in detail in the body of the chapter.

8.2 Scientific/Industrial Chemical Societies/Sections of Societies Established before 1907

8.2.1 The Class of Mathematics and Natural Sciences of the Hungarian Academy of Sciences (MTA)

From the original six classes of the Hungarian Academy of Sciences (*Magyar Tudományos Akadémia*, MTA) founded in 1825 there remained only five after the Compromise (1867) as the Classes of Mathematics and of Natural Sciences merged. Its members were, of course, exclusively corresponding and regular members of MTA. It had 20 chemist members, out of whom three were, at different periods, vice-presidents of the Academy.[3] Its journal, *Mathematikai és*

---

[1] For future reference, see the list of abbreviations.
[2] We will not deal here with the Trade Union of Chemical Workers, which was essentially an institution safeguarding the workers' interests in the chemical industry and other institutions wherever chemists/workers of the chemical industry were working.
[3] See Szabadváry and Szőkefalvi-Nagy (1972), 215–216.

Természettudományi Füzetek (Brochures of Mathematics and Natural Sciences), called from 1861 *Mathematikai és Természettudományi Közlemények* (Communications in Mathematics and Natural Sciences), was brought into being by Antal Csengery.[4] It dealt with, besides mathematics, all the branches of sciences that constituted MTA3 as did the scientific sessions of the class.

8.2.2 *Királyi Magyar Természettudományi Társulat* (KMTT) *Kémiai és Ásványtani Szakosztálya* (KAS) – (Section of Chemistry, later Chemistry and Mineralogy of the Royal Hungarian Society for Natural Sciences)

According to the objectives of the Hungarian Academy of Sciences, the following text was enacted in 1827: "The Hungarian Society of Scholars strives to improve the national language in all the branches of sciences and arts."[5] In the activities of MTA, cultivation of the Hungarian language and humanities was prevailing in the nineteenth century. Thus the other learned society, the Royal Hungarian Society for Natural Sciences (*Királyi Magyar Természettudományi Társulat* – KMTT), acted in many respects as the Academy of Natural Sciences. It seems that by 1841 the élan of the Reform period ripened the demand for acquainting the wider public with natural sciences. In 1841, the *Magyar Orvosok és Természetvizsgálók Vándorgyûlései* (Itinerary Congresses of Hungarian Physicians and Naturalists) were founded. It was at their first soirée that the assembled scholars brought into being the Royal Hungarian Society for Natural Sciences.[6]

During the first 50 years of the society's existence the sessions of the different branches of natural sciences were not held separately, thus "a physicist was obliged to listen to a lecture on phloristics, while a botanist possibly had to follow a long theoretical deduction in physics".[7] In 1887, university professor and member of MTA József Fodor,[8] then secretary of the society, formulated as follows: "I hardly can imagine a society which would be able to promote and vividly develop zoology, botany, physics, chemistry, biology, astronomy and anthropology; however, as I see it, separate societies for the scientific purposes of zoology, botany, chemistry, physics, *etc.*, could well and successfully work."[9] As a result of these ideas the short-lived and soon-dissolved *Magyar Chemiai*

---

[4] A. Csengery (1822–1880) was a politician, journalist, economist, historian and member of MTA. See Új magyar Lexikon (1959), **1**, 501; Gombocz (1941), 108.

[5] Vázlatok (1881), p. 3.

[6] Szabadváry and Szőkefalvi-Nagy (1972), 216–217.

[7] Gombocz (1941), 218.

[8] J. Fodor (1843–1901) graduated from Pest University of Sciences as M.D. in 1865. First he was assistant at the Department of Forensic Medicine. His scientific career started in 1868. In 1869 he became private docent, then went abroad for studies in various European countries. In 1872 he was appointed professor by the Department of Forensic Medicine of recently founded Kolozsvár University of Sciences, and in 1874 first professor of the Department for Public Health, which he had to organize and install. He might be considered the pioneer of domestic public hygiene. See Lambrecht (1997).

[9] See Gombocz (1941), 219.

Társulat (Society of Hungarian Chemists) was launched, which should not be confused with the Hungarian Chemical Society. Professor of chemistry Béla Lengyel,[10] secretary in 1891, wrote in that year as follows: "Owing to our small numbers we should not strive to found as many specialised societies as possible, along with related journals, because this would lead to dispersing our forces."[11] Finally, it was professor of the Budapest Technical University (BTU) and member of MTA Lajos Ilosvay[12] who suggested the founding of sections that later became serious centres of the individual branches of sciences.

The Section for Chemistry and Mineralogy came into being on 5th January 1892. Its first presidents were university professor and member of MTA Károly Than,[13] perhaps the greatest dignitary in chemistry of nineteenth-century Hungary, and geologist and mineralogist József Szabó,[14] while Lajos Ilosvay acted as assistant secretary. In 1893 it was decided to found an independent journal, published from 1895 under the name of *Magyar Chemiai Folyóirat* (Hungarian Journal of Chemistry). Its first editor was chemist and pharmacist professor Lajos Winkler.[15] All of its authors were outstanding chemists, university professors and members of MTA.

---

[10] B. Lengyel Sen. (1844–1913) performed his studies first at the Technical University, then at Pest University of Sciences. After becoming assistant professor with K. Than, he was appointed professor of chemistry for Debrecen Agricultural School. After a longer study sojourn in Heidelberg with Bunsen, he obtained his Ph.D. from there. Then he returned to Budapest, where finally he was appointed head of the department of chemistry II of the University of Sciences, a post he held until his death. He dealt, in the first place, with analyses of mineral waters. He was elected correspondent member of MTA in 1876, and regular member in 1889. From its foundation (1882) until his death he was editor of the journal "*Mathematikai és Természetudományi Értesítő*" (Communications in Mathematics and Natural Sciences) published by MTA. See Szőkefalvi-Nagy, Z. (1982); Szabadváry (1997a).

[11] Lengyel (1891).

[12] L. Ilosvay (1851–1936) graduated from Budapest University of Sciences in pharmacy, and obtained a degree as secondary school teacher and a Ph.D. in philosophy. After becoming assistant professor, he obtained a two-year scholarship, which he spent in Heidelberg with Bunsen, in Paris with Berthelot and in Munich with Baeyer. Returning home, he became head of the department for general chemistry of József Technical University in 1882. In 1891 he was elected correspondent, in 1905 regular member and from 1916 until his death president of MTA. He introduced the first ion specific reagent in analytical chemistry for the detection of nitrates, and wrote the first book of organic chemistry in the Hungarian language (1905). Szőkefalvi-Nagy (1978).

[13] K. Than (1834–1908) was a most versatile professor of chemistry at Budapest University of Sciences; besides many scientific honorary offices he was a member of MTA, of great influence on the development of chemistry in Hungary. See Szabadváry (1997b).

[14] J. Szabó (1822–1894) was a student of philosophy and the law at Pest University of Sciences, and later graduated from Selmecbánya Mining Academy. Later he also graduated in law. He obtained his Ph.D. in 1850. After various jobs in mineralogy he was appointed professor of the subject at Pest University of Sciences in 1862. In 1867–1868 he was dean, in 1883 rector of the university. He was a founding member of the Hungarian Geological Society, and from 1872 its president. He was also vice-president of KMTT. In 1858 he was elected corresponding member of MTA, in 1867 regular member and from 1870 secretary of MTA3. His activities in the literature of geology, mineralogy, petrography and chemistry were considerable. See Dudich (1997).

[15] L. Winkler (1863–1939) graduated from Budapest University of Sciences as a pharmacist (1892). In 1889 he obtained his Ph.D. in pharmacy. After various university jobs, he became director of the Chemical Institute No.1 of Budapest University of Sciences in 1909. He became corresponding member of MTA in 1892 and regular member in 1922. He published over 200 papers in the fields of analytical chemistry and analysis of pharmaceuticals. See Szabadváry (1975).

Table 8.1 The increase in the number of the members of the *Királyi Magyar Természettudományi Társulat* (Royal Hungarian Society for Natural Sciences, KMTT). (Source: Gombocz [1941].)

Year	Number of members[a]
1841	134
1867	659
1868	804
1869	1658
1870	2228
1875	4432
1880	5380
1885	5780
1890	7173

[a] The figure for 1841 represents the members who joined at the foundation of the society, the rest for the respective ends of the years.

In the first two years of the Section's existence, all the branches of chemistry were dealt with but only a few papers were presented on mineralogy. Most of the papers concerned organic chemistry but physical chemistry, biochemistry and agricultural chemistry were also given attention. A separate commission was brought into being for the enhanced control of food. The Hungarian proposals to the International Committee of Establishing Atomic Weights were submitted by Károly Than in 1900.[16] The members visited plants and institutions in common, and they organized festivities in honour of Kekulé and van 't Hoff.

For 15 years Lajos Ilosvay was the president of the Section for Chemistry and Mineralogy, followed in this office by the famous organic chemist and professor of Budapest Technical University, Géza Zemplén.

The number of the section's members is not known as such data are only available for the membership of the whole society. The increase in the number of members from the foundation from 1841 to 1890 is shown in Table 8.1.[17] It can be seen that after a slow growth during the first years of the society's existence, the number of members rose to a considerable figure. This was due to the fact that the society started popularizing science (from 1866 onwards). It is interesting to note that from 1868 to 1869 the number of the members doubled. The same is true for the years 1870 to 1875. By 1890 the number of the members was more than 50 times the figure of 1841.

All the above activities carried out by the section for many decades went in parallel with those of the Hungarian Chemical Society. The Hungarian Society of Natural Sciences after reorganization still exists today.

---

[16] Gombocz (1941), 261.
[17] Compiled by the author from Gombocz (1941), pp. 14–15, 39, 78, 132–133.

8.2.3 Vegyészeti Gyárosok Országos Egyesülete (Association of Chemical Industrialists, VEGYOE)

The Association of Chemical Industrialists (*Vegyészeti Gyárosok Országos Egyesülete*, VEGYOE) was founded in 1904, two years after the foundation of the Federation of Hungarian Industrialists (*Gyáriparosok Országos Szövetsége*).[18] According to the records of industrial development of 1904 the number of chemical companies was about 190, the number of workers employed in the chemical industry was over 20 000 and the value of its production amounted to 84 million Crowns. The association was meant to be an advocate of the interests of the chemical industry in the fields of customs policy, traffic matters, government contracts and social policy.

The first step in public was completed in May 1904. Ágoston Kohner, president of the board of *Hungária műtrágya, kénsav és vegyipari Rt.* (*Share Company Hungária for Fertilizers, Sulfuric Acid and Chemical Industry*) convened a meeting in the matter on 3rd May, where some representatives of different branches of the chemical industry were invited. An appeal was sent off to the Hungarian chemical industrialists. The signatories expressed their opinion to the effect that an organization similar to the professional association founded in Austria as far back as 1878 could also be successful in Hungary. Forty-seven Hungarian chemical companies sent representatives to the statutory meeting. In his opening address, Ágoston Kohner mentioned that the boom of the German chemical industry was due in large part to the fact that it had an outstanding organization at its disposal in representing its interests. In the more modest Hungarian conditions he considered establishing the directives for the development of the Hungarian chemical industry as one of the most urgent tasks. The statutes then adopted were modified in 1910.

The Association had four levels of membership: ordinary, funding, honorary and corresponding members. It carried out its work in sections always deeply engaged in the problems of the promotion of industry, trade regulations, state control and the competition of state-owned factories.

The number of members shows a slow but continuous development until the outbreak of the First World War (Table 8.2).[19] In the ten years analysed the number of members rose three-fold. It must be noted that the members were, in this case, not individuals but companies.

Questions of social policy and workers' problems were dealt with by the Association from the employers' side. From time to time the Association gave an extensive evaluation of the development of Hungary's chemical industry. Using this and similar information the Association tried to influence the further development of the industrial companies, steering them towards, instead of unnecessary competition, existing market demands. It supported the training of chemists and chemical engineers.

---

[18] See Vámos (1996).

[19] Data compiled by the author from the annual reports of the Association of Hungarian Chemical Industrialists for the period 1904–1914. The reports were published yearly as separate leaflets by Pesti Lloyd Társaság, Budapest.

Table 8.2   Number of members of the Association of Hungarian Chemical Industrialists. Source: Gombocz (1941).

Year	Number of members
1904	47
1907	84
1909	116
1912	150
1914	155

The Association founded in 1904 had, from October 1906, its own journal, *Vegyészeti Lapok* (Chemical Gazette), which bore the subtitle "Official Bulletin of the Association". In the opening announcement in Issue No. 1 of Volume 1 (20th October 1906), the reason for its being published is given by a quotation from the 1905 memorandum of the Association: "The prosperity of a chemical factory depends, to an extent much greater than that of any other branch, on the level of the theoretical and practical knowledge of its technical director." Therefore: " . . . the country that cannot grant its chemists a proper living for their studies and education will never have an independent, self-relying chemical industry and chemical technology." Further on, the announcement makes it clear that chemists and the industry are dependent on each other and should, therefore, cooperate.[20]

Among the original publications there were scientific papers, exclusively from the field of applied chemistry, and also review articles. These reviews and papers were often written by well-known scientists or experts. The column "Economical News" contained information from abroad and from Hungary, the "Technical News" published information on the lives of other associations; sometimes the latter column also contained personal news (*e.g.* appointment of university professors). The "Professional Literature" informed about books that had appeared in Hungary or abroad. The "Patents" recommended, through patent agents, patents issued in Hungary or abroad for utilization. The section "Industrial Copyright Protection" dealt with Hungarian patent legislation. In the "Trivia" part, readers' letters often appeared and also the editor's replies. Practically every issue published advertisements of chemists seeking jobs.

On the whole, the journal looks very colourful and interesting, even to today's readers. It always gives ample publicity to questions of interest to the community of chemists. Among others, as far back as in Volumes 1 and 2, the polemics about the training of technical chemists (chemical engineers) in Hungary and about the possibilities of Hungarian chemists obtaining domestic jobs was trailing on.[21] Remarks to the discussion were added, not only by

---

[20] *Vegyészeti Lapok* (1906), **1**, 1.
[21] Requinyi, see *Vegyészeti Lapok* (1907), **2**, 1 and 6.

domestic experts but also by those working in foreign countries.[22] This shows that the gazette was well read abroad as well as in Hungary.

8.2.4   Section of Chemical Engineering of the *Magyar Mérnök- és Építészegylet* (Association of Hungarian Engineers and Architects, MÉ)

The Association of Hungarian Engineers and Architects (MÉ) was founded in 1867 with the aim "to promote the financial and intellectual interests of the cause of technology". Sections existed within the Association right from the beginning; the Section of Chemical Engineering was brought into being in 1906.[23] The statutory meeting was convened by Tamás Kosutány:[24] "The Chemical Section within the Association of Engineers has been founded in order to be able to duly represent the professional interests and discuss the arising scientific and practical problems as well as to summon the colleagues to join the Association". Their objective was "that chemical engineers should have a representation, and that every Hungarian factory should employ Hungarian chemists".[25]

It was made a duty of the section's meetings to discuss the following problems: the draft of chemists' tariffs; the use of the title "chemical engineer" and engineers' tariffs, the recognition of the right of private institutes of chemical analysis to give expertise, the countrywide control of nutriments and the modifications of the law of quality. With the increase in the number of chemical engineers, the number of the section's members increased as well: from 30, at the beginning, to 208 by 1914.

MÉ got in touch with VEGYOE and sent questionnaires to the Hungarian engineers employed in Hungarian chemical factories. The results of the survey were published under the title "Memorandum on the education, in factories, of mechanical and chemical engineers, further mining and metallurgical engineers, and their employment by domestic industrial companies".

The greatest achievement of the section, during the period investigated, was the organization of the 1st National Congress of Hungarian Chemists in

---

[22] Bartal, see *Vegyészeti Lapok* (1906) **1**, 3; Dely, see *Vegyészeti Lapok* 1907, **2**, 84.

[23] Fábián, see *A Nehézvegyipari Kutatóintézet Közleményei* (1989).

[24] T. Kosutány (1848–1915) first graduated from Keszthely Agricultural School in 1869, then continued his studies at the faculty of philosophy of Budapest University of Sciences and finished them as scholarship holder at Halle University. In 1871 he became assistant professor at Magyaróvár Agricultural Academy. He defended his doctoral thesis, written in German language, in Leipzig (1873). From 1884 he was head of department in Magyaróvár and head of the Chemical Experimental Station established in the same town. In 1903 he was appointed director of the Budapest National Institute of Chemistry and the Central Experimental Station of Chemistry, posts he held until his death. He was invited lecturer at the Department of Agricultural Chemical Technology of József Technical University (1903–1908). He was an excellent teacher and a fertile author of scientific books and papers. His interests focused around tobacco, agricultural distilleries, wine yeasts, fertilizers and feeds. In 1894 he was elected corresponding member of MTA. See Móra (1997).

[25] *A Magyar Mérnök- és Építészegylet Heti Értesítője* (1906), 24–29.

1910.[26] This congress was an outstanding display of the development of chemistry in Hungary. Chemical engineer, university professor, secretary, vice-president, and later president Béla Lengyel spoke about the role of Hungarian chemical engineers as follows: "We are, in the first place, a state of lawyers and therefore everybody, who possesses ambitions to participate in dealing with, and improving the fate of, our country and our nation, lives in the erroneous belief that it is only the law degree that qualifies to it."

The call for participation, as formulated by royal assistant chemist Gyula Halmi and issued on 10th October 1910, started as follows: "In Hungary the conditions with respect to the specialists in chemistry are still quite adverse. Our industry is small and is developing but slowly; moreover, the extremely narrow space that offers itself for chemical work, is – not to a slight part – occupied by foreign chemists." Further he complained that in public offices, even though there was an urgent need for them, the number of chemists' posts was not being increased; promotion of the numerous chemists in the civil service was highly restricted and – compared to that of officials of other professions – was scanty and slow. No less distressing was the situation in the fields of scientific work and technical literature. He remarked that, astonishingly, the youth was still seeking chemistry as a profession to a greater extent than required. They would like to restore the balance of supply and demand in this respect. One of the main reasons of the country's backwardness lay, according to Halmi, in the unconcern of the Hungarian society of chemists. There was still no healthy, lively and bustling life or cooperation among colleagues; chemists working in the fields of science and industry did, so to speak, not even know of each other's existence.[27]

This call for participation, which remained by no means unnoticed by the Hungarian society of chemists or related trades,[28] implicitly was a justification of the foundation of the Hungarian Chemical Society.

8.3 *Magyar Kémikusok Egyesülete* (the Hungarian Chemical Society, MKE)

8.3.1 The Establishment of the Hungarian Chemical Society, Arguments Justifying It and Its Initiators

The Hungarian Chemical Society (*Magyar Kémikusok Egyesülete*, MKE) was established in Budapest on 27th June 1907. Prior to that event, a critical survey of the institutions related to chemists and chemistry was published. It stated that the Class of Mathematics and Natural Sciences of the MTA was working "entirely secluded from the world"; the Section of Chemistry and Mineralogy

---

[26] Of course, in its organization all the other chemical societies or associations took part as well.
[27] Halmi (1911), 7.
[28] The First Congress of Hungarian chemists had 221 participants. Not all of them were, of course, members of MKE as it was a joint achievement of all the societies dealing with chemistry at the time.

of the Royal Hungarian Society for Natural Sciences, although successful in supporting the development of chemistry and chemical literature, was not able to extend its functions beyond that point; the Association of Chemical Industrialists was dealing, in the first place in its own interests, with the economic issues of the branches of chemical industry; while the Section of Chemical Engineering of the MÉ was representing the interests of only some of the chemists. Therefore, the need arose to create an institution representing equally the interests of the profession and the progress of science.[29]

According to the requirements sketched above, the aim of the society was to improve the social life and professional spirit of chemists as well as safeguarding their interests. In order to achieve its aim, the society was going to keep up premises, found a library, subscribe to professional journals, promote the publishing activities of its members and organize lectures and excursions; it was going to enter relationships with societies and bodies of similar aims and protect their activities and, as a body, join other societies as a member. In problems related to chemistry or chemists it would offer its services to the Government and attentively follow the pertinent activities of legislation. In questions related to the profession it would provide information to members and authorities. As soon as its financial power would allow, it would establish insurance and widows' and orphans' funds for its members within the frame of the society, in the form of societies with separate statutes.[30]

Two years after its foundation, at the first regular general assembly (28th March 1909), the social interests of the chemists were formulated in a very explicit way. Among others, employment of foreign chemists should be eliminated, the problems of salaries, modes of employment, pensions and accident insurances of chemists working for public offices should be settled and the interests of chemists should be duly represented in Parliament and ensured by legislation. University and college as well as secondary school chemistry education should be improved. The spelling of Hungarian chemical names should be unified and the tariffs of analyses and other work carried out by chemists should be duly settled.[31]

The founding president of the society was professor of the Kolozsvár (Kluj) University, Rudolf Fabinyi;[32] industrialist and chemist Adolf Kohner[33] was elected deputy president, chemist Sándor Kalecsinszky vice-president[34] and

---

[29] Halmi (1908), 281.

[30] László (1910).

[31] Halmi (1907), 15.

[32] R. Fabinyi (1849–1920) was, after studies in Budapest, assistant to Károly Than and, on longer sojourn abroad at the institutes of Wislicenus, Baeyer and Bunsen, appointed professor of chemistry at Kolozsvár University, an office he held until his death. He was elected corrrespondent and regular member, respectively, of MTA in 1895 and 1915. See Szabadváry (1997b) and Móra (1999).

[33] A. Kohner (1866–1937) came from a family of industrialists and was himself a shareholder of a factory producing sulfuric acid and fertilizers. He had obtained a doctor's degree in chemistry from Berlin university. Vámos (1996).

[34] S. Kalecsinszky (1854–1913) had been assistant to K. Than and B. Lengyel, and later became chemist at the Institute of Geology. In Kenyeres (1967).

Gyula Halmi secretary. Three university professors were elected honorary members,[35] and the famed professor Károly Than[36] honorary president. The board had 12 members.[37]

8.3.2 Members, Executives and Board; Numbers, Background and Positions

According to the statutes, the society had four kinds of members:

a) Honorary members, elected by the general assembly (upon recommendation of the board);

b) Funding members who endowed a fund, in the amount set by the board, at the point of the establishment of the society or later;

c) Regular members. The society was open to every chemist of honest character who:

1) participated in the foundation of the society;
2) graduated from the faculty of chemistry of a technical university or obtained a degree in chemistry as main speciality from the faculty of philosophy (of the University of Sciences);
3) was in possession of a secondary or commercial school teacher's diploma in chemistry;
4) was employed, as assistant professor, by any department of chemistry of a university or technical university; or
5) had been working as a chemist for an industrial company for three years.

d) Supporting member. Any person interested in chemistry could be a supporting member.

Regular and supporting members were admitted by the board, upon recommendation by two regular members, and passed by simple majority in secret ballot.

Funding members constituted an important group of the society's membership as – besides the membership fees – they provided for its financial background. The donations, together with their amounts, were regularly reported on the columns of the society's journal and thanks were simultaneously expressed.

At the point of establishment the society had 45 members.[38] In 1909 the society had 68 regular members and 1 funding member; the board meeting of

---

[35] The three professors (all members of MTA) were V. Wartha, L. Ilosvay and B. Lengyel. V. Wartha (1844–1914) was professor at Budapest Technical University, the first head of the department for chemical technology, a post he held for 42 years until his retirement in 1912. He was a member of MTA. It was his procedure that enabled the porcelain factory Zsolnay (Pécs) to manufacture its world famous objects with eosin enamel. See Móra in Nagy (1997). For L. Ilosvay and B. Lengyel see notes 9 and 11, respectively.

[36] *cf.* note 13.

[37] *cf. Vegyészeti Lapok* (1907), **2**, 14.

[38] *Ibid.*

7th February 1910 reported 91 members.[39] By the time of the second regular general assembly the simultaneous admittance of 55 new members raised the number of members to 146.[40] By 1911 the number of regular members rose to 180.[41] By March 1911 (board meeting) the number of members rose to 194. Among the new members there were, besides individual persons, institutions and factories such as the Nobel Dynamite Factory, the Military School Pécs, the Agricultural Experimental Station Mosonmagyaróvár and the High School of Sciences in Kassa (today Košice, Slovakia).[42]

In the account of the society given by the secretary general at the third regular general assembly, it was stressed that the number of members rose, in four years, from 40 to 200, among whom 72 were from the capital.[43] Later on, the numerical increase in members was not given but the newcomers – all highly qualified and renowned participants in Hungarian chemical public life – were all listed.

The total number of persons with a degree in chemistry was, at that time, around 400. Thus, it can be said that about half of the chemists in the country were members of MKE.

The Hungarian Chemical Society was open to a wide range of professions besides those of chemists and chemical engineers. However, most members belonged to those two trades. For instance, the board meeting of 26th May 1914 admitted six new members, four of them chemical engineers, of whom one was also an industrialist.[44] One of the new members was director of an industrial experimental station and one – a student of Budapest Technical University – a supporting member.

The distribution of members according to the list of participants of the first National Congress of Hungarian Chemists (Budapest, 5th–8th November 1910) is shown in Table 8.3.[45] Although the 221 participants of the congress were, as already said, not all members of the society, the data may give an insight into the stratification of the part of the society interested in chemistry at the time.

The majority of the participants were chemists (in Hungary "chemists" means graduates from the two universities of sciences) and chemical engineers. Together they formed 36% of the participants. Another significant group was formed by factory owners and factory directors (16.8%). A third important group was that of companies, university departments and societies (12%).

The educational background and position on the labour market of the members of the society's executive committee is shown in Table 8.4.[46] By 1911 one attorney at law joined the executive committee.

---

[39] *Magyar Kémikusok Lapja* (1911), **2**, 26.

[40] *Magyar Kémikusok Lapja* (1910), **1**, 69.

[41] *Magyar Kémikusok Lapja* (1911), **2**, 26.

[42] *Magyar Kémikusok Lapja* (1911), **2**, 32.

[43] *Magyar Kémikusok Lapja* (1911), **2**, 51.

[44] E. Wolf (1886–1947) graduated from Budapest Technical University as chemical engineer. He was one of the founders of the renowned Hungarian pharmaceutical factory Chinoin. See Fábián (2005).

[45] Compiled by the author from Halmi (1911), 239–244.

[46] *Magyar Kémikusok Lapja* (1910), **1**, 79.

Table 8.3 Distribution of the participants of the first National Congress of Hungarian Chemists (6th November 1910), according to educational background and/or position on the labour market. Compiled by the author from Halmi (1911).

Groups of participants[a]	Number	% of total
Chemists (51) + chemical engineers (29)	80	36
Professors + teachers: university (11); academy/college (4); senior lecturers (2); assistants (13); private docents (3); secondary school teachers/directors (6)	39	17.7
Factory owners + industrialists (13); factory directors, director generals (24)	37	16.8
Pharmacists	2	0.9
Patent attorneys	3	1.4
Society secretaries	2	0.9
Mechanical engineers + other engineers	7	3.2
Directors of industrial experimental stations	7	3.2
Others	17	7.7
Companies, university departments, societies *etc.*	27	12.2
Total	221	100.0

[a] The figures in brackets indicate the numbers of participants in the subgroups.

Table 8.4 Educational background and position on the labour market of the executive committee of the society. (Source: *Magyar Kémikusok Lapja* (1910), **1**, 79.)

Office held in the society	Educational background	Position on the labour market
President	Chemist	University professor
Deputy president	Chemist	Factory owner
Vice-president 1	Chemist	Chief chemist of Royal Institute
Vice-president 2	Chemist	University professor
Vice-president 3	Chemist	Factory owner
Secretary general	Chemical engineer	Sworn royal forensic chemist
Secretary 1	Chemist	Chief engineer-chemist of state railways
Secretary 2	Pharmacist	University private docent
Treasurer	Chemist	Sworn forensic chemist
Auditor	Chemist	Professor at Academy of Commerce & Technical University
"Sergeant-at-Arms"	Chemist	Chief chemist of Capital
Librarian	Chemist	Royal assistant chemist
Editor-in-chief of MKL	Chemist	Sworn royal forensic chemist

The board consisted of about one third university professors, one third chemists (graduated from universities of sciences) holding various jobs and one third factory owners or directors, and a few company representatives.

8.3.3 Events Organized by MKE

According to the statutes, the general assembly had to take place yearly in March. The board had to have its sessions every three months; besides the board members the editor of the official journal was also invited. From that time on, when MKE had its own premises, it had a library, and "Chemists' Days" were held regularly. At the beginning, the books of the library came exclusively from donations. As can be read in an advertisement: "We accept any book pertinent to our profession or a related trade, be it new, old, obsolete or fragmentary."[47] As a result, not only private persons but also companies, even those from abroad, donated books to the library.[48] By 1911 the library became so plentiful that the 600 volumes at its disposal could be lent to the members. At that time the society subscribed to 40 journals, out of which 10 came from abroad.[49] The library also had a librarian, László Szathmáry, one of the first historians of chemistry in Hungary. He wrote a comprehensive work under the title "Hungarian alchemists".[50]

The social life of chemists was promoted by keeping open, and at their disposal, the premises in 14–16 Kossuth Lajos Street every day from 4 p.m. The second Wednesday of every month was "Chemists' Day", allowing members to meet members of the board.[51] At the board meeting of 15th December 1910, President Adolf Kohner suggested organizing – similarly to other societies – special meetings once a month, at which papers were to be presented.[52]

Many members of the leading bodies (Executive Committee, honorary members and board members) and also other members of the society gave papers at the scientific meetings. The papers covered a wide range of the branches of chemistry; the most frequent ones being those dealing with problems of agriculture, food control and analysis and sometimes technology. This can be easily understood if we consider that Hungary was called, at that time, "the pantry of the monarchy."

---

[47] *Magyar Kémikusok Lapja* (1910), **1**, 19.
[48] Every issue of MKL enumerated the freshly donated books and the donors. *E.g.* in 1911 books were obtained from Farbwerke vorm. Meister Lucius Brüning, Höchst; from two Hungarian professors (one of them sent his own work); the greatest gift came form MKL itself: seven books in the German language. See *Magyar Kémikusok Lapja* (1911), **2**, 111.
[49] The titles of the foreign journals were: *Chemiker Zeitung*, Cöthen; *Zeitschrift für angewandte Chemie*; *Chemisches Zentralblatt*; *Österreichische Chemiker-Zeitung*, Wien; *Zeitschrift für Elektrochemie*; *Zeitschrift für Zuckerindustrie und Landwirtschaft*, Wien; *Seifensieder Zeitung*, Augsburg; *Technische Neuerungen*; *Technische Monatshefte*; *Experiment Station Record*, USA.
[50] L. Szathmáry (1880–1944) was a chemical engineer, at a time working for Badische Anilin und Sodafabrik, Germany. After returning to Hungary, he first was senior lecturer at Budapest Technical University then, from 1915 until his death, professor at Budapest Academy of Commerce. See Kenyeres, (1969).
[51] *Magyar Kémikusok Lapja* (1910), **1**, 36.
[52] *Magyar Kémikusok Lapja* (1911), **2**, 26.

8.3.4   Relations of MKE with the Surrounding Society

MKE was, in general, not involved in political issues. However, in 1908 a letter was sent to all the members of the Government, in which the society wanted to lay down the chemists' situation in the regulations to be issued about the rights of civil servants.[53] In the letter it was claimed:

1. Secret records on civil servants, in use until then, should be abolished as they were unjust and humiliating, moreover not in keeping with the views of the age;
2. Unpaid jobs should be abolished;
3. All civil servants should be able to apply for being entitled, after 35 years of service, to a full pension, like those working for the State Railways or the Capital. According to statistical data, working with poisonous gases and vapours can attack the respiratory organs of chemists. Thus it would be fair and justified if chemists – both civil servants and teachers – were entitled to a full pension after 30 years of service;
4. The promotion of chemists in civil service should be equal to that of those who had graduated in law.

On other occasions the society participated in the preparation of a law draft on patents as far as the rights of employees in relation to their patented inventions were concerned.[54]

It can be seen that the political involvement of the society was always related to the welfare not only of their members, but also of the whole society of Hungarian chemists.

The society followed, with great interest, the training of chemists at Technical University and at the universities of sciences of Budapest and Kolozsvár (Cluj, today in Romania). Reports were given, *e.g.* on a visit of students of Belgrade University to the laboratory of general chemistry of Technical University;[55] on the intolerable conditions at the department of general and theoretical chemistry of Budapest University of Sciences: 600 students had enrolled but the lecture hall could hold only 300. Thus the professor was compelled to give his lectures in a room where there was no possibility for presenting experiments.[56] Information was given on the curriculum of the faculty of chemistry of the Technical University in the second semester of the academic year 1911–1912;[57] on the election of the rector and the faculty deans of Technical University for the academic year 1913–1914, whereby Imre Szarvassy[58] was elected dean of

---

[53] Szabadváry and Szőkefalvi-Nagy (1972).

[54] *Magyar Kémikusok Lapja* (1910), **1**, 24, 30.

[55] *Magyar Kémikusok Lapja* (1911), **2**, 49.

[56] *Magyar Kémikusok Lapja* (1911), **2**, 32.

[57] *Magyar Kémikusok Lapja* (1912), **3**, 6.

[58] I. Szarvassy (1872–1942) graduated from Budapest Technical University as chemical engineer and obtained his Ph.D. from Budapest University of Sciences. He became head of the department of electrochemistry at Technical University in 1905, a post he held until his death. He was a member of MTA. He dealt with many fields of chemical technology and electrochemistry. See Kenyeres (1969).

the faculty of chemistry.[59] In 1914 the society decided to offer prizes to the students preparing the best doctoral theses at the universities of sciences of Budapest and Kolozsvár and at Technical University.[60]

The official journal of the society also informed about appointments of professors and other teaching staff of universities and colleges. Information was also given about courses delivered[61] or to be delivered in the near future.[62]

Royal chief chemist Henrik Géza Donáth gave a detailed comparison, on the columns of the official journal of the society, of the training of food chemists in Hungary and abroad, whereby he pointed out the deficiencies of pertinent domestic training and legislation.[63]

Social issues were dealt with, among others, by giving space – on the columns of the journal – to announcements of vacancies, on the one hand, and advertisements of chemists looking for jobs, on the other.

8.3.5 The Relations of MKE with Other Professional Societies at the National and International Level

MKE established and maintained friendly relations with quite a number of other professional societies. The intention to do so was explicitly formulated in the opening address given by court counsellor Adolf Kohner,[64] co-president of the society, at the regular general assembly on 5th June 1914: "While we are organizing – in the service of scientific progress – lectures and meetings, are trying to establish direct contacts, moreover are maintaining a journal and have brought into being a library, we are seeking . . . the friendly and understanding contacts and co-operation with the bodies and societies involved."[65]

The society regularly informed, on the pages of its official journal, about the meetings of the Maths and Natural Sciences Section of the MTA, even if the papers presented were dealing with mathematics, geology, physics or astronomy.[66] The meetings dealing with chemistry (*e.g.* on 18th June 1914) were, in general, not treated in a more detailed way: in most cases, only the names of the speakers and the titles of the papers were given.[67]

In a similar way reports were given on the scientific meetings and general assemblies of the Section of Chemistry and Mineralogy of KMTT,[68] of the MÉ,[69] of the Chemical Society of Kolozsvár (*Kolozsvári Chemikus*

---

[59] *Magyar Kémikusok Lapja* (1912), **3**, 6.

[60] *Magyar Kémikusok Lapja* (1914), **5**, 11.

[61] *Magyar Kémikusok Lapja* (1912), **3**, 15: information about a training course in "scientific microscopy", organized by university professor and bacteriologist Hugó Preisz (1860–1940), where the lecturers were scientists from abroad.

[62] *Magyar Kémikusok Lapja* (1911), **2**, 49: information about a training course for distillers, in which they might obtain a licence for managing a distillery.

[63] *Magyar Kémikusok Lapja* (1911), **2**, 10–11, 59, 70.

[64] *cf.* note 33.

[65] *Magyar Kémikusok Lapja* (1914), **5**, 73.

[66] *E.g. Magyar Kémikusok Lapja* (1911), **2**, 23 24, 131; (1912), **3**, 23; (1914), **5**, 19, 69.

[67] *Magyar Kémikusok Lapja* (1914), **5**, 69.

[68] *E.g. Magyar Kémikusok Lapja* (1912), **3**, 23; (1914), **5**, 14, 37, 60, 69, 131.

[69] *E.g. Magyar Kémikusok Lapja* (1910), **1**, 13, 41, 75, 159.

Társaság)[70] and of the *Verein Österreichischer Chemiker.*[71] At the meeting of 21st March 1914 of the *Verein Österreichischer Chemiker*, the Hungarian chemical engineer Gyula Szilágyi[72] presented a paper "On the historical development of distillery management".[73]

Occasionally short accounts were given on the activities of other associations too, *e.g.* on the Session of the National Industrial Association (*Országos Iparegyesület*) of 7th May 1914, where the new products of the Ganz factory were presented;[74] on the 25th Congress of German Naturalists and Physicians in Karlsruhe (24th–28th September 1914), where 28 papers of interest to chemists were presented and professor Rudolf Fabinyi,[75] president of the Hungarian Chemical Society, gave a paper on the colorimetric determination of morphine and colchicine.[76]

It must be considered a friendly gesture that upon request of the society of the Hungarian Refrigerating Industry (*Magyar Hűtőipari Egyesület*), the board of MKE decided to put their premises at the disposal of the former society's board meeting convened for 24th November 1914. It was also decided, at the same board meeting of MKE, that it would act in a similar way whenever a fellow society was in need of their premises.[77]

The material available on these early years of the society's activities does not contain any hints at its international relations.

8.3.6 MKE's Journal, Its Contents and Intended Audience

Two and a half years after its foundation, MKE launched the *Magyar Kémikusok Lapja* (Hungarian Chemists' Journal, MKL), which served, at the same time, as its official bulletin. The first issue appeared on 14th January 1910 (Figure 8.1). In his opening address, the editor-in-chief, Dr Zsigmond Neumann, wrote as follows:

"The Hungarian chemists, dispersed not only in the country but nearly all over the world, have heard or known – by chance – very rarely of each other. However, as in every family the brotherly feeling attracts the members of the family towards each other – around the family hearth – thus in the big family of *Hungarian chemists* there was a desire to *found a common hearth*. This desire remained, however, only a wish for a long

---

[70] *E.g. Magyar Kémikusok Lapja* (1910), **1**, 51, 65, 92, 159; (1914), **5**, 42.
[71] *E.g. Magyar Kémikusok Lapja* (1911), **2**, 6, 14, 30, 60, 69, 131; (1912), **3**, 23.
[72] Gy. Szilágyi (1860–1924) graduated from Budapest Technical University as chemical engineer, then acquired his Ph.D. from Budapest University of Sciences. He first was a private docent, later extraordinary professor at Technical University. He was an expert in matters of distilleries, breweries and vinegar manufacture. He founded his own experimental station and became a sworn forensic expert. See Kenyeres (1969).
[73] *Magyar Kémikusok Lapja* (1914), **5**, 42.
[74] *Magyar Kémikusok Lapja* (1914), **5**, 69.
[75] *cf.* note 32.
[76] *Magyar Kémikusok Lapja* (1914), **5**, 124.
[77] *Magyar Kémikusok Lapja* (1914), **5**, 11.

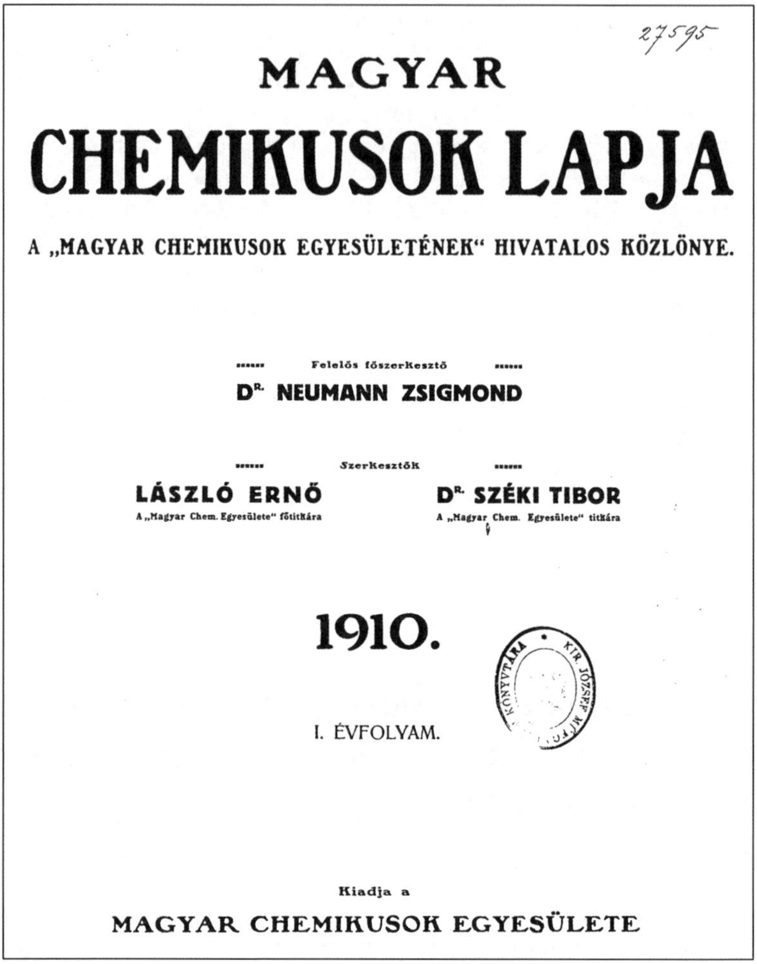

MAGYAR

CHEMIKUSOK LAPJA

A „MAGYAR CHEMIKUSOK EGYESÜLETÉNEK" HIVATALOS KÖZLÖNYE.

Felelős főszerkesztő

Dᴿ NEUMANN ZSIGMOND

Szerkesztők

LÁSZLÓ ERNŐ   **Dᴿ SZÉKI TIBOR**

A „Magyar Chem. Egyesülete" főtitkára   A „Magyar Chem. Egyesülete" titkára

1910.

I. ÉVFOLYAM.

Kiadja a

MAGYAR CHEMIKUSOK EGYESÜLETE

Figure 8.1   Title page and first page of the first volume of *Magyar Kemikusok Lapja*, 1910.

time and – owing to their small numbers and, perhaps to their lacking strength – they had to content themselves with the possibility to take shelter, like orphans, in the circle of richer and bigger families."

The editor-in-chief reminds that the workers of every other trade had their independent and self-contained body. It was just the chemists that had to tolerate that anybody might sneak into their ranks, anybody might call themselves chemists and not only usurp that title but often make use of it in practical life in an unauthorized way, even if they did not know their right hand from their left in chemistry. This had been and still was of great harm to the profession. In the last decades, however, the number of chemists had considerably increased in Hungary. Thus the time was considered ripe for establishing

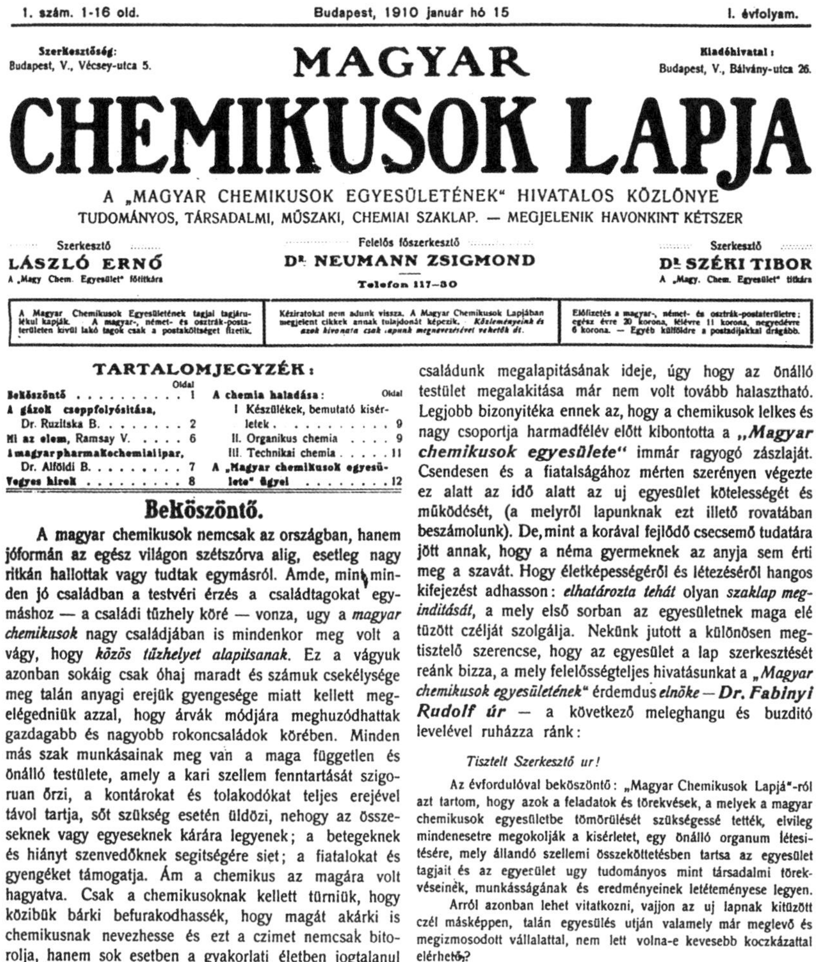

1. szám. 1-16 old.　　　　　Budapest, 1910 január hó 15　　　　I. évfolyam.

Szerkesztőség:　　　　　　　**MAGYAR**　　　　　　　Kiadóhivatal:
Budapest, V., Vécsey-utca 5.　　　　　　　　　　　　　　　Budapest, V., Bálvány-utca 26.

CHEMIKUSOK LAPJA

A „MAGYAR CHEMIKUSOK EGYESÜLETÉNEK" HIVATALOS KÖZLÖNYE
TUDOMÁNYOS, TÁRSADALMI, MŰSZAKI, CHEMIAI SZAKLAP. — MEGJELENIK HAVONKINT KÉTSZER

Szerkesztő　　　　　　　Felelős főszerkesztő　　　　　　　Szerkesztő
LÁSZLÓ ERNŐ　　　　DR NEUMANN ZSIGMOND　　　　DR SZÉKI TIBOR
A „Magy Chem. Egyesület" főtitkára　　　　Telefon 117-30　　　A „Magy. Chem. Egyesület" titkára

A Magyar Chemikusok Egyesületének tagjai tagjárulékul kapják. A magyar-, német- és osztrák-posta-területen kívül lakó tagok csak a postaköltséget fizetik. | Kéziratokat nem adunk vissza. A Magyar Chemikusok Lapjában megjelent cikkek annak tulajdonát képezik. Közleményeink és azok kivonata csak »apunk megnevezésével vehetők át. | Előfizetés a magyar-, német- és osztrák-postaterületre; egész évre 20 korona, félévre 11 korona, negyedévre 6 korona. — Egyéb külföldre a postadíjakkal drágább.

Beköszöntő.

A magyar chemikusok nemcsak az országban, hanem jóformán az egész világon szétszórva alig, esetleg nagy ritkán hallottak vagy tudtak egymásról. Amde, mint minden jó családban a testvéri érzés a családtagokat egymáshoz — a családi tűzhely köré — vonza, ugy a *magyar chemikusok* nagy családjában is mindenkor meg volt a vágy, hogy *közös tűzhelyet alapítsanak*. Ez a vágyuk azonban sokáig csak óhaj maradt és számuk csekélysége meg talán anyagi erejük gyengesége miatt kellett megelégedniük azzal, hogy árvák módjára meghuzódhattak gazdagabb és nagyobb rokoncsaládok körében. Minden más szak munkásainak meg van a maga független és önálló testülete, amely a kari szellem fenntartását szigoruan őrzi, a kontárokat és tolakodókat teljes erejével távol tartja, sőt szükség esetén üldözi, nehogy az összeseknek vagy egyeseknek kárára legyenek; a betegeknek és hiányt szenvedőknek segitségére siet; a fiatalokat és gyengéket támogatja. Ám a chemikus az magára volt hagyatva. Csak a chemikusoknak kellett türniük, hogy közibük bárki befurakodhassék, hogy magát akárki is chemikusnak nevezhesse és ezt a czimet nemcsak bitorolja, hanem sok esetben a gyakorlati életben jogtalanul értékesitse, ha mindjárt annyi köze is volt a chemiához, mint a tyuknak az ábécéhez. Nagy kárára és szégyenére volt és talán van is még ez a karunkra. De volt-e nekünk elég izmos, erőteljes testületünk, a mely e métélyt karunktól távol tudta volna tartani? Sajnos, a tények azt bizonyitják, hogy nem volt. Ha azok a

családunk megalapitásának ideje, úgy hogy az önálló testület megalakitása már nem volt tovább halasztható. Legjobb bizonyitéka ennek az, hogy a chemikusok lelkes és nagy csoportja harmadfélév előtt kibontotta a *„Magyar chemikusok egyesülete"* immár ragyogó zászlaját. Csendesen és a fiatalságához mérten szerényen végezte ez alatt az idő alatt az uj egyesület kötelességét és működését, (a melyről lapunkban ezt illető rovatában beszámolunk). De, mint a korával fejlődő csecsemő tudatára jött annak, hogy a néma gyermeknek az anyja sem érti meg a szavát. Hogy életképességéről és létezéséről hangos kifejezést adhasson: *elhatározta tehát* olyan *szaklap meginditását*, a mely első sorban az egyesületnek maga elé tüzött czélját szolgálja. Nekünk jutott a különösen megtisztelő szerencse, hogy az egyesület a lap szerkesztését reánk bizza, a mely felelősségteljes hivatásunkat a „Magyar chemikusok egyesületének" érdemdus *elnöke — Dr. Fabinyi Rudolf úr* — a következő meleghangu és buzditó levelével ruházza ránk :

Tisztelt Szerkesztő ur!

Az évfordulóval beköszöntő : „Magyar Chemikusok Lapjá"-ról azt tartom, hogy azok a feladatok és törekvések, a melyek a magyar chemikusok egyesületbe tömörülését szükségesse tették, elvileg mindenesetre megokolják a kisérletet, egy önálló organum létesitésére, mely állandó szellemi összeköttetésben tartsa az egyesület tagjait és az egyesület ugy tudományos mint társadalmi törekvéseinek, munkásságának és eredményeinek letéteményese legyen.

Arról azonban lehet vitatkozni, vajjon az uj lapnak kitüzött czél másképpen, talán egyesülés utján valamely már meglevő és megizmosodott vállalattal, nem lett volna-e kevesebb koczkázattal elérhető?

A földből kikelő növény első hajtásából vajmi nehéz a növény minőségére következtetni. Faji jellege csak később válik biztosan meghatározhatóvá s mennél tovább halad fejlődésében, annál határozottabban válik külön egyéb növénytársaitól s érvényre juttatja faji jellegét, bármenemű talajban növekedjék s bármilyen viszonyok közzé is kerüljön. Fejlődésének folyamatában a döntő, az elhatározó, a tő életképessége, nedvkeringése, mig az irányt, a melyben . . .

Figure 8.1 (continued)

---

an independent body. Consequently, an enthusiastic and large group of chemists decided to found the Hungarian Chemical Society. Later on, in due course, it decided to launch a professional journal serving, in the first place, the goals set by the society to itself.[78]

---

[78] *Magyar Kémikusok Lapja* (1910), **1**, 1.

The journal dealt, in a retrospective way, with the foundation of MKE and its early activities.[79] We may learn that the idea of launching a journal was first put forward at the first regular general assembly of the society held on 28th March 1909, when secretary Gyula Halmi raised the question, "Why do we still not have a professional chemical journal that would meet our demands in every respect and really would be the professional journal of the whole trade of Hungarian chemists?" Then, somewhat later, "The valuable chemical papers appear exclusively abroad or in academic or other official publications hermetically closed to the public, so that their real values will be recognized perhaps by the archaeologists of future ages only."[80] As a result the president suggested launching a journal, an idea unanimously approved, in principle, by the Assembly. Some months later, on 15th June 1909, at the society's second board meeting, president Kohner expressed his wish that the journal should be launched by 1st October of the same year as a weekly.[81] The fourth board meeting (20th October 1909) decided that the title of the journal should be *Magyar Chemikusok Lapja* (Hungarian Chemists' Journal, *MKL*), and that about half of the space should be devoted to scientific issues, the other half to social ones. Upon the president's proposal the board decided that the authors of original papers should be accorded royalties. The president estimated the appearance of advertisements in the journal as permissible. Members of the society should obtain their copies free of charge as premium of the membership dues.[82]

The contents of the first issue of the first volume of the journal read as follows:

Opening address
Liquefaction of gases, by Dr B. Ruzitska
What is an element? by V. Ramsay
Hungarian pharmacochemical industry, by Dr B. Alföldy
Miscellaneous news
Progress of chemistry:
    I. Apparatuses, experiments for presentation
    II. Organic chemistry
    III. Technical chemistry (= chemical technology)

8.3.7   Affairs of MKE[83]

Thus it can be seen that right from the beginning, as had been decided by the board of the society, the journal published original articles – not only by

---

[79] This is of great importance from the aspect of the reconstruction of its history. As it is, MKE had to change premises four times since the end of the Second World War, thus much of its original documentation has been lost.

[80] *Magyar Kémikusok Lapja* (1910), **1**, 15.

[81] *Magyar Kémikusok Lapja* (1910), **1**, 16.

[82] *Ibid.*

[83] *Ibid.*

domestic authors but also by foreign authors – as well as information of different kinds. The column bearing the title "Progress of chemistry" gave reviews of mainly foreign but sometimes also Hungarian literature published elsewhere, whereby the branches of chemistry were covered alternately, a given branch always bearing the same Roman number. The section *"Vegyes hírek"* (Miscellaneous news) was much variegated.

Very soon, still in Vol. I, the number of informative sections was completed by one bearing the title "Literature",[84] which reviewed both books and papers from journals recently published, and another one about societies and associations,[85] which gave accounts on the activities of fellow societies. Ample space was given to issues in economics in the column on Economics.[86] Advertisements occupied sometimes half, and sometimes whole, pages. Advertisements of chemists looking for jobs were free of charge.[87] The column *"A 'Magyar Chemikusok Egyesülete' ügyei"* (Affairs of the Hungarian Chemical Society) gave detailed accounts on the general assemblies and board meetings of MKE but also on the new books the library obtained, and the journals to which it subscribed.[88] The structure of the journal did not change essentially during the period investigated.

In order to give the reader an idea of the sections on Economics (a) and "Miscellaneous" (b) some items taken from various issues shall be quoted. In the Economics section of Vol. I we can read, among others, about the manufacture of paper from corn stems, exports, imports, and prices of rubber and tyres, mineral oil, sulfur, boric acid, soap and rapeseed oil;[89] about favours granted by the state to companies, about nitric acid production in Chile, manufacture of sulfuric acid in Russia, Russian factories manufacturing rubber goods, the mineral oil market in Germany and the USA and transport of chemicals;[90] the "Miscellaneous" section contained information *e.g.* about the 25-year anniversary of service of an assistant professor charged of wine falsification using saccharine, the establishment of a new chemical experimental station in the county town Szentes, the shortage of chemists in China, the vigorous development of industry and education in that country and about a graduate chemical engineer who opened up a grocery in a county town, as he was not able to live on the salary granted him in his trade.[91]

It can be seen that *MKL* was, during the period investigated, a very interesting and many-sided professional journal. It was, in the first place, intended to meet the demands of the members of MKE. However, it can be said that through its versatility it served not only the society of chemists in Hungary but also the country's chemical industry and its companies.

---

[84] *E.g. Magyar Kémikusok Lapja* (1910), **1**, 24, 40, 49, 160, a.s.o.

[85] *E.g. Magyar Kémikusok Lapja* (1910), **1**, 19, 41, 65, 73, 92, a.s.o.

[86] *E.g. Magyar Kémikusok Lapja* (1910), **1**, 62, 143, a.s.o.

[87] *E.g. Magyar Kémikusok Lapja* (1910), **1**, 79.

[88] *E.g. Magyar Kémikusok Lapja* (1910), **1**, 24, 54, 62; (1911), **2**, 32, a.s.o.

[89] *Magyar Kémikusok Lapja* (1910), **1**, 32.

[90] *Magyar Kémikusok Lapja* (1910), **1**, 143.

[91] *Magyar Kémikusok Lapja* (1910), **1**, 79.

8.3.8   MKE's Attitude Towards Policies of Expansion, Monopolization or Professional Autonomy

As already said before, MKE was an open society. Although its statutes explicitly defined the sphere of its membership, in practice – as it could be seen – it adopted even institutions and industrial companies among its members. As the last – in chronological order – of four societies directed towards chemistry, it did not harbour intentions of monopolizing the trade, nor was the question of professional autonomy raised by its leaders. On the contrary, it strived for maintaining friendly relations with the other "chemical" societies that preceded it in time; moreover it extended this friendly attitude even towards societies of foreign countries. On one occasion only, at the third regular general assembly of the society (11th April 1911) did the president of the society, Professor Fabinyi, raise the issue of monopolization: " . . . I cannot be averse to the assumption that all the objects and tasks that our society wishes to reach and solve could be reached and solved easier, safer, and within a shorter time, if the Hungarian chemists' fractions – now distributed among several societies, and working separately – would act united, *viribus unitis* . . . "[92] At the end of the page, a footnote by the editor marked by an asterisk said in small characters that the president's above idea had not met with the approval of the general assembly.

8.4   Concluding Remarks

On the example of the five chemical societies that existed in the last two centuries in Hungary we can see that the situation of chemistry and chemical societies in Hungary was very specific. The reasons for the evolvement of the special Hungarian situation were, among others, the following:

- At the Hungarian Academy of Sciences the positions of philology and literary sciences were the strongest as part of the national aspiration to change the country's situation within the Habsburg Empire or become independent of the Empire. That is why the community of natural sciences was looking for other forums as well;
- In the Hungarian chemical industry a great number of German and Austrian chemists were employed. *E.g.* in 1900, 32% of the technical staff in the Hungarian industry were foreigners, on average, while in the chemical industry 64% of the employees were not Hungarian. It was the societies that had to encourage the employment of Hungarian chemists.

Three out of the five societies were established as late as at the beginning of the twentieth century. Until then only sections of academic societies were dealing with the subject. This shows that the vigorous growth of the discipline started

---

[92] *Magyar Kémikusok Lapja* (1911), **2**, 51.

but about 100 years ago, in contrast to the great boom Hungary underwent in other fields of economy, science and culture from 1867 onwards.

Each of the societies, as could be seen, had – in spite of their similarities – their own specific profiles shaped interestingly enough by the same leading personages, Hungarian scientists and university professors of the nineteenth and the first half of the twentieth century. All these persons were working in the societies to prepare the great boom in the chemical, and in particular in the pharmaceutical, industry of the country in the interwar period.

The Hungarian Chemical Society probably would not have come into being without the experience gained by the preceding societies. MKE survived two world wars and two revolutions, several economic crises and several restructurings of the chemical industry that forms its financial background.

Abbreviations

Abbreviation	Hungarian name (English translation)
KMTT	*Királyi Magyar Természettudományi Társulat* (Royal Hungarian Society for Natural Sciences)
KMTT KAS	*A Királyi Magyar Természettudományi Társulat Kémiai és Ásványtani Szakosztálya* (Section for Chemistry and Mineralogy of KMTT)
MKE	*Magyar Kémikusok Egyesülete* (Hungarian Chemical Society, MKE)
MKL	*Magyar Kémikusok Lapja* (Hungarian Chemists' Journal)
MÉ	*Magyar Mérnök- és Építészegylet* (Association of Hungarian Engineers and Architects)
MÉ VMSZ	*Magyar Mérnök- és Építészegylet Vegyészmérnöki Szakosztálya* (Section of Chemical Engineering of MÉ)
MTA	*Magyar Tudományos Akadémia* (Hungarian Academy of Sciences)
MTA3 oszt.	*A Magyar Tudományos Akadémia III. osztálya (Matematika és Természettudomány)* (Class III of MTA (Mathematics and Natural Sciences))
VEGYOE	*Vegyészeti Gyárosok Országos Egyesülete* (Association of Chemical Industrialists)

References[93]

A Magyar Mérnök- és Építészegylet Heti Értesítője (Weekly Bulletin of the Hungarian Association of Engineers and Architects) 1905–1910, 24–29.
A Nehézvegyipari Kutatóintézet Közleményei (1989), **19**, 145–150.

---

[93] In some cases the articles do not have an author and/or title. Therefore, only the journal is given and details of the reference are in the notes.

Bartal, A. (1906), Vegyészeink kiképzéséről [On the training of our chemists], *Vegyészeti Lapok* 1, 3.

Dely, G. (1907), A vegyészek helyzete és kiképzése Amerikában [Situation and training of chemists in America], *Vegyészeti Lapok* 2, 84.

Dudich, E. (1997), Szabó József, in Nagy, F. (ed.), *Magyar Tudóslexikon A-Zs*, BETTER-MTESZ-OMIKK, Budapest, 741–742.

Fábián, É. (1980), A Magyar Mérnök- és Építészegylet Vegyészmérnöki Szakosztályának története 1906-tól 1944-ig [History of the Section of Chemical Engineering of the Association of Hungarian Engineers and Architects, 1906–1944], *A Nehézvegyipari Kutatóintézet Közleményei* 19, 145–150.

Fábián, É. (2005), Az ő szellemében dolgozunk tovább [We shall go on working in his spirit], *Tanulmányok a természettudományok, a technika és az orvoslás történetéből* 12, 59–63.

Gombocz, E. (1941), *A Királyi Magyar Természettudományi Társulat története 1841–1941* [History of the Royal Hungarian Society for Natural Sciences 1841–1941], Királyi Magyar Természettudományi Társulat, Budapest.

Halmi, Gy. (1907), *Vegyészeti Lapok* 1, 14, 15.

Halmi, Gy. (1908), *Vegyészeti Lapok* 1, 281.

Halmi, Gy. (1911), *A magyar vegyészek első kongresszusa* [First Congress of Hungarian Chemists], A Pesti Lloyd Társulat Nyomdája, Budapest.

Kenyeres, Á. (ed.) (1967), *Magyar Életrajzi Lexikon* [Hungarian Biographical Encyclopaedia], Akadémiai Kiadó, Budapest, Vol. 1, item "Kalecsinszky Sándor", 840.

Kenyeres, Á. (ed.) (1969), *Magyar Életrajzi Lexikon* [Hungarian Biographical Encyclopaedia], Akadémiai Kiadó, Budapest, Vol. 2, items "Szarvassy Imre", 709; "Szathmáry László", 714; "Szilágyi Gyula", 775.

Lambrecht, M. (1997), Fodor József, in Nagy F. (ed.), *Magyar Tudóslexikon A-Zs.*, BETTER-MTESZ-OMIKK, Budapest, 302–303.

László, E. (1910), A Magyar Kémikusok Egyesületének ügyei [Affairs of the Hungarian Chemical Society], *Magyar Kémikusok Lapja* 1(1), 12.

Lengyel, B. (1891), A titkár jelentése az 1891. évről [Secretary's Report for the Year 1891], in E. Gombocz, *A Királyi Magyar Természettudományi Társulat története 1841–1941* Királyi Magyar Természettudományi Társulat, Budapest, 1941, 220.

Magyar Kémikusok Lapja [Journal of Hungarian Chemists].

Móra, L. (1974), 'Sigmond Elek. Magyar Vegyészeti Múzeum, Várpalota.

Móra, L. (1997), Kosutány Tamás, in Nagy, F. (ed.), *Magyar Tudóslexikon A-Zs.*, BETTER-MTESZ-OMIKK, Budapest, 489–490.

Móra, L. (1997), Wartha Vince, in Nagy, F. (ed.), *Magyar Tudóslexikon A-Zs.*, BETTER-MTESZ-OMIKK, Budapest, 860.

Móra, L. (1999), *Fabinyi Rudolf élete és kora (1849–1920)* [The life and time of Fabinyi], Technika Alapítvány, Budapest.

Requinyi, G. (1907), A műszaki vegyészek képzéséről [On the training of chemical engineers], *Vegyészeti Lapok* 2, 1, 6.

Szabadváry, F. (1975), *Winkler Lajos*, Akadémiai Kiadó, Budapest.

Szabadváry, F. (1997), Fabinyi Rudolf, in Nagy, F. (ed.), *Magyar Tudóslexikon A-Zs.*, BETTER-MTESZ-OMIKK, Budapest, 278.

Szabadváry, F. (1997a), Lengyel Béla, in Nagy, F. (ed.), *Magyar Tudóslexikon A-Zs.*, BETTER-MTESZ-OMIKK, Budapest, 534–535.

Szabadváry, F. (1997b), Than Károly, in Nagy, F. (ed.), *Magyar Tudóslexikon A-Zs.*, BETTER-MTESZ-OMIKK, Budapest, 804–805.

Szabadváry, F. and Szőkefalvi-Nagy, Z. (1972), *A kémia története Magyarországon* [History of Chemistry in Hungary], Akadémiai Kiadó, Budapest.

Szőkefalvi-Nagy, Z. (1959), *Ilosvay Lajos*, Akadémiai Kiadó, Budapest, 1978.

Szőkefalvi-Nagy, Z. (1982), *Lengyel Béla*, Akadémiai Kiadó, Budapest.

Új magyar Lexikon (1959) [New Hungarian Encyclopaedia], Akadémiai Kiadó, Budapest.

Vámos, É. K. (1996), Egy vegyész gyáros mint a művészetek mecénása [A chemist and industrialist as Maecaenas of arts], in *Tanulmányok a természettudományok, a technika és az orvoslás történetéből* 3, MTESZ, Budapest, 39–43.

Vámos, É. K. (1996), Contributions to the history of the Association of Hungarian Chemical Industrialists, *Technikatörténeti Szemle* 22, 303–321.

Vázlatok a Magyar Tudományos Akadémia félszázados történelméből [Sketches from the 50-year history of the Hungarian Academy of Sciences] (1881), Akadémiai Kiadó, Budapest.

Vegyészeti Lapok [Chemical Gazzette].

CHAPTER 9

THE NETHERLANDS: Keeping the Ranks Closed: The Dutch Chemical Society, 1903–1914

ERNST HOMBURG

9.1 Introduction

The founding of chemical societies in Europe took place in three "waves". First, between about 1750 and the 1820s, local chemical societies were founded, such as dinner clubs, student clubs, local societies focusing on lecturing and collective experimentation, and reading circles.[1]

During a second wave, between about 1840 and 1870, national science-oriented chemical societies were founded. Examples are the chemical societies of London, Paris, Berlin and St. Petersburg. While those of Paris and St. Petersburg, initially, had a purely scientific orientation, the societies of London and Berlin also tried to be a platform on which science and industry could meet.

After about 1875, as a third wave, national profession-oriented chemical societies were established. Examples are the *Society of Public Analysts* (1874), the *Institute of Chemistry* (1877), the *Verein analytischer Chemiker* (Association of Analytical Chemists) (1878), the *Society of Chemical Industry* (1881), the *Association des Chimistes de Sucrerie et de Distillerie de France et des Colonies* (Association of Chemists of the Sugar Works and Distilleries of France and its Colonies) (1883), the *Deutsche Gesellschaft für angewandte Chemie* (German Society for Applied Chemistry) (1887; later *Verein deutscher Chemiker* [Association of German Chemists]), the *Association Belge des Chimistes* (Belgium

---

[1] Bolton (1902); Voornaamste chemische vereenigingen (1903/1904); Russell, Coley and Roberts (1977).

Creating Networks in Chemistry: The Founding and Early History of Chemical Societies in Europe
Edited by Anita Kildebæk Nielsen and Soňa Štrbáňová
© The Royal Society of Chemistry 2008

Chemists Association) (1887) and the *Syndicat Central des Chimistes et Essay-eurs de France* (Central Syndicate of Chemists and Assayers of France) (1890).

All three waves will be discussed below in relation to the social history of Dutch chemistry. Compared to other European chemical societies the Dutch Chemical Society (*Nederlandsche Chemische Vereeniging* [NCV]) was founded rather late, in 1903. But earlier, local chemical societies existed. To a certain extent, the founding of a national society still belonged to the "third wave" of applied-chemical and profession-oriented societies, because, as we will see, chemists working in the (applied) analysis of foodstuffs and industrial products took the lead. But other organizational models – such as the influential *Deutsche Chemische Gesellschaft*, with many Dutch members – were available as well, and for a number of reasons already at its founding date the NCV had a hybrid character that combined elements of a science-oriented society with those of a profession-oriented one.

In this chapter I will try to explain why the creation of the Dutch chemical society occurred relatively late, and why it took on a hybrid character from the start. We will show that the founding of the NCV was the result of a carefully negotiated equilibrium between different groups of chemists, and also that initial opposition within academic circles had to be overcome. In order to understand these peculiarities of the Dutch case, it is important to first tell more about the institutional and social development of chemistry in the Netherlands up to 1903. In the course of this historical sketch the mutual relations and conflicts between the different groups of chemical practitioners will become clear, and also why the NCV was not founded earlier.

9.2   The First Dutch Chemical Societies

Just like in several other European countries, chemistry as a teaching subject was first institutionalized during the seventeenth century in medical faculties of the Dutch universities. Next to that, there were several so-called 'athenea' that prepared for university studies. In the eighteenth century almost all universities and athenea had chairs of chemistry, although often in combination with botany or medicine.[2] After the Napoleonic wars, only three out of five universities were continued by the national government – Leiden, Groningen and Utrecht – and in the course of the nineteenth century only two athenea survived – Amsterdam and Deventer. In 1876 the *Atheneum Illustre* of Amsterdam was raised to university status. Outside the universities chemistry was practiced mainly by pharmacists, just like abroad.

In the course of the eighteenth century, chemistry and other natural sciences became fashionable subjects. All over Europe scientific societies were founded, and the Netherlands were no exception. These societies were of different kinds. A few of them were of national, or even international, importance. Next to that there were numerous societies of a local nature. In these societies "chemistry",

---

[2] Snelders (1993), 173–177; Snelders (1986).

in the form of lectures, demonstration experiments or chemical investigations, was frequently on the agenda. The same was true of societies of different types, such as dinner clubs, student societies and societies of competent experimentalists. In the Netherlands, especially between 1770 and 1820, dozens of these societies were founded. Visser counted 78 scientific societies founded before 1808, and Snelders, who focused more narrowly on local societies devoted to the natural sciences, counted 45 societies established between 1748 and 1850.[3]

Although in many of these societies chemical topics were discussed, in a few instances the term "chemistry" even entered into its name. In the literature quite some attention has been paid to eighteenth-century "chemical societies" in Scotland, England and the United States.[4] The Netherlands offer examples of the same phenomenon. In 1767 the pharmacist, chemist and physician Willem de Loos founded a *Chemisch Gezelschap* (Chemical Society) in Rotterdam, in which he gave courses on chemistry of a strongly entertaining nature. After De Loos' death these lessons were continued for a few years by the then well-known lecturer and physician Leonard Stocke (*cf.* Table 9.1).[5]

Far better known is the group of four friends who, at the initiative of J. R. Deiman and A. Paets van Troostwijk, united themselves in about 1790 as the *Scheikundig Gezelschap* (Chemical Society) in Amsterdam. They were able and creative experimenters, who contributed significantly to the acceptance and diffusion of the theories of Lavoisier. They founded their own journals, published frequently, and belonged without doubt to the emerging international scientific community of chemists of their days. In about 1802, this Society of Dutch Chemists, as it is most commonly called, ceased to be active as a group.[6]

In 1801 six students at the University of Groningen founded a student society aimed at doing experiments, once a week, and at collectively buying instruments and books. This society, soon named the *Natuur- en Scheikundig Genootschap* (Society of Natural and Chemical Sciences), flourished and was gradually transformed between 1801 and 1810 into a larger society in which lectures and, later, courses were given, and experiments performed, both for demonstration and for scientific purposes. In 1826 membership of the society had grown to 154. After a merger with a Groningen society of naturalists, in 1830, the word "Scheikundig" was dropped from the name, but chemical experiments and lessons continued to be on the agenda for the rest of the nineteenth century.[7] In 1817, the town of Deventer followed with its own *Natuur- en Scheikundig Genootschap*, modelled after the *Natuurkundig Gezelschap* of Utrecht, which in its turn was organized along similar lines to the Groningen society after 1810. These three societies were dominated by the professors of the

---

[3] Visser (1970); Snelders (1983).
[4] Averley (1986); Kendall (1942); Miles (1959).
[5] Lieburg (1978), 134–138.
[6] Snelders (1980).
[7] Swinderen (1826); Wiese (2001).

Table 9.1 Local chemical societies in the Netherlands, 1767–1890.

Founding date	End date	Town	Dutch name	English translation
1767	1771 (1787[a])	Rotterdam	*Chemisch Gezelschap*	Chemical Society
1790	1802	Amsterdam	*Scheikundig Gezelschap*	Chemical Society
1792	1798	Delft	*Gezelschap van Beminnaaren der Scheikunde*	Society of Lovers of Chemistry
1801	1830 (present[a])	Groningen	*Natuur- en Scheikundig Genootschap*	Society of Natural and Chemical Sciences
1817	After 1917	Deventer	*Natuur- en Scheikundig Genootschap*	Society of Natural and Chemical Sciences
1820	1840 (1879[a])	Nijmegen	*De Scheikunde Gehuldigd*	Doing Honour to Chemistry
1822	About 1886	Leiden	*Pharmaceutisch en Chemisch Gezelschap "Concordia"*	Pharmaceutical, and Chemical Society "Concordia"
1824	1881	The Hague	*Schei-, Kruit- en Artsenijmengkundig Leesgezelschap*	Chemical, Botanical and Pharmaceutical Reading Circle
1835	1878	Amsterdam	*Scheikundig Leesgezelschap*	Chemical Reading Circle,
Before 1842	1862	Amsterdam	*Chemisch-Pharmaceutisch Leesgezelschap*	Chemico-Pharmaceutical Reading Circle
1846	1857	Nijmegen	*Vereeniging tot Beoefening van Practische Scheikunde*	Association for the Cultivation of Practical Chemistry
1890	Present	Delft (with nationwide membership)	*Technologisch Gezelschap*	Technological Society

[a] After 1771, 1830 and 1840, the societies at Rotterdam, Groningen and Nijmegen, respectively, continued to exist as more general "societies for the cultivation of natural sciences". The Groningen society still exists today.

local universities and atheneum, and the substantial collections of instruments were used both by the societies and in university teaching.[8]

Except for the *Scheikundig Gezelschap* at Amsterdam, all these "chemical" societies mentioned so far had an extremely heterogeneous membership. Some of the members, such as pharmacists and professors of chemistry, physics and botany, were involved in chemistry and other sciences on a more or less professional basis. Others, including numerous vicars and clergymen, were interested in science for physico-theological reasons. For the large majority of

---

[8] Goeij (1993); Cittert-Eymers (1977).

the members, going to weekly or monthly evening lectures probably was only a form of sociability in which members of the (upper) middle classes just had to take part.

9.3   The Rise of Scientific Pharmacy

After 1818, chemical societies of a very different nature, with a far more homogeneous membership, emerged. These were the *Scheikundige Leesgezel-schappen*, or *Chemisch-Pharmaceutische Leesgezelschappen* (chemical or chemico-pharmaceutical reading circles, or reading societies), the membership of which consisted almost exclusively of pharmacists. The first of these was the reading circle *De Scheikunde Gehuldigd* (Doing Honour to Chemistry), founded in the town of Nijmegen in 1820. Pharmacists in other towns followed: Leiden (1822), The Hague (1824) and Amsterdam (1835) (Table 9.1).[9]

The founding of these societies in the 1820s and 1830s was one of the consequences of new medical acts that had come into force after the founding of the Kingdom of the Netherlands in 1813. In 1818 the licensing practice of pharmacists was reorganized. From then on, chemistry played a far greater role in the examinations. As a consequence, the chemical training of future pharmacists had to be raised to a higher level. In 1823, Dutch government decided that towns could erect so-called Clinical Schools for the training of future pharmacists, surgeons and obstetricians. The first of these schools was founded in Haarlem in 1825, and between 1825 and 1828 other towns followed. In towns where a Clinical School was lacking, the chemico-pharmaceutical reading circles played a crucial role. They maintained a common biological garden, both for botanical teaching and for the supply of medicinal plants, they organized chemical lectures for pharmaceutical apprentices and they bought books and subscribed to scientific journals.

When we try to summarize the social situation of Dutch chemistry around 1840, the time when the *Chemical Society of London* was founded, it is important to note that of the two groups which played the crucial role in the case of the London society – professional chemical analysts and consultants, and younger academics[10] – the first one was almost non-existent in the Netherlands, and the second one was tiny. The Dutch chemical industry was too small for a career in consultancy, and the absence of government regulations in the field of food control and sanitation led to the same result. Also prospects for chemistry teaching in the Netherlands were lacking, until a school reform in 1863 dramatically changed the situation.

In theory, the local chemical societies could have become nuclei of a national society. But in the reading circles chemistry was just a subject of study, and not the core of the professional identity of the participants. They were pharmacists in the first place. When Dutch government in 1841 installed an official commission to prepare pharmaceutical reforms, consisting of medical doctors only,

---

[9] Snelders (1983); Bierman (1988), 119–123; Wittop Koning (1948), 39–41.
[10] Bud (1980), 35–67, 92; Bud (1991).

Dutch pharmacists became furious that not a single pharmacist had been appointed into the commission. In Amsterdam, the *Scheikundig Leesgezelschap* and the *Chemisch-Pharmaceutisch Leesgezelschap* joined forces and decided to found a national pharmaceutical society to protect and promote the professional interests of Dutch pharmacists. Early in 1842 this *Nederlandsche Maatschappij ter Bevordering der Pharmacie* (NMBP) (Dutch Society for the Advancement of Pharmacy) came into being. During the following years several local chemical and pharmaceutical societies were integrated into the national organization. At the time of the foundation of the NCV (1903) the Dutch pharmaceutical society had become a well-organized and influential body, with about 480 members (out of a total of about 680 apothecaries).[11]

The role played by the Amsterdam chemical and chemico-pharmaceutical reading circles in the founding of a national pharmaceutical society perfectly illustrates the social position of Dutch chemistry in the early 1840s. Chemistry was closely linked to pharmacy and medicine, but was not a profession in its own right. During the first two years of the NMBP, the Amsterdam pharmacists had been quite liberal with respect to membership of the new society. Everybody interested in the advancement of pharmacy could join. As a result there were several manufacturers of chemicals, medical doctors and chemists and druggists among the early members. But in 1845 it was decided that full membership should be restricted to pharmacists who owned an apothecary shop. Since that date, even true pharmacists who had passed the provincial examinations, and who therefore were entitled to call themselves "apothecaries", but who did not have their own shop, could only become associate members.[12]

9.4   Reforming the Educational and Scientific Institutions, 1863–1887

In 1863 Dutch government issued a new law on secondary education, which modernized the Dutch educational system, but at the same time kept the traditional class structure of society intact, or perhaps even reinforced it. Together with the reforms of the medical profession of 1865, and the law on higher education of 1876, a binary educational system was created that stayed in force unchanged until 1917. These educational reforms had a great impact on the emergence of the Dutch chemical profession, as well as on the boundaries that would divide the different factions inside that professional group.

On the one hand, a new type of secondary school was created, called the *Hogere Burger School* (HBS) (School for the Higher Middle Classes), where mathematics and natural sciences were taught intensely during five years

---

[11] Stoeder (1891), 382–384, 391, 410; Wittop Koning (1948), 13–28, 77–78, 80, 164–165, 204–208, 218–229; Wittop Koning (1986), 259, 273–274; *Toekomst der gegradueerden* (1936), 231.
[12] Wittop Koning (1948), 24–25. Contrary to England, but similar to Germany, in Dutch the terms pharmacist and apothecary are virtually identical, and are used in this paper as synonyms. Since 1818 though, "apothecary" was the official term protected by law.

(chemistry only the last three years), and which prepared for the Polytechnic College at Delft. On the other hand there was the gymnasium, where Latin and Greek were taught, that prepared for the university. But there were escape routes. At the end of the nineteenth century, several chemistry students with HBS backgrounds decided to study at a Dutch university – which was allowed, as long as they didn't take part in examinations – but to submit their dissertations at a foreign university where a gymnasium examination was not required. The Swiss universities at Basel, Bern and Zürich, and the universities of Marburg and Heidelberg in Germany, were among the favourites.[13]

The introduction of the HBS in 1863 led to growing numbers of students who were better prepared for the Polytechnic College and the science faculties of the universities than before. It also boosted the labour market for teachers tremendously. Between 1863 and 1870 no less than 32 HBS schools were founded in the country, which meant, in principle, that there were 32 new teaching positions available in chemistry.[14] This was probably not much less than the total number of academic chemists in Holland in 1863.[15]

The laws on higher education of 1876 rounded off this period of reforms. For chemistry three aspects were crucial: the creation of a specific doctorate in chemistry; the introduction of pharmacy as an academic study, preparing both for the pharmaceutical exams based on the law of 1865 and for the doctorate in pharmacy, created in 1876; and the raising of the Amsterdam Atheneum to university status.

After the appointment of the young J. Henry van 't Hoff at the University of Amsterdam in 1877, he developed a flourishing research school in the then just-emerging field of physical chemistry. When van 't Hoff went to Berlin, in 1895, H. W. Bakhuis Roozeboom succeeded him. Until far into the twentieth century, physical chemistry in the Netherlands was completely dominated by chemists who had studied with van 't Hoff or Bakhuis Roozeboom at Amsterdam. At the same time, A. P. N. Franchimont succeeded in creating at Leiden University a very successful research school in organic chemistry. Franchimont and his pupils monopolized that field. All professors of organic chemistry appointed after 1877 were either students of Franchimont, or students of his students.[16]

The changes in the Dutch educational system and the growing role of original research at the universities, which partly resulted from that, also called for reforms of the scientific institutions. Several new scientific journals were founded, in which chemists could publish the results of their research: the *Archives Néerlandaises des Sciences Exactes et Naturelles* founded in 1865, the *Maandblad voor Natuurwetenschappen* (Natural Sciences Monthly), starting in 1870 and dominated by the Amsterdam physico-chemical school, and the

---

[13] Leden der Nederl. Chem. Vereeniging (1913/1914); Bolton (1901); Wittop Koning (1986), 284–287.
[14] Bartels (1963); Mandemakers (1996).
[15] In 1863 there was no specific academic degree in chemistry. But if one counted the number of Dr.Phil.s who defended a chemical dissertation, the number would not be much greater than 32. Between 1840 and 1868 G. J. Mulder supervised 22 dissertations. The number of chemical doctorates at Leiden and Groningen was far less. For Mulder, see Snelders (1993), 102.
[16] Cohen (1912); Kerkwijk (1934).

Recueil des Travaux Chimiques des Pays-Bas, founded 1882 by Franchimont and his colleagues in organic chemistry, W. A. van Dorp, S. Hoogewerff, E. Mulder and A. C. Oudemans.[17]

Although van 't Hoff and Franchimont certainly met during the bi-annual meetings of the *Nederlandsch Natuur- en Geneeskundig Congres* (Dutch Natural Science and Medical Congress; founded in 1887) and at meetings of other national bodies, this should not lead to the idea that there was frequent communication between the research groups at Leiden and Amsterdam. On the contrary, it seems that there was a deep rift between these two segments of the Dutch chemical community, which is confirmed by the founding history of the NCV (see below). Van 't Hoff and his students focused on physical chemistry, went to their own meetings within the *Genootschap voor Natuur-, Genees- en Heelkunde* (Society for Natural Sciences, Medicine and Surgery) in Amsterdam and published in the *Archives Néerlandaises*, the *Maandblad voor Natuurwetenschappen* and the *Zeitschrift für physikalische Chemie*. Contrary to this, Franchimont and his followers focused on organic chemistry and published in the *Receuil*.[18]

9.5   The State of the Chemical Profession around 1890

At this stage, it is good to reflect for a second time on the prospect that a national chemical society would be founded. How was the social situation of Dutch chemistry around 1890 viewed in that light? After 1841, science-oriented national chemical societies had been founded in many countries, and a number of profession-oriented chemical societies had followed in Britain, Germany, France and Belgium. What were the chances for similar organizations in the Netherlands? To answer that question, we should take a look at the different segments of the profession. In the Netherlands, apart from the pharmacists, four groups of chemical practitioners were the most important ones in those days: (1) the well-established university professors; (2) the young academics; (3) the teachers at the HBS, at professional schools and at some gymnasia; and (4) the industrial chemists. From all four groups a not insignificant number were members of the *Deutsche Chemische Gesellschaft* (DCG) – between 1868 and 1892 at least 92 of them.[19]

From the well-established academics few impulses for founding a national chemical society could be expected. They had plenty of possibilities to meet their fellow academics within the Royal Academy of Sciences, or within other prestigious national societies; they also were a member of the DCG, and of local societies such as the *Genootschap voor Natuur-, Genees- en Heelkunde* at Amsterdam.[20] Moreover, as the examples of Franchimont and van 't Hoff have shown, they were controlling their own networks of former students and

---

[17] Berkel (2002); Snelders (1991); Berkel (1991), 25–34; Cohen (1912), 125, 238–239; Kerkwijk (1934), 58–61.

[18] Snelders (1993), 110–111, 136; Cohen (1912), 24–71; Kerkwijk (1934), 5–6, 47–56.

[19] *Verzeichniss der Mitglieder.*

[20] Jorissen, Reicher and Rutten (1902/1903b).

journals, and often had little to gain by sharing power and making compromises within the framework of a national chemical body. Only when the national society could act as an effective vehicle to control the discipline and to promote certain specific scientific ideas, as the example of Wurtz shows in the case of France, it would be attractive for a well-established professor to support the founding of such a society.[21] In the case of the Netherlands, the robust equilibrium between the organic segment and the physical segment of the academic chemical community made such a possibility an illusion. Moreover, the Netherlands were lacking a true metropolis, such as London, Paris or Berlin, of which the local society could take the lead in the formation of a national society. In the egalitarian Netherlands a similar move by, for instance, the Amsterdam chemists would not have been acceptable for the other centres of learning.

Also from the point of view of sociability – which, for instance, in Copenhagen had motivated the founding of a chemical society – the established professors had little to gain from a national chemical society.[22] In many cases they were embedded quite well inside local or national elites, and moved in circles in which other important professors of science and medicine operated, such as national scientific societies and local dinner clubs.

For young academics, the situation was different. They were not inside the prestigious national scientific bodies, and the establishment of a national chemical society could certainly be attractive to them. Nevertheless, there were so many alternative circles in which they operated that there was no pressing need for them to found such a society. Many of them were members of the DCG, and within their institutes there often were informal gatherings where scientific topics could be discussed. In Amsterdam, for instance, apart from the monthly meetings of the *Genootschap van Natuur-, Genees- and Heelkunde*, van 't Hoff, his assistants and students held regular evening meetings where papers were presented.[23] And for establishing trans-local, national contacts, the *Natuur- en Geneeskundig Congres* worked rather well.

From the secondary school teachers even fewer impulses could be expected for founding a national chemical society. In those days, the position of teachers at the HBS and gymnasium was quite good. They were well paid, and fulfilled important roles within the local cultural elites. Many teachers were crucial figures within the local scientific societies of their towns. In view of their professional interest they often had joined the *Vereeniging van Leeraren aan Inrichtingen van Middelbaar Onderwijs* (Society of Teachers Connected to Institutions for Secondary Education), founded in 1867.[24] And for keeping informed about their discipline, quite a few of them had joined the DCG, or read the national and international journals to which their local societies had a subscription. So, from the point of view of sociability, and from a scientific point of view, the importance of the network of local scientific societies in the

---

[21] Fell and Rocke (2007), Rocke (2001), 216–220.
[22] Kildebæk Nielsen in this volume.
[23] Jorissen and Reicher (1912), 45–46.
[24] *Feestnummer* (1917).

Netherlands should not be underestimated. In 1903 there were still 12 *Natuurkundige Genootschappen* (Societies for the Natural Sciences) active in the country, several of them more than a century old.[25] Probably the existence of these local societies has retarded the founding of national disciplinary-oriented societies in the country.

The last group to be discussed is the industrial chemists. Before 1890, they were hardly connected to the three groups discussed above. All three previous groups consisted of chemists with a (Dutch) university education. But there were hardly any university chemists working in industry. At that time the chemical industry of the country was still quite small, so there were not many industrial positions available to chemists. Moreover, the few factory owners that employed chemists strongly preferred chemists from German universities. Chemistry at the Dutch universities was considered to be too theoretical, too academic and not well tuned to solving practical problems. Also alumni from the Polytechnic College of Delft worked in industry, but before 1890 their numbers were not greater than the numbers of chemists with a German university background.[26] Finally, a last group of chemists in industry was the *Chemiker* – the German term was in common use for them – who worked in the Dutch beet sugar industry. Many of them had been trained at the German sugar schools at Brunswick and Berlin. After about 1890, similar sugar schools were established inside the Netherlands, and from then on *Chemiker* for the sugar industries of the Netherlands and the Dutch Indies could be recruited from these schools at Amsterdam, Wageningen, Delft, Dordrecht and Deventer.[27]

Common to all these kinds of industrial chemists is that connections to the Dutch university chemists were mostly lacking, either because of the lower level of the schools where they had been trained, or because of the fact that they had been trained and educated abroad. Because of this, and because of the diverse nature of their backgrounds, industrial chemists before 1890 often were in rather isolated positions, and therefore the chance that they would unite to form a national society of technical chemists seems to have been rather small.

As a result of this review of the professional situation of 1890, we can conclude that for some segments of the profession (university professors and the school teachers) pressing reasons to found such a society were lacking, whereas for other segments (industrial chemists) the possibilities to do so were limited. After 1890 though, this situation would change. Two specific groups of chemists rose to prominence, and they would become important nuclei for the founding of the Dutch chemical society: the (academic) chemists working in private laboratories and in agricultural experimental stations, and the techno-logists of the Polytechnic College of Delft.

[25] *Honderdjarig bestaan* (1901), 4, 56.
[26] Verbong and Homburg (1994).
[27] Bakker (1989), 193–232.

9.6 The Rise of Private Laboratories and Agricultural Stations

When, after 1876, growing numbers of chemistry students started to populate the university laboratories, it became increasingly clear that the stream of graduates could not be absorbed by the labour market for school teachers. A large unemployment crisis did not occur though, because at the same time new employment options emerged.[28]

One of these labour markets was agriculture. When fertilizer use expanded, there was a growing need to analyse the quality and the amounts of nutrients of the commercial fertilizers. In 1877 Dutch government established the first official Rijkslandbouwproefstation (State Agricultural Experimental Station) at Wageningen, and from 1889 onwards four additional experimental stations followed where several academic chemists found employ.[29]

In the same period, demand for chemical analyses also grew in other sectors of society, partly because of the growth of (long-distance) transport and trade, and partly because new methods of chemical analysis created that demand. As a result, several laboratories were founded for the analysis of foodstuffs and articles of commerce. Before 1900 most of these laboratories were privately owned. In 1880 there were about five, in 1890 about ten, in 1900 thirty, while in 1909 the absolute maximum of 52 privately owned laboratories was reached. After that year, the establishment of government laboratories caused a steep decline.[30] In these laboratories numerous academic chemists found a job. As will be shown below, several of them – organized around the *Tijdschrift voor toegepaste scheikunde en hygiène* (Journal for applied chemistry and hygienics) – played a pivotal role in the formation of the Dutch Chemical Society.[31]

The laboratories were not the exclusive realm of academic chemists. Also technologists from Delft Polytechnic and, especially, pharmacists found employ in private analytical laboratories. In 1889, the Pharmaceutical Society (NMBP) started a campaign to stipulate that the members of the society would connect an analytical laboratory to their pharmacy shop. The analysis of drinking water and foodstuffs was considered to be a domain on which pharmacists could and should be active successfully. An investigation executed by the NMBP the same year showed that of 145 Dutch municipalities in which there was a pharmacy shop, there were 58 in which the local pharmacist also analysed foodstuffs.[32]

During the following decades this area of activity certainly became a contested terrain between chemists and pharmacists, and numerous boundary conflicts resulted from it. When in late 1898 the new pharmaceutical laboratory of Leiden University was opened, the professor of pharmacy and toxicology,

---

[28] On the labour marker for chemists, see: *Toekomst der gegradueerden* (1936), 168–195; Jorissen (1901/1902); Snelders (1997), 32–34.

[29] Knuttel (1928); Homburg (2004), 27–30.

[30] Vledder, Homburg and Houwaart (1999–2000); Raalte (1928).

[31] Snelders (1993), 137.

[32] Wittop Koning (1948), 43–45.

Hendrik Wijsman, held a lecture in which he emphasized the important role that pharmacists should play in the analysis of foodstuffs, fertilizers and numerous other commodities. Shortly afterwards, he was heavily attacked by C. A. Lobry de Bruyn, professor of organic and pharmaceutical chemistry at Amsterdam, who argued in a long letter in the *Pharmaceutisch Weekblad* (Pharmaceutical Weekly) that pharmacists should fully concentrate on their pharmacist shops, and not enter domains such as the analysis of foodstuffs, for which chemists had been trained far more thoroughly and extensively. In the weeks that followed a true polemic was fought in the columns of the *Pharmaceutisch Weeklad*.[33]

9.7   The Rise of the Chemical Engineers from Delft

During the same years, the position of Dutch academic chemists was also threatened by another group, that of the so-called "technologen" (technologists) – later chemical engineers – from Delft Polytechnic College.[34] After 1885, the new professor of (organic) chemistry, Bas Hoogewerff, who had studied in Germany, restructured the curriculum at Delft completely. As well as (future) factory owners and managers, from now on industrial chemists were trained at Delft as well, and the number of chemical courses was enlarged considerably. That the curriculum was heavily tilted towards chemistry was emphasized in 1905, when the Polytechnic College was elevated to the rank of a Technical University. On that occasion, the name of the technology department was changed into the department of chemical technology, and the title "technoloog", for its alumni, was changed into "scheikundig ingenieur" (chemical engineer).[35]

In 1890 students of the Delft technology department, supported by Hoogewerff, founded the *Technologisch Gezelschap* (Technological Society) (*cf*. Table 9.1), with the aim of intensifying the contacts between industry and the technology department at Delft. *Via* excursions, public lectures and membership of their society by industrialists, the Delft students tried to improve the industrial labour market for technologists. During the 1890s the labour market for these "chemical engineers" of the Delft school developed satisfactorily, and student numbers grew accordingly – from 25 in 1890, to 139 in 1905. In the last year the chemical engineering department at Delft had more chemistry students than all four Dutch universities together. The *Technologisch Gezelschap* also prospered. It grew from 23 ordinary members (students) and 13 external members (alumni) in 1891, to 80 members and 140 external members in 1905. In both years about 30 industrialists were extraordinary members. As a professional organization of chemical engineers, it was in fact the first Dutch professional society in the chemical sphere.[36]

---

[33] Lobry de Bruyn (1899a and 1899b); Wijsman (1899). *cf*. Jorissen (1901/1902).
[34] On the history of Delft Polytechnic College and the training of technologists/chemical engineers in particular, see Eijdman Jr. (1906); Nieuwenburg (1955).
[35] On the profession of chemical engineer, see also Hasselt (1913).
[36] *Eeuwboek* (1989), 20–21, 26–33, 46–53.

During the 1890s Dutch industrialization intensified, and labour markets for engineers expanded in all areas. It was a decade of turmoil for the Dutch engineering profession. Between 1895 and 1900 negotiations took place between the old established *Koninklijk Instituut van Ingenieurs* (KIVI) (Royal Institute of Engineers), founded in 1847, and the societies of mechanical and electrical engineers, founded in 1889 and 1895, respectively, with the aim of forming one powerful organization of Dutch engineers, with sections for the different branches of engineering. As a result, in 1899 a merger took place between KIVI and the two other societies.[37]

During the discussions on the formation of disciplinary sections within the KIVI, understandably the idea was also put forward to form a section on chemical engineering under the larger umbrella of the engineering society. In 1897, the technologists H. Baucke and A. Vosmaer – who in 1903 both would become council members of the Dutch Chemical Society – were active on that front. But their attempts failed, and it would not be until 1948 that a section on chemical technology within KIVI was founded.[38]

In the meantime, tensions grew between the technologists from Delft and the chemists from the universities. The more Hoogewerff steered the technology curriculum in a chemical direction,[39] the more the Delft alumni and the academic chemists were at crossed purposes. In theory, the Delft technologists were educated for industry, and the academic chemists for scientific positions and for the teaching profession. In practice though, the labour markets were not neatly divided that way. Technologists were appointed as chemists in agricultural experimental stations, in private laboratories, and around 1900 also increasingly as teachers at professional and secondary schools. It is not very surprising, therefore, that chemists from the universities became quite upset when Dutch government in March 1903 announced that the Polytechnical College at Delft would become a Technical University.[40]

In June 1903, chemistry students at the University of Amsterdam sent a petition to Dutch parliament, in which they protested against the elevation of the Polytechnical College to university status. They feared heavier competition from the Delft alumni on all their labour markets, and on the labour market for teachers – their traditional home market – especially. Until early 1904, a heated boundary conflict between the university professors of chemistry and their colleagues from Delft would follow, in which the role of university chemists in industry and the role of chemical engineers in teaching were the key issues. In the end, the university professors could not turn the tide, and in 1905 the Polytechnical College at Delft was transformed into a Technische Hoogeschool, with all the same rights as a university.[41]

---

[37] Lintsen (1980), 97–99, 205–210, 226–243, 326–330.

[38] Vosmaer (1897); Baucke (1908).

[39] *cf.* Eijdman (1906).

[40] For similar debates in Germany, see Fischer (1897 and 1898); Burchardt (1978).

[41] For this debate, see Somsen (1998).

9.8   Founding the *Nederlandsche Chemische Vereeniging* (NCV)

Just before these clashes between Delft and the universities intensified, two former students of van 't Hoff and a chemical engineer who had studied at Delft sent a circular letter to at least 600 chemical practitioners in the Netherlands. The circular summarized a draft version of the by-laws of a nationwide chemical society which they intended to establish, and contained an appeal to visit a meeting which would be held in The Hague on 15th April 1903.[42]

The circular and the meeting of 15th April had been preceded by several years of careful preparation. Already in the winter of 1896–1897 Willem P. Jorissen, who in October 1896 had received his Amsterdam chemical doctorate, and Jan Rutten, who in the same year had graduated from Delft Polytechnic College, had discussed the importance of founding a national chemical society during long evening walks in Rotterdam. Both had moved only recently to Rotterdam, and were cut off from the inspiring intellectual atmospheres of their former schools. Rutten was employed as a chemist in a phosphate fertilizer works, and Jorissen worked in the private analytical laboratory of Dr Bonno van Dijken. Although both Rutten and Jorissen left Rotterdam rather soon – in 1898 and 1900 respectively – they developed a life-long friendship during the 20 months they spent together.

During their discussions, they were confronted inevitably with the question of how to define the membership of the new society. Should it unite all chemical practitioners – so all who did chemical investigations, or had a profound knowledge of chemistry? If so, then this would mean that it would include all academic chemists, chemical engineers and pharmacists. Or should it be limited to those devoted to applied chemistry and sanitation – just like Jorissen and Rutten themselves – for the simple reason that many of the academic chemists were already members of one or more scientific societies? These were complicated questions, and before drawing some final conclusions, both friends decided first to compile a list of all Dutch chemical practitioners, and to publish that list, together with practical information that would be useful to any chemist (such as tables of boiling points, molecular weights, *etc.*), in the new *Scheikundig Jaarboekje* (Chemical Annual) founded by them. Perhaps a published address list of Dutch chemists would stimulate a sense of community among them, or, at least, it would be a very useful tool to arrive at a more complete list later, as well as a necessary prerequisite to communicate with the entire community – as the example of the circular shows.[43]

When the *Scheikundig Jaarboekje* was published, in late 1898 or early 1899, it contained a list with no less than 650 names. Next to Jorissen and Rutten, three other editors had helped in preparing the book: B. A. van Ketel, a pharmacist, bacteriologist and chemist who had studied at the University of Amsterdam,

---

[42] Jorissen, Reicher and Rutten (1902/1903c). For the early history of the Dutch Chemical Society, see Snelders (1993), 169–172 and (1997), 23–37.

[43] Reicher (1928); Olivier (1928).

and who at that time owned a commercial laboratory for the analysis of food, drinking water and, his specialty, cinchona bark; H. C. Prinsen Geerligs, who also had studied in Amsterdam, and then headed the Experimental Station of the West-Java sugar industry at Tegal; and, last but not least, van 't Hoff's assistant for many years Lodewijk T. Reicher, who in 1893 had left the university laboratory to take charge of the chemical laboratory of the town of Amsterdam, where as part of the Municipal Medical Service, foodstuffs, drinking water and other matters of public hygiene were analysed. They obviously were all recruited by Jorissen, who had studied together with them in the laboratories of van 't Hoff and his colleague Gunning at Amsterdam. In 1902 Reicher would become the *dritte im Bunde* when he joined Jorissen and Rutten in founding the Dutch Chemical Society.[44]

The publication was well received. A second edition of the yearbook appeared in 1901, and the third edition of 1902 even widened its scope to Flanders. The head of the municipal laboratory of the town of Ghent, A. J. J. Vandevelde, now joined the group of editors, and the title of the yearbook changed to *Scheikundig Jaarboekje voor Nederland, België en Nederlands-Indië* (Chemical Annual for the Netherlands, Belgium and the Dutch Indies).

Jorissen's publishing activities were not confined to the yearbook. In 1897, his employer Van Dijken took the initiative to publish a journal devoted to applied chemistry and public hygienics – the *Tijdschrift voor toegepaste scheikunde en hygiène* – and asked Jorissen to join him as editor. In those years the analysis of foodstuffs, the fight against adulterations and the monitoring of water sources attracted much attention, and a growing number of public and, especially, private laboratories was active in that field. Reicher and Henry's brother H. J. van 't Hoff became the first collaborators of the journal, and later Reicher replaced Van Dijken as editor. During the six years of its existence, Jorissen and Reicher succeeded in attracting a growing number of collaborators, mostly from Holland, but also from Belgium.[45]

Together with the *Scheikundig Jaarboekje*, the *Tijdschrift voor toegepaste scheikunde* played a crucial role in preparing the ground for founding a nation-wide chemical society. The *Tijdschrift* acted as the mouthpiece of the founders of the new society. From May 1902 onwards a lively debate on the necessity and nature of a future society was initiated on the pages of the journal, from which we can learn that there were opposing views on the viability of such an enterprise, and on the course it should take. The discussion was opened by the Flemish chemist A. J. J. Vandevelde, who emphasized that the new society should be devoted mainly to applied chemistry and public hygienics, and therefore should have a similar scope to the *Association Belge des Chimistes*, founded in 1887, during its early years.[46] In June, Rutten responded that he agreed with the focus on applied chemistry, but not on public hygienics (which

---

[44] Jorissen and Reicher (1912), 65–70; Cohen (1912), 175–181.
[45] *Tijdschrift* (1897/1898), (1902/1903).
[46] Vandevelde (1901/1902a and 1928); Deelstra and Van Tiggelen (2003).

already had its own organization), and he even proposed to found one unified society of Dutch and Belgium chemists.[47]

A completely different approach was put forward by Vosmaer, who in August 1902 returned to his old ideal of 1897 to found a chemical section within the engineering society KIVI. In his view, no separate chemical society should be founded at all, but both chemists and chemical engineers should be united under the KIVI umbrella.[48]

Before November 1902 though, probably under the influence of Jorissen, Rutten and Reicher, the idea of a far broader society gained ground, which would not be restricted to applied chemistry, but would also include academic chemists and pharmacists. The example of Germany had shown, Rutten argued, that also a broad society could serve as a good platform on which science and industry could meet. It was this approach that finally gained the upper hand. The result was a broad society in which the equilibrium between the different factions was guarded carefully. It was also proposed that the existing local chemical societies and chemical reading circles should become local sections within the new national chemical society.[49]

As a result of their mutual consensus on a broad chemical society, Jorissen, Reicher and Rutten drafted a proposal which contained the basic principles of the scope, aims and working rules of the new society, and sent this text in 1902 to all professors of chemistry and pharmacy of the four Dutch universities and the Polytechnic College.[50] When in November 1902 these draft statutes were made public for the first time – in the *Tijdschrift* – Jorissen and his friends clearly tried to enrol the joint group of professors for their purposes. "Until now", they wrote, "we received replies from all [professors], except two, with remarks and comments, as solicited. From these replies it appeared to us, that *in general* there is sympathy for our plan among the professors."[51]

Clearly, the professors were not over-enthusiastic. In fact, most of them, as we will argue below, were rather sceptical about the future of the society, and some of them even opposed it. Despite this head wind, the *Nederlandsche Chemische Vereeniging* (NCV) was founded. During the meeting of 15th April, chaired by Jorissen, a provisional council was formed. The composition of that council illustrates wonderfully both the scope the new society would have and the care with which the equilibrium between the different constituencies would be guarded: it consisted of two chemists (Jorissen and Reicher), two professors (Ernst Cohen and F. A. H. Schreinemakers), two pharmacists (Alide Grutte-rink and Jan J. Hofman) and three chemical engineers (Rutten, H. Baucke and A. Vosmaer). The young professor of physical and inorganic chemistry at Utrecht, Ernst Cohen, who had studied in Amsterdam with van 't Hoff, was elected the first president. A clear statement was made by the election of a woman (Grutterink) into the council: despite Cohen's (later?) ambivalence with

---

[47] Rutten (1901/1902a). Also see Vandevelde (1901/1902b).
[48] Vosmaer (1901/1902).
[49] Rutten (1901/1902b); Vandevelde (1902/1903); Jorissen, Reicher and Rutten (1902/1903a).
[50] Snelders (1997), 28; Reicher (1928), 346.
[51] Jorissen, Reicher and Rutten (1902/1903a) and (1902/1903b).

respect to women in science, women would be welcome in the new society, and could even become officers.[52]

That Jorissen and Reicher took the lead in the establishment of a broad chemical society, and not one devoted to applied chemistry alone, was closely related to their roles as the two editors of the *Tijdschrift voor toegepaste scheikunde en hygiëne*. They wanted to give that journal a stronger economic basis by making it the journal for the members of a chemical society, and by transforming it into a more general chemical journal that would attract a larger readership. A chemical society that would unite all chemists of the country would serve that purpose far better than a society of a more restricted nature.

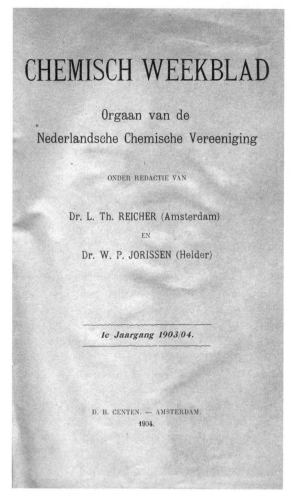

Figure 9.1   Title page and first page of the first volume of *Chemisch Weekblad*, 1903/04.

---

[52] Rutten (1902/193); Snelders (1993), 172, 204; Cohen (1905); Bosch (1994), 524.

CHEMISCH WEEKBLAD.

Orgaan van de Nederlandsche Chemische Vereeniging.

ONDER REDACTIE VAN

Dr. L. TH. REICHER (Amsterdam) en Dr. W. P. JORISSEN (Helder).

Uitgever: D. B. CENTEN, Amsterdam.

Agent voor Ned. Indië: H. VAN INGEN, *Soerabaia.*

*Het auteursrecht van den inhoud van dit Blad wordt verzekerd volgens
de Wet van 28 Juni 1881, Staatsblad No. 124.*

| N°. 1. | Amsterdam, 3 October 1903. | 1ᵉ Jaargang. |

INHOUD : Prof. Dr. ERNST COHEN, Toespraak gehouden op de eerste Algemeene Vergadering der Nederlandsche Chemische Vereeniging te Utrecht den 4ᵈᵉⁿ Juli 1903. — Prof. J. F. EIJKMAN, Over eenige gehydreerde cyclische koolwaterstoffen. — Dr. A. J. SALM, Eenige mededeelingen uit de kliniek te Bordeaux. — Nederlandsche Chemische Vereeniging. — Correspondentie.

L. S.

Dit Weekblad treedt in de plaats van het „Tijdschrift voor toegepaste scheikunde en hygiëne". Het zal in de eerste plaats opnemen oorspronkelijke mededeelingen, zoowel zuiver chemische als zoodanige, die betrekking hebben op toepassingen der chemie op de techniek, de hygiëne, het levensmiddelen-onderzoek, enz. Daarnaast zullen er in verschijnen opstellen, die een overzicht geven van den stand der wetenschap betreffende een of ander onderdeel der chemie.

Ook korte mededeelingen van algemeen belang, chemisch-industrieele berichten, personalia en boekaankondigingen, zullen o. a. opgenomen worden.

DE REDACTIE.

**Toespraak, gehouden op de eerste Algemeene Vergadering *)
der Nederlandsche Chemische Vereeniging te Utrecht
den 4den Juli 1903.**

DOOR

Prof. Dr. ERNST COHEN.

Mijne Heeren,

Nu gij uit alle deelen des lands hier zijt samengekomen ter bijwoning van de eerste vergadering onzer zoo juist tot stand ge-

*) Een verslag dezer vergadering is opgenomen in de Augustus-aflevering van het „Tijdschrift voor toegepaste scheikunde en hygiëne".

Figure 9.1   (Continued)

Their proposal to transform the monthly *Tijdschrift* into the more general *Chemisch Weekblad* therefore was part and parcel of their initiative to create a national chemical society. The decision to publish a weekly instead of a monthly had been taken after strong pressure by the publisher of the *Tijdschrift*, D. B. Centen, who therefore, as a non-chemist, also put his mark on the nature of the new society. In September 1903 the last issue of the *Tijdschrift* appeared, and on 3rd October 1903 the first issue of the *Chemisch Weekblad* (Figure 9.1) came out as the official periodical of the *Nederlandsche Chemische Vereeniging* (NCV), although the ownership of the journal remained with the publisher. Jorissen and Reicher also became the editors of the new journal.[53]

---

[53] Centen (1902/1903); Reicher and Jorissen (1903/1904); Jorissen (1928); Reicher (1928); Centen and Van Crans (1928); Linden (1953), 545–547.

There can be no doubt that the policies of the editors and publisher of the journal to a large degree determined the nature of the new chemical society.

9.9   The First Members of the *Nederlandsche Chemische Vereeniging*

On 4th July 1903 the first general assembly took place and the NCV was formally established. The provisional council was re-elected without any change, and after long discussions on many details the statutes were drawn up. Thereby the original proposal that the society would be watchful of "the interests of Dutch chemists, at home, in the colonies, and abroad" was changed into "the interests of its members", which circumvented any future discussion on who was a chemist. Next to taking care of the interests of its members, the society was supposed to "pay attention to the prosperity and development of chemistry and of the chemical and related industries in the Netherlands and the Dutch Indies".[54] That summer, the society counted 149 original members, who had become members just by proposing themselves. After the statutes had come into force, one could only become a member after nomination by two members, a three-week term in which objections could be raised, followed by a decision by the council.[55]

This procedure to become a member of the society emphasized the principle that the society should be open to all those engaged in chemical practice and/or devoted to the promotion of chemistry. No definition of a chemist was given, and no specific degree or diploma was required! In practice though, the society was far less open than its statutes suggested. The mechanism of cooptation proved to be quite an effective means, as we shall see, for hindering the membership of chemical practitioners without higher degrees. Also the founding process of the society, in which obviously the personal networks of Jorissen, Reicher, Rutten and Cohen played a crucial role, resulted in a society with a far less heterogeneous membership than would have been possible on the basis of the statutes. This becomes strikingly clear when we analyse in detail the group of the 149 founding members that became members before 4th July 1903 (see Tables 9.2–9.4). That analysis will throw light on the frontiers that separated members of the society from outsiders, as well as on the boundaries that divided different segments within the chemical community itself.[56]

Tables 9.2 and 9.3 early show that the vast majority of the original NCV members was highly educated. At least 63% had studied at a Dutch or foreign university (Table 9.3), and almost half of the members even had obtained the doctorate (66), or were preparing for it (6) (Table 9.2). Most of the 19 members

---

[54] Notulen (1902/1903).

[55] Leden (1902/1903).

[56] For biographical data, degrees, *etc.* I have primarily consulted the Adreslijst (1899) and (1901), and the Leden der Nederl. Chem. Vereeniging (1904/1905), (1913/1914) and (1915/1916) (these last two editions are more precise with respect to diplomas and degrees, and their dates). Additional sources were Bolton (1901); *Album Academicum* (1913); Kroon (ed.) (1925); *Album Promotorum* (1963); *Delftse dissertaties* (1992).

Table 9.2   Highest degrees of the 149 original members of the NCV.

Highest Degree	Number
Dr (= doctorate)	66
Chem. drs. (= master degree)	6
Chem. cand. (= bachelor degree)	7
T. (= chemical engineers from Delft)	38
Ap. (= pharmacists)	28
Other	4
Total	149

Table 9.3   Universities where the 149 original members of the NCV studied.

University	Number	Percentage
Amsterdam	31	19
Utrecht	19	12
Leiden	11	7
Groningen	9	6
Delft	42	26
Foreign university	30	19
Unknown	19	12
Total	161	

N.B. The total number exceeds 149 because several chemists or pharmacists studied at more than one institution.

Table 9.4   Professional affiliation of the 149 original members of the NCV.

Professional Affiliation	Number
University professor	5
Assistant at a university or professional school	11
Teacher Professional School (agriculture; trade; veterinary medicine; navy)	8
Teacher Gymnasium or HBS	15
Chemist in private (analytical) laboratory and/or consultant	16
Chemist in government laboratory	9
Chemist in experimental agricultural station	11
Industrial chemist	30
Pharmacist	28
Student	7
Unknown	9
Total	149

of which the *alma mater* is unknown were pharmacists, and the majority of them – namely all that received their degree after 1876 – will have finished a university course as well. So, certainly more than 63% of the members had obtained a university degree. A second group were the chemical engineers (42) who had studied at the Polytechnic College at Delft, and four of them had obtained a doctorate at a foreign university later (*cf.* Tables 9.2 and 9.3). The pharmacists (42, including 14 that had obtained the doctorate) formed a third

Table 9.5 NCV members compared to all Dutch chemists, according to educational background (1903).

Education	All Dutch chemists (rough estimate)	Dutch NCV members	Percentage NCV members of that educational category
University chemists (Dr and chem. drs.)	205	53	26%
Chemical engineers	170	42	25%
Pharmacists with laboratories	120	42	35%
Non-academic chemists (mainly sugar chemists)	165	4	2–3%
(Advanced) students	40	7	18%
Total	700	148	21%

NB: One founding NCV member was a foreigner, so there were 148 Dutch members.

group. So, all three segments were of the same order of magnitude. One can therefore conclude that the aim of the founders of the society to get chemists, chemical engineers and pharmacists all on board was realized successfully (see also Table 9.5 below). Except for two female pharmacists, all founding members of the society were male.[57]

When we leave the seven students out of consideration, only four NCV members did not have an academic degree of some kind. Two of them, C. M. F. van Rossum and J. Weinberg, worked in the sugar industry, and therefore might have studied at one of the Dutch or foreign sugar schools. The third, J. H. Aberson, had studied at the University of Amsterdam, but did not have an official academic title. He had obtained a licence to become a teacher and, in 1903, taught at the Agricultural College at Wageningen. As a result, there was only one member, the manufacturer of nitrogen fertilizers Ch. E. H. Boissevain, of whom we know with certainty that he had no chemical training at all. We therefore can safely conclude that the open membership made possible by the statutes only existed on paper. In practice the NCV at its founding date was a society that united persons with a high level of training in chemistry. This is strikingly confirmed by Table 9.5. Of all non-academic chemists listed by Jorissen and Rutten in their lists of Dutch chemists, who mainly worked in the sugar industry in the Dutch Indies, only about 2–3% became a founding member of the NCV. In comparison to England, for instance, in the Netherlands there was a far larger social and intellectual gulf between the academic chemists and the chemists with middle-rank training. Also later the numerous sugar *Chemiker* in the Netherlands and the Dutch Indies remained largely

---

[57] In 1912 there were 40 women among a total of 260 chemical engineering students at Delft. See Hasselt (1913), 41–42. Among students of pharmacy the percentage of women students was far higher. Between 1889 and 1934 more than 30% of the graduates was female. See *Toekomst der gegradueerden* (1936), 234.

outsiders to the society; and the same was true for laboratory assistants and middle-rank chemical technicians in the chemical industry.[58] This situation was formalized in 1921, when new statutes came into force. From then on, only those who had an academic degree could become ordinary members.[59]

That the early NCV was not a well-balanced representation of the Dutch community of chemical practitioners as a whole is also shown clearly when we look at the schools and universities where the founding members of the society had studied. Institutions that the founders of the society came from were over-represented: Delft, Amsterdam and Utrecht accounted for more than 80% of the members whose (Dutch) educational background is known (in this calculation I excluded the foreign universities; because almost all those who studied abroad also studied at one of the Dutch institutions).[60] Rutten (Delft), Jorissen and Reicher (Amsterdam) and the first president of the society, Cohen (Utrecht), who had studied in Amsterdam, obviously had used their personal networks to recruit members into the new society. As a result, the early NCV was dominated by chemical engineers from Delft and chemists that came from the Amsterdam school of van 't Hoff – to which Jorissen, Reicher and Cohen belonged. Organic chemists from the Leiden school of Franchimont were clearly under-represented.

This conclusion is confirmed when we focus on the professors that were founding members of the societies. Of the 13 Dutch professors of chemistry and pharmacy in 1903 only 5 were founding members of the NCV (Table 9.4): all professors of Utrecht (Cohen, Van Romburgh, Wefers Bettink) joined the society, and two professors of physical chemistry of Leiden (Van Bemmelen, Schreinemakers), whom Jorissen knew from the time he was an assistant in Leiden in the 1890s. Of the professors of organic chemistry, only Cohen's close colleague Van Romburgh became a founding member. When Ernst Cohen later looked back on the establishment of the NCV, he emphasized the negative attitude of most of his professorial colleagues with respect to the new society: some doubted its viability, others were afraid that it would become a kind of trade union, and those from Delft did not want to be together with the university professors in one organization. Only after long talks could he convince some of his colleagues to enter the NCV. Van Romburgh, in another commemorative publication, emphasized specifically that the organic chemists connected to the journal *Recueil* saw no reason whatsoever to found a second scientific journal in a small country such as the Netherlands, and therefore were opposed to the founding of the NCV and its journal the *Chemisch Weekblad*. In the following years, the successful growth of the society and the quality of the *Chemisch Weekblad* meant that in the years after 1903 most Dutch chemistry professors overcame their initial resistance and entered the NCV: Eijkman (Groningen) in 1903, Aronstein (Delft), Hoogewerff (Delft) and Lobry de

---

[58] *cf.* Jorissen (1901/1902); Barger (1928).

[59] Linden (1953), 522.

[60] Delft and Amsterdam were also the largest breeding school of chemists, but the chemistry department of Utrecht, in terms of student numbers, was smaller than Leiden. See Jorissen (1901/1902), 46; Hondius Boldingh (1909), 10–11; Snelders (1997), 33.

Bruyn (Amsterdam) in 1904, and, finally, Holleman (Groningen/Amsterdam) in 1905 – all of them, except Aronstein, professors of organic chemistry. In 1920 the NCV took over the *Recueil*, which meant that the cold war between the two journals came to an end. Franchimont himself, though, never became an ordinary member of the NCV. In 1912 he was made an honorary member, an offer he obviously could not refuse, after Nobel laureates such as van 't Hoff (1908), Arrhenius (1909), Van der Waals (1912) and Lorentz (1912) had received the same mark of honour.[61] By this elegant gesture the ranks definitely had been closed.

The subtle boundary between physico-chemical van 't Hoff school and the organic school of Franchimont was not the only dividing line which separated the early members of NCV from other segments of the Dutch chemical community. When we look to professional affiliations of the founding members (Table 9.4), it is evident that the vast majority – accounting for more than 70% – worked in the sphere of applied chemistry. Of the 140 members whose jobs are known, 36 worked in private, agricultural and state analytical laboratories, 30 in industry, 28 in pharmacy and 8 were teachers at schools of an applied nature. Only 31 members, who were connected to universities and secondary schools, can considered to be pure chemists. Obviously, a large portion of the founding members of the NCV was recruited from the subscribers to the *Tijdschrift voor toegepaste scheikunde en hygiene*, which served especially the interests of the analytical laboratories and the pharmacists.

The question of to what degree the NCV was a faithful representation of the Dutch chemical community has still another dimension, namely what percentage of that community joined the society at its foundation. It is not easy to give a conclusive answer. If we take a broad definition of the chemical community – one that includes the sugar *Chemiker*, the laboratory assistants and middle-rank chemical technicians – then we are confronted with the fact that exact numbers of those three last groups are not known. Nevertheless, during the early years of the NCV Jorissen and Rutten still included many non-academic chemists in their lists of chemists. On the basis of those lists we can make the reasonable guess that there were about 700 Dutch chemists in 1903.[62] This number included non-academic chemists (about 165), students who had done the halfway "candidaats" examination (a kind of bachelor examination) (about 40), as well as chemical engineers (about 165) and pharmacists with chemical laboratories (about 120, out of a total of about 680 pharmacists) (see Table 9.5). Also included were about 130 chemists working in the Dutch

---

[61] Cohen (1928), 339; Romburgh (1928); Linden (1953), 523.

[62] Different estimation methods gave different results. From a detailed analysis of the lists of 1899 (550 names) and 1914 (882 names), using also information from the report *Toekomst der gegradueerden*, interpolation gave the estimate of about 650 chemists in 1903. Counting of the lists of 1901 (670 names) and 1905 (840 names) produced an estimate of 750 chemists in 1903 *via* interpolation. These numbers for 1901 and 1905 are probably too high, because of double counting of persons with two last names, such as Bakhuis Roozeboom, who were listed twice, and NCV members in the Dutch Indies, who were also listed twice (in the membership list, and in the list of chemists in the Indies). In all calculations for 1903 I have assumed that there were about 700 Dutch chemists in the Netherlands, its colonies and abroad. See Adreslijst van H. H. Scheikundigen (1899); Adreslijst van scheikundigen (1901); Leden der Nederl. Chem. Vereeniging (1904/1905) and (1913/1914); *Toekomst der gegradueerden* (1936), 169–177. *cf.* Snelders (1993), 169.

Table 9.6   NCV members compared to non-members in 1913/1914, according to educational background.

Education	NCV members	Non-members	Total	Percentage NCV members of that educational category
University chemists (Dr and chem. drs.)	184	99	283	65%
Chemical engineers	155	142	297	52%
Pharmacists with laboratories	99	18	117	85%
Non-academic chemists (mainly sugar chemists)	44	71	115	38%
(Advanced) students	26	44	70	37%
Total	508	374	882	58%

Indies and about 20 chemists working abroad. So, there were about 550 chemists in the Netherlands, including the other professionals mentioned.[63]

Of the 149 founding members of the NCV, 5 worked in the Dutch Indies and 4 abroad. Of these last chemists, only one was a true foreigner: A. J. J. Vandevelde of Ghent. On the basis of these figures one can conclude that about 21% of all Dutch chemists were founding members of the NCV, and 27% of all the chemists living in the Netherlands. This certainly was a very good start for the new society. Table 9.5 strongly confirms that there was a good equilibrium within the NCV between academic chemists, chemical engineers and pharmacists. About 25% of all academic chemists and chemical engineers joined the society at its foundation. Of the pharmacists with chemical laboratories, listed by Jorissen and Rutten, 35% joined the society. This was about 18% of all pharmacists with apothecary shops in the Netherlands at that time (about 680). Almost none of the non-academic chemists joined the society. Later that situation would change (see Table 9.6).

9.10   The First Ten Years of the Society

After its start in July 1903 the membership figures of the NCV continued to grow. On 1st January 1905 the society counted 225 members, and half a year later 286. On 1st January 1914 a figure of 575 members had been reached.[64] In

---

[63] Table 9.4 is based on Adreslijst van H. H. Scheikundigen (1899), Leden der Nederl. Chem. Vereeniging (1913/1914) and the report *Toekomst der gegradueerden* (1936), 169–177. There are discrepancies between the lists of chemists and the report: the report gives higher figures for the number of chemical engineers (97 versus 56 in 1899, and 319 versus 297 in 1913/1914), probably because the report also took alumni of the Polytechnic College into account that left the school without graduation. As the lists give only (academic) titles, these alumni were counted by me as non-academic chemists. I cannot explain why the report gives lower figures for the university chemists compared to the list (124 versus 160 in 1899, and 250 versus 283 in 1913/1914). In Tables 9.4 and 9.5 I have taken the average of the figures from the report and the lists, as far as the university chemists and the chemical engineers are concerned. Figures for the other categories are not given in the report.

[64] *CW* 2 (1905), 457; Linden (1953), 522.

that year there were about 325 Dutch chemists and chemical engineers outside the chemical society. So, of the about 900 chemists working in the Netherlands, the Dutch Indies and abroad, almost two-thirds were members of the NCV. After the First World War the NCV succeeded in attracting an even larger share of the Dutch chemical community.[65]

In Table 9.6 a detailed analysis of the list of chemists of 1913/1914 is presented. With respect to the interpretation of the figures one should take into account that the listing of pharmacists and non-academic chemists among the non-members was far less complete than in the lists from the years 1899–1905. So, there were certainly more non-academic chemists than 115 in the Netherlands and the Dutch Indies. Nevertheless, it is striking that the number of non-academic chemists within the NCV grew from 4 (2.6% of all members) in 1903 to 44 (8.6%) in 1913/1914. Also the estimate that 85% of the pharmacists with laboratories joined the society is too high, because Jorissen and Rutten had skipped many pharmacists from the list of non-members (as indicated by a footnote to that list). Table 9.6 confirms, however, the picture sketched above of a society with three strong constituencies: the university chemists were the largest group, the chemical engineers were also present in large numbers – although only about half of the total number of chemical engineers joined the NCV (many decided to become KIVI members instead) – and the pharmacists formed a third large group.

In the initial growth of membership during the first few years, an active and personal approach by members of the council played a large role. All of them recruited mainly among their own constituencies. For instance, during 1904 and 1905, after they had become members themselves, the Delft professors Aronstein and Hoogewerff succeeded in persuading at least 24 chemical engineers to become members of the NCV.[66] Similarly, the pharmacists Hofman and Grutterink during 1903–1905 recruited more than 18 new pharmacists to the NCV. And Rutten, Reicher, Cohen and Jorissen recruited about an equal number of university chemists.[67] Obviously, for some years at least, the initial equilibrium between the different constituent groups of the society was well preserved.

In order to understand why the new society made such a head start, one should analyse in more detail what the NCV could offer its members. According to the (draft) statutes of May and July 1903 the chemical society had the following aims:

1. to organize meetings on theoretical and technical subjects, and to discuss the interests of its members;
2. to publish a (weekly) journal;

---

[65] Leden der Nederl. Chem. Vereeniging (1913/1914) and (1915/1916); Homburg and Palm (2004), 8.
[66] Hissink (1928). Also Van Hasselt stipulated that, as a rule, advanced students from Delft should become NCV members. See Hasselt (1913), 39.
[67] CW 1 (1903/1904) and 2 (1905), passim.

3. to found a consultation office;
4. to found local sections and reading societies;
5. to influence government policy, as far as is related to the interests of its members;
6. to offer good contracts for collective insurances to its members;
7. to financially support chemical research;
8. to organize excursions;
9. to standardize analytical-chemical methods and commercial tariffs;
10. to organize courses for teachers during holidays; and
11. to cooperate with similar societies.[68]

This list nicely illustrates the hybrid character of the new society. Some aims, for instance the organization of meetings, the publication of a journal, the founding of local sections and the support given to research, fitted perfectly well in a science-oriented chemical society. Others, such as the erection of a consultation office, the conclusion of insurance contracts, the standardization of tariffs and taking action towards government, were typical for profession-oriented chemical societies. To realize all these activities, each member paid a membership fee of Dfl. 7.50 each year, and Dfl. 2.50 for each visit to a general meeting.[69] Initially there were no industrial sponsors, but that changed during the 1910s. In 1912 eight chemical companies sponsored the society, in 1915 thirteen, and in 1918 the sponsorship by the industry was enlarged considerably – from then on about 50 companies gave support. In 1918 and 1919 Dfl. 26 500 was raised in order to permit the acquisition of the journal *Recueil* and to enlarge the size of the *Chemisch Weekblad*. Also an industrial section was added to the NCV weekly from 1923 onwards.[70]

Although all aims were formally of equal importance, in practice some aims related to professional interests were hardly realized, despite the fact that a few commissions within the society discussed these aims during many years. As a result, no consultation bureau was established, no collective insurance contracts were concluded and excursions hardly took place. It certainly belongs to the irony of history that these professional roles of the NCV, which had been strongly emphasized at the start of the society, did not get off the ground.[71]

From the very start of the society its first president, Ernst Cohen, took care to get the more scientific-oriented chemists on board, and also the publishing of the *Chemisch Weekblad* generated a dynamics of its own, away from its previous applied orientation, because it was aimed to be a general chemical journal for chemists of all sorts.[72]

During the first few years, two activities of the NCV certainly were the most important ones: the organization of (general) meetings, and the publishing of

---

[68] Concept-statuten (1902/1903); Notulen (1902/1903).
[69] Notulen (1902/1903), 378.
[70] Linden (1953), 524, 546–547, 550, 554; Leden der Nederl. Chem. Vereeniging (1915/1916), 393; Olivier (1928), 328; Kruyt (1928).
[71] Linden (1953), 521.
[72] Cohen (1903/1904).

the *Chemisch Weekblad*. Compared to several other chemical societies – for instance its Belgian sister-organization with its many sections and numerous meetings – the social dimension of the NCV was very modest. During the period under consideration, at first only two general meetings a year were held, and from 1908 onwards three meetings a year. It was not until 1923 that local chemical reading circles and societies, which each often had ten or more meetings a year, were integrated into the NCV. Nevertheless, the two or three yearly general meetings were later considered to be important events, and offered good occasions on which the NCV members could meet. Dinners, excursions to industries and music performances were standard ingredients of the general meetings, and offered ample opportunity for the NCV members to intensify personal contacts. During the first few years of the society, though, the numbers of attendants to these meetings were only modest.[73]

Papers presented at the general meetings covered all aspects chemistry, both pure and applied, as well as topics that related to the interests of the profession, and speakers were from academia as well as from practice. On 16th July 1904, for instance, the academics C. H. Wind (Utrecht) and F. M. Jaeger (Amsterdam) gave purely scientific expositions on the role of electrons in chemistry and on crystal chemistry, respectively; whereas Ko Hissink (Agricultural Station, Goes) and G. van Iterson Jr. (Delft) gave papers on more applied topics, such as the investigation of soils and denitrification. And, to give another example, on 18th July 1914 purely scientific, applied chemical and professional topics all were presented: H. J. Prins (industrial research) lectured on the fundamentals of catalysis, J. Dekker (Colonial Institute) gave a paper on tanning agents and O. de Vries (Experimental Station for Tobacco, Java) discussed the labour market of chemists in the Dutch Indies.[74]

The presentation of Van Iterson's paper on denitrification by bacteria was not by accident. During the whole period 1903–1914 there were several papers on microbiology presented at the general meetings of the NCV, due to the strong position of microbiology within the department of chemical technology at Delft and the chemistry department of the Agricultural School at Wageningen (which acquired university status in 1918). The general meetings kept their broad character until 1916 when, for the first time, the meeting was split into two sections: one on general chemistry and one on applied chemistry.[75]

This leaves no doubt that the publication of the *Chemisch Weekblad* was the most important activity of the NCV by far, and probably also the prime reason for chemists to become members. Jorissen and Reicher were competent editors, who were able to transform the *Tijdschrift voor toegepaste scheikunde en hygiene*, with 384 pages in 1902–1903, into the much more voluminous *Chemisch Weekblad*, with 1046 pages in 1903–1904 and 836 pages in 1905.

[73] Olivier (1928), 326; Holleman (1928); Linden (1953), 536, 552, 565–566.
[74] Wind (1903/1904); Jaeger (1903/1904); Hissink (1903/1904); Iterson Jr. (1903/1904); Dekker (1914); Prins (1914); Vries (1914).
[75] Meerburg (1928).

Apart from articles on applied chemistry, which had also appeared in the *Tijdschrift*, now also papers on purely scientific topics were published – *e.g.* on organic chemistry by Eijkman, on physical chemistry by van 't Hoff and Arrhenius, on the phase rule by Schreinemakers and P. A. Meerburg – together with news of the NCV, articles on the history of chemistry by Cohen and Jorissen, economic news, book reviews, news of research done at state laboratories and on the meetings of other scientific societies, and obituaries, even of non-members, which played an important role in strengthening the sense of community among the chemists. Many articles in the 1903–1904 volume were written by members of the council, obviously because it was not easy to fill an issue every week. But in 1905 most of these problems were solved. While a number of different authors were involved, the character of the journal also moved increasingly in the direction of pure science, without becoming dominantly pure though. During the entire period under investigation in all volumes a fair equilibrium between pure and applied chemistry, and news for the profession, was maintained.[76]

That the NCV lost its initial applied character during the first few years of its existence is also illustrated by the composition of the council, and by the election of its presidents especially. Although professors of chemistry had played no role at all during the embryonic phase of the society, during the first 13 years it was only professors that presided over the NCV: Cohen 1903–1904, Aronstein 1905–1907, Holleman 1908–1909, Hoogewerff 1910–1912 and, again, Cohen 1913–1915. Not until 1916 was a non-academic president elected: A. Lam, the director of the municipal laboratory of Rotterdam. The fact that also organic chemists such as Holleman and Hoogewerff presided over the NCV shows that they had overcome their initial hesitations, and had chosen to put their own, scientific, stamp on the NCV. All professors mentioned were from the universities of Amsterdam, Utrecht and Delft. It would not be until 1966(!) that a professor from Leiden became president and, until at least 1985, professors from Groningen never obtained that position. So, the bias that the NCV had at its start, seems to have persisted for a long time.[77]

Next to these two core activities – the publication of a journal and the organization of two or three national meetings a year – many other initiatives were taken by the NCV. I will not give a full overview, but will briefly discuss a few of the most important ones.[78]

In 1903 a Library Committee was inaugurated, which took on the task of making bibliographical overviews of all chemical periodicals and chemical books present in the Dutch university libraries and in several other large libraries in the Netherlands. In 1914 the journal and books of more than 30 libraries were listed. These bibliographies were published in the *Chemische*

---

[76] Snelders (1997), 23–25; Olivier (1928), 325.
[77] Snelders (1997), 221; Homburg and Palm, ed. (2004), 333; Linden (1953), 560–562.
[78] See Olivier (1928), 326–327; Linden (1953), 524–528, 549, 552.

Jaarboekjes, together with codes indicating in which library a journal or book could be found.[79]

The national chemical bibliography certainly was useful for all NCV members, both scientific and professional, but it was not a professional activity in the strictest sense. That professional activities were also undertaken by the NCV, though, shows the example of a committee, installed in 1904, that should advise about the tariffs for chemical investigations. In 1906 this committee published a list of tariffs that chemists should ask for their work. The next year this list was officially accepted by the NCV. It stayed in force until 1915, when a new committee revised the list.

On several occasions, the NCV wrote addresses to parliament or government on issues that were relevant to the chemical profession. An early example is an address from February or March 1905 on the new law on Higher Education. This was, as we have seen above, a very sensitive issue that divided the professors of the universities and the Polytechnic College at Delft. The NCV, therefore, steered a middle course, and included those issues in the address on which the professors of both types of institutions agreed: the HBS should give access to university studies in chemistry and pharmacy, and chemistry students should be allowed to shift halfway through their studies from a university to Delft, and *vice versa*.[80]

As the founding history of the NCV has shown, the society was very good at keeping the balance between different interest groups. During large parts of its history this strategy was followed, which resulted, for instance, in coalitions with other scientific societies. A good example is the domain of food analysis, which was a contested terrain between the pharmacists and the chemists. In 1905, after an extensive debate within the NCV on whether the German example should be followed to create a specific degree for food analysts (*Nahrungsmittel-Chemiker*), the society concluded that no special degree was needed, and that chemists and pharmacists should continue to work in this field as before.[81] Three years later the NCV joined forces with the Pharmaceutical Society NMBP. A joint commission was installed that would organize conferences about food chemistry every two years, in which both chemists and pharmacists would participate. This activity was successfully continued for many decades.

Another joint activity was a united address to Dutch government in 1912 on the rules with respect to patent agents, together with the engineering society KIVI. A joint commission was installed that prepared alternative regulations for patent agents. These were sent to Dutch government in 1916, but the action did not have the desired effect.[82]

Later, from 1917 onwards, the NCV played a very important role in the organization of examinations for laboratory assistants and analysts. But that activity falls outside the area of discussion.

---

[79] *cf.* also Reicher (1903/1904); Baucke (1903/1904a) and (1903/1904b); Verslag van de vergadering (1914); Gorter (1951).
[80] Hissink (1905).
[81] *CW* 2 (1905), 441, 460–463.
[82] *cf.* Verslag van de werkzaamheden (1914).

9.11   Conclusion

Why was the Dutch Chemical Society founded relatively late, and how can we explain the hybrid nature of the society? These were the questions posed at the start of this chapter. I hope to have shown that there were several reasons for the late establishment of the NCV. In the first place, there was an extensive network of local scientific societies in the Netherlands that also paid attention to chemistry. This made it less urgent for academic chemists to found a national discipline-oriented society. Secondly, the Netherlands did not have a dominant metropolitan centre, and therefore there was no town whose local chemical society could develop into a national one. The fact that there were also tensions between the Amsterdam chemical school and the Leiden school was an extra reason why the initiative for a national society could not come from academic circles. The third, and final, reason for the late emergence of the NCV was the relatively late industrialization of the Netherlands, and the late expansion of applied chemical research. Only in the 1890s, when agricultural experimental stations and analytical laboratories flourished, and the number of chemical engineers grew rapidly, were the conditions fulfilled for the establishment of a profession-oriented chemical society.

Indeed, it was the chemists from the private and governmental analytical laboratories who took the initiative to establish a national chemical society. Within these laboratories there was a close cooperation between academic chemists, chemical engineers and pharmacists, and the founders of the NCV included all three groups in their project from the start. Also important was the necessity to give a strong economic basis to the journal of the new society. The leading role of the editors of that journal in the founding of the NCV makes it plausible that they played a crucial role in broadening the society from a profession-oriented organization to a hybrid society in which both academic chemistry and professional interests could play a role. They took the decision to ask a professor of chemistry to become the first chairman of the NCV, and, after this start, gradually also other professors of chemistry and pharmacy became involved in the NCV.

A true unification of the chemical profession did not take place. Until after the Second World War, the different groups within the society – the university chemists, the pharmacists and the chemical engineers – were always visible. But mostly the NCV succeeded in letting them cooperate harmoniously. The society was always good at making coalitions with other societies and professional groups, such as the pharmacists (in the field of foodstuffs), the physicists and the engineers, and in embracing local groups and groups of specialists (such as the biochemists, the photochemists, experts in ceramics, *etc.*). Around 1980 the society was again opened to non-academic chemists, who had been excluded from membership in 1921. As a result of this new inclusive strategy the society became one of the largest chemical societies of Europe. With more than 10000 members it is now in third place, after the British and the German societies.

Abbreviations

ap.	apotheker (= pharmacist)
cand.	candidaat (= chemistry student with bachelor degree)
chem.	chemisch (= chemical)
CJ	*Chemisch Jaarboekje*
CW	*Chemisch Weekblad*
DCG	*Deutsche Chemische Gesellschaft*
Dr	doctor (= chemist with Ph.D.)
drs.	doctorandus (= chemist with master degree)
HBS	Hogere Burger School
KIVI	Koninklijk Instituut van Ingenieurs
NCV	*Nederlandsche Chemische Vereeniging*
NMBP	*Nederlandsche Maatschappij ter Bevordering der Pharmacie*
SJ	*Scheikundig Jaarboekje*
t.	technoloog (= technologist = chemical engineer)
TTSH	*Tijdschrift voor toegepaste scheikunde en hygiène*

References

Adreslijst van H. H. Scheikundigen (1899), *Scheikundig Jaarboekje* (hereafter: *SJ*) **1**, 163–196.

Adreslijst van Scheikundigen (1901), *SJ* **2**, 196–233.

Album Academicum van het Atheneum Illustre en van de Universiteit van Amsterdam (1913), R. W. P. de Vries, Amsterdam.

Album Promotorum der Rijksuniversiteit Utrecht 1815–1936 (1963), E. J. Brill, Leiden.

Averley, Gwen (1986), The 'Social Chemists': English Chemical Societies in the Eighteenth and Early Nineteenth Century, *Ambix* **33**, 99–128.

Bakker, Martijn S. C. (1989), *Ondernemerschap en vernieuwing. De Nederlandse bietsuikerindustrie 1858–1919*, NEHA, Amsterdam.

Barger, George (1928), De chemische opleiding in Nederland en Engeland, *Chemisch Weekblad* (hereafter: *CW*) **25**, 362–364.

Bartels, A. (1963), *Een eeuw middelbaar onderwijs 1863–1963*, J. B. Wolters, Groningen.

Baucke, H. (1903/1904a), Boekenlijst voor het Chemisch Jaarboekje, *CW* **1** (1903/1904), 544–549, 562–568, 598–602, 614–620, 639–645, 740–741.

Baucke, H. (1903/1904b), Verslag van de Bibliotheek-Commissie, *CW* **1** (1903/1904), 1024–1026.

Baucke, H. (1908), Iets over het ontstaan der Ned. Chem. Vereeniging, *CW* **5** (1908), 295–299.

Berkel, K. van (1991), Het Genootschap als Spiegel van twee eeuwen wetenschapsgeschiedenis in Nederland, in K. van Berkel, M. J. van Lieburg and H. A. M. Snelders, *Spiegelbeeld der Wetenschap. Het Genootschap ter Bevordering van Natuur-, Genees- en Heelkunde*, Erasmus, Rotterdam, 11–58.

Berkel, Klaas van (2002), De Hollandsche Maatschappij, de Archives Néerlandaises en de Nederlandse natuurwetenschap rond 1870, in *Geleerden en leken. De wereld van de Hollandsche Maatschappij der Wetenschappen 1840–1880*, Haarlem and Rotterdam, 59–81.

Bierman, Annette Irene (1988), *Van artsenijmengkunde naar artsenijbereidkunde. Ontwikkelingen van de Nederlandse farmacie in de negentiende eeuw*, Rodopi, Amsterdam.

Bolton, Henry Carrington (1901), *A Select Bibliography of Chemistry 1492–1897. Second Supplement. Section VIII – Academic Dissertation*, Smithsonian Institution, Washington (also: Kraus Reprint, New York, 1967).

Bolton, Henry Carrington (1902), *Chemical Societies of the Nineteenth Century*, Smithsonian Institution, Washington.

Bosch, Mineke (1994), *Het geslacht van de wetenschap. Vrouwen en hoger onderwijs in Nederland 1878–1948*, SUA, Amsterdam.

Bud, Robert Franklin (1980), *The Discipline of Chemistry: The Origins and Early Years of the Chemical Society of London*, Ph.D. thesis, University of Pennsylvania.

Bud, Robert (1991), The Chemical Society – a glimpse at the foundations, *Chemistry in Britain*, 230–232.

Burchardt, Lothar (1978), Die Ausbildung des Chemikers im Kaiserreich, *Zeitschrift für Unternehmensgeschichte*, **23**, 31–53.

Centen, D. B. (1902/1903), L. S., *Tijdschrift voor toegepaste scheikunde en hygiène* (hereafter: *TTSH*) **6**, 384.

Centen, D. B. and van Crans, P. J. (1928), Naschrift, *CW* **25**, 375–376.

Chemisch Jaarboekje voor Nederland, België en Nederl. Indië, tevens jaarboekje der Nederlandsche Chemische Vereeniging, **5** (1904/1905)–**13** (1915/1916) (= *CJ*).

Chemisch Weekblad **1** (1903/1904)–**11** (1914) (= *CW*).

Cittert-Eymers, J. G. van (1977), Het Natuurkundig Gezelschap te Utrecht 1777–1977, in *NG 200: Natuurkundig Gezelschap te Utrecht 1777–1977*, Natuurkundig Gezelschap, Utrecht, 39–82.

Cohen, Ernst (1903/1904), Toespraak, gehouden op de eerste Algemeene Vergadering der Nederlandsche Chemische Vereeniging te Utrecht, *CW* **1**, 1–7.

Cohen, Ernst (1905), Feminisme en exacte wetenschap, *CW* **2**, 349–367.

Cohen, Ernst (1912), *Jacobus Henricus van 't Hoff. Sein Leben und Wirken*, Akademische Verlagsgesellschaft, Leipzig.

Cohen, Ernst (1928), Niet-uitgesproken rede ter gelegenheid van het 25-jarig bestaan der Ned. Chemische Vereeniging (1903–1928), *CW* **25**, 337–342.

Concept-statuten van de Algemeene Nederlandsche Chemische Vereeniging (1902/1903), *TTSH* **6**, 316–320.

Deelstra, Hendrik and Van Tiggelen, Brigitte (2003), Edouard Hanuise: Président-fondateur de l'Association des chimistes en Belgique, *Chimie nouvelle* **81**, 21–24.

Dekker, J. (1914), Over looistoffen, *CW* **11**, 719–725.

2000 Delftse dissertaties, een bibliografie (1992), Bibliotheek Technische Universiteit Delft, Delft.

1e Eeuwboek van het Technologisch Gezelschap te Delft, opgericht 15 december 1890, 1890–1990 (1989), Technologisch Gezelschap, Delft.

Eijdman Jr., F. H. (1906), De scheikunde aan de Polytechische School, in H. H. R. Roelofs Heyrmans, ed., *Gedenkschrift van de Koninklijke Akademie en van de Polytechische School 1842–1905*, J. Waltman, Delft, 265–278.

Feestnummer uitgeg. bij de viering van het 50-jarig bestaan van de Vereeniging van leeraren bij het middelbaar onderwijs, 1867–1917 (1917), Versluys, Amsterdam.

Fell, Ulrike and Rocke, Alan (2007), The Chemical Society of France in its Formative Years, 1857–1914: Disciplinary Identity and the Struggle for Unity, Chapter 5 in this volume.

Fischer, Ferd. (1897), *Das Studium der technischen Chemie an den Universitäten und Technischen Hochschulen Deutschlands und das Chemiker-Examen*, Vieweg, Braunschweig.

Fischer, Ferd. (1898), *Chemische Technologie an den Universitäten und Technischen Hochschulen Deutschlands*, Vieweg, Braunschweig.

Goeij, Carel de (1993), Het Deventer Natuur- en Scheikundig Genootschap en de opleving van de natuurwetenschap in Nederland in de tweede helft van de negentiende eeuw, *Deventer Jaarboek* 22–44.

Gorter, A. (1951), *Inventaris van periodieken op chemisch en verwant gebied aanwezig in Nederlandse bibliotheken*, D. B. Centen, Amsterdam.

Hasselt, Rolf van (1913), Chemisch ingenieur, in C. Beets and Rolf van Hasselt, *De fabrieks-ingenieurs. Chemisch ingenieur*, C. Morks, Dordrecht, 23–52.

Hissink, D. J. (1903/04), Grondonderzoek, *CW* **1**, 681–688.

Hissink, D. J. (secretary of NCV) (1905), Address NCV to Parliament, *CW* **2**, 297–298.

Hissink, D. J. (1928), De Nederl. Chem. Vereeniging tijdens mijn secretariaat, *CW* **25**, 352–353.

Holleman, A. F. (1928), De Nederlandsche Chemische Vereeniging in 1908 en 1909, *CW* **25**, 335–336.

Homburg, Ernst, with cooperation by Arjan van Rooij (2004), *Groeien door kunstmest: DSM Agro 1929–2004*, Verloren, Hilversum.

Homburg, Ernst and Palm, Lodewijk (2004), Grenzen aan de groei – groei aan de grenzen: enkele ontwikkelingslijnen van de na-oorlogse chemie, in Homburg, Ernst and Palm, Lodewijk, ed., *De geschiedenis van de scheikunde in Nederland 3: De ontwikkeling van de chemie van 1945 tot het begin van de jaren tachtig*, Delft University Press, Delft, 3–18.

Het honderdjarig bestaan van het Natuurkundig Genootschap te Groningen, gevierd op 1 en 2 maart 1901 (1901), Gebroeders Hoitsema, Groningen.

Hondius Boldingh, G. (1909), *De maatschappelijke waarde van ons Hooger Onderwijs in de scheikunde*, Hoorn.

Iterson Jr., G. van (1903/1904), Over denitrificatie, *CW* **1**, 691–699.

Jaeger, F. M. (1903/1904), Kristalchemische bijdrage, *CW* **1**, 657–658.

Jorissen, W. P. (1901/1902), De vooruitzichten onzer aanstaande scheikundigen, *TTSH* **5**, 46–48.

Jorissen, W. P. (1928), Ter inleiding, *CW* **25**, 321–323.

Jorissen, W. P. and Reicher, L. Th. (1912), *J.H. van 't Hoffs Amsterdamer Periode 1877–1895*, C. de Boer, Den Helder.

Jorissen, W. P., Reicher, L. Th. and Rutten, J. (1902/1903a), Eene Nederlandsche Chemische Vereeniging, *TTSH* **6**, 65–68.

Jorissen, W. P., Reicher, L. Th. and Rutten, J. (1902/1903b), Naschrift, *TTSH* **6**, 68–70.

Jorissen, W. P., Reicher, L. Th. and Rutten, J. (1902/1903c), Chemische Vereeniging, *TTSH* **6**, 224.

Kendall, James (1942), Some eighteenth-century chemical societies, *Endeavour* **1**, 106–109.

Kerkwijk, C. P. van (1934), *Antoine Paul Nicolas Franchimont 1844–1919*, Ph.D. thesis, Leiden.

Knuttel, D. (1928), De Rijkslandbouwproefstations, *CW* **25**, 370–371.

Kroon, J. E., ed. (1925), *Album Studiosorum Academiae Lugduno-Batavae MDCCCLXXV-MCMXXV*, A.W. Sijthoff, Leiden.

Kruyt, H. R. (1928), Herinneringen, *CW* **25**, 342–344.

Leden der Nederl. Chem. Vereeniging; 1e Opgaaf; 2e Opgaaf, *TTSH* **6** (1902/1903), 351–352, 379–380.

Leden der Nederl. Chem. Vereeniging; Adreslijst van Nederlandsche chemici, niet-leden der Nederl. Chem. Vereeniging; Nederlandsch Oost-Indië (1904/1905), *Chemisch Jaarboekje* (hereafter: *CJ*) **5**, 209–254.

Leden der Nederl. Chem. Vereeniging; Adreslijst van Nederlandsche chemici, niet-leden der Nederl. Chem. Vereeniging (1909/1910), *CJ* **10**, 311–354.

Leden der Nederl. Chem. Vereeniging; Adreslijst van Nederlandsche chemici, niet-leden der Nederl. Chem. Vereeniging (1910/1911), *CJ* **11**, 315–365.

Leden der Nederl. Chem. Vereeniging; Adreslijst van Nederlandsche chemici, niet-leden der Nederl. Chem. Vereeniging (1913/1914), *CJ* **12**, 398–474.

Leden der Nederl. Chem. Vereeniging; Adreslijst van Nederlandsche chemici, niet-leden der Nederl. Chem. Vereeniging (1915/1916), *CJ* **13**, 393–471.

Lieburg, M. J. van (1978), De geneeskunde en natuurwetenschappen binnen de Rotterdamse genootschappen uit de 18e eeuw, *Tijdschrift voor de Geschiedenis der Geneeskunde, Natuurwetenschappen, Wiskunde en Techniek* **1**, 14–22, 124–143.

Linden, T. van der (1953), Vijftig jaren Nederlandse Chemische Vereniging, *CW* **49**, 519–566.

Lintsen, H. W. (1980), *Ingenieurs in Nederland in de negentiende eeuw*, Nijhof, Den Haag.

Lobry de Bruyn, C. A. (1899a), Chemici en pharmaceuten, *Pharmaceutisch Weekblad* **36**(37) (14 January), 2–3.

Lobry de Bruyn, C. A. (1899b), Chemici en pharmaceuten, *Pharmaceutisch Weekblad* **36**(45) (15 March).

Mandemakers, Kees (1996), *HBS en gymnasium. Ontwikkeling, structuur, sociale achtergrond en schoolprestaties, Nederland, circa 1800–1968*, Stichting beheer IISG, Amsterdam.

Meerburg, P. A. (1928), Mijn secretariaat, *CW* **25**, 354–355.

Miles, Wyndham D. (1959), The Columbian Chemical Society, *Chymia* **5**, 145–154.

Nieuwenburg, C. J. van (1955), De opleiding tot scheikundig ingenieur, in Kamp, A.F. ed., *De Technische Hogeschool te Delft 1905–1955*, Staatsuitgeverij, 's-Gravenhage, 246–255.

Notulen van de Algemeene Vergadering der Nederlandsche Chemische Vereeniging (1902/1903), *TTSH* **6**, 370–379.

Olivier, S. C. J. (1928), Na vijf-en-twintig jaren, *CW* **25**, 323–329.

Prins, H. J. (1914), Over wederzijdse activering, *CW* **11**, 784–790.

Raalte, A. van (1928), De ontwikkeling van het keuringsdienstwezen in Nederland, *CW* **25**, 368–370.

Reicher, L. Th. (1903/1904), De tijdschriftenlijst in het Chemisch Jaarboekje, *CW* **1**, 519–522.

Reicher, L. Th. (1928), Het tot stand komen van de Nederlandsche Chemische Vereeniging, het Chemisch Weekblad en het Scheikundig Jaarboekje (later "Chemisch Jaarboekje"), *CW* **25**, 344–347.

Reicher, L. Th. and Jorissen, W. P. (1903/1904), L. S., *CW* **1**, 1.

Rocke, Alan (2001), *Nationalizing Science. Adolphe Wurtz and the Battle for French Chemistry*, MIT Press, Cambridge, MA.

Romburgh, P. van (1928), Herinneringen aan de 25-jarige en iets meer, *CW* **25**, 349–351.

Russell, C. A., Coley, N. G. and Roberts, G. K. (1977), *Chemists by Profession. The origins and Rise of The Royal Institute of Chemistry*, Open University Press, Milton Keynes.

Rutten, J. (1901/1902a), Letter to the editors, *TTSH* **5**, 315–317.

Rutten, J. (1901/1902b), Letter to the editors, *TTSH* **5**, 379–380.

Rutten, J. (1902/1903), Algemeene Nederlandsche Chemische Vereeniging, *TTSH* **6**, 287–288.

Scheikundig Jaarboekje **1** (1899), **2** (1901) (= *SJ*).

Scheikundig Jaarboekje voor Nederland, België en Nederlandsch-Indië **3** (1902), **4** (1903) (= *SJ*).

Snelders, H. A. M. (1980), *Het Gezelschap der Hollandsche Scheikundigen. Amsterdamse chemici uit het einde van de achttiende eeuw*, Rodopi, Amsterdam.

Snelders, H. A. M. (1983), De natuurwetenschappen in de lokale wetenschappelijke genootschappen uit de eerste helft van de negentiende eeuw, *De Negentiende Eeuw* **7**(2), 102–122.

Snelders, H. A. M. (1986), Chemistry at the Dutch universities: 1669–1900, *Mededelingen van de Koninklijke Academie voor Wetenschappen, Letteren en Schone Kunsten van België. Klasse der Wetenschappen* **48**(4), 59–75.

Snelders, H. A. M. (1991), Het Genootschap en de beoefening van de natuurwetenschappen sinds 1870, in van Berkel, K., van Lieburg, M. J. and Snelders, H. A. M., *Spiegelbeeld der Wetenschap. Het Genootschap ter Bevordering van Natuur-, Genees- en Heelkunde*, Erasmus, Rotterdam, 119–156.

Snelders, H. A. M. (1993), *De geschiedenis van de scheikunde in Nederland: Van alchemie tot chemie en chemische industrie rond 1900*, Delftse Universitaire Pers, Delft.

Snelders, H. A. M. (1997), *De geschiedenis van de scheikunde in Nederland 2: De ontwikkeling van chemie en chemische technologie in de eerste helft van de twintigste eeuw*, Delftse Universitaire Pers, Delft.

Somsen, Geert J. (1998), Selling science: Dutch debates on the industrial significance of university chemistry, 1903–1932, in Travis, Anthony S. *et al.*, ed., *Determinants in the Evolution of the European Chemical Industry, 1900–1939. New Technologies, Political Frameworks, Markets and Companies*, Kluwer Academic, Dordrecht, 143–168.

Stoeder, W. (1891), *Geschiedenis der Pharmacie in Nederland*, D. B. Centen, Amsterdam.

Swinderen, Th. van (1826), *Het vijfentwintigjarig bestaan van het Natuur- en Scheikundig Genootschap te Groningen, plegtig gevierd op den 1 maart 1826*, J. Oomkens, Groningen.

De toekomst der gegradueerden. Rapport van de Commissie ter bestudering van de toenemende bevolking van universiteiten en hogescholen en de werkgelegenheid voor academisch gevormden (1936), J. B. Wolters, Groningen and Batavia.

Tijdschrift voor toegepaste scheikunde en hygiène **1** (1897/1898), (1902/1903) (= *TTSH*).

Vandevelde, A. J. J. (1901/1902a), Verbond voor de belangen der toegepaste scheikunde en hygiene, *TTSH* **5**, 268–269.

Vandevelde, A. J. J. (1901/1902b), Letter to the editors, *TTSH* **5**, 317.

Vandevelde, A. J. J. (1902/1903), Eene Nederlandsche Chemische Vereeniging, *TTSH* **6**, 17–18.

Vandevelde, A. J. J. (1928), De Nederlandsche Chemische Vereeniging 25-jarig, *CW* **25**, 347–349.

Verbong, G. P. J. and Homburg, E. (1994), Chemische kennis en de chemische industrie, in Lintsen, H.W. *et al.*, ed., *Geschiedenis van de techniek in Nederland. De wording van een moderne samenleving, 1800–1890*, vol. 5, Walburg Pers, Zutphen, 242–269.

Verslag van de vergadering van der Bibliotheek-Commissie, gehouden te Leiden den 17den januari 1914 (1914), *CW* **11**, 114–115.

Verslag van de werkzaamheden der Commissie, *etc.* (1914), *CW* **11**, 265–278.

Verzeichniss der Mitglieder der Deutschen chemischen Gesellschaft, 1868–1888, 1892, 1901 (separate brochures, often bound with the *Berichte der Deutschen chemischen Gesellschaft*).

Visser, R. P. W. (1970), De Nederlandse geleerde genootschappen in de achttiende eeuw, *Dokumentatieblad Werkgroep 18e Eeuw* **7**, 7–18.

Vledder, Ingrid, Homburg, Ernst and Houwaart, Eddy (1999–2000), Particulier laboratoria in Nederland, *NEHA-Jaarboek voor de economische-, bedrijfs- en techniekgeschiedenis* **62**, 249–290 and **63**, 104–142.

Voornaamste chemische vereenigingen, (1903/1904), *CW* **1**, 630–633.

Vosmaer, A. (1897), De fusie der technische Vereenigingen, *De Ingenieur* **12**, 126.

Vosmaer, A. (1901/1902), Letter to the editors, *TTSH* **5**, 379.

Vries, O. de (1914), De werkkring der chemici in Ned. Indië, *CW* **11**, 725–728.

Wiese, Kees, ed. (2001), *Een spiegel der wetenschap: 200 jaar Koninklijk Natuurkundig Genootschap te Groningen*, Profiel Uitgeverij, Bedum.

Wind, C. H. (1903/1904), Electronen, *CW* **1**, 664–680.

Wittop Koning, D. A. (1948), *De Nederlandsche Maatschappij ter Bevordering der Pharmacie 1842–1942*, D. B. Centen, Amsterdam.

Wittop Koning, D. A. (1986), *Compendium voor de Geschiedenis van de Pharmacie van Nederland*, De Tijdstroom, Lochem.

Wijsman, H. P. (1899), Chemici en pharmaceuten, *Pharmaceutisch Weekblad* **36**(41) (15 February).

CHAPTER 10

NORWAY: A Group of Chemists in the Polytechnic Society in Christiania. The Norwegian Chemical Society, 1893–1916

BJØRN PEDERSEN

10.1 Setting the Norwegian Scene

A chemical society was founded in Norway fairly late compared to other countries in Europe. We have to look at the history of Norway to understand why. In the nineteenth century, Norway was a country at the outskirts of Europe fighting for its independence with a population of less than one million in 1800.

In 1380, a king common to both Denmark and Norway formed the union Denmark-Norway, and in the next 434 years Denmark became the dominating part in the union. There was no university in Norway before 1811 so Norwegians who wanted to study at an advanced level had to go to the University of Copenhagen or to one of the German universities. The only higher institution founded in Norway by the King of Denmark was the Det Kongelige Bergseminarium (the Royal Mountain School) founded in 1757 and connected to Kongsberg sølvverk (the silver company at Kongsberg) founded in 1624.

In an effort to strengthen the ties to Denmark a Norwegian university was founded by King Frederik VI in 1811, but as a result of the Napoleonic war Denmark lost Norway to Sweden in 1814. In the union with Sweden, Norway was more independent, but full independence was first obtained in 1905 when the union was broken.

Creating Networks in Chemistry: The Founding and Early History of Chemical Societies in Europe
Edited by Anita Kildebæk Nielsen and Soňa Štrbáňová
© The Royal Society of Chemistry 2008

Despite the union with Sweden the cultural ties to Denmark remained strong. In the golden age in Norway from about 1850 to 1914 science blossomed at the university in the capital Christiania and an increasing number of chemical engineers were educated at technical schools in Norway and in Germany.[1] Chemistry slowly became an established profession and not only a part of the education of pharmacists, physicians and teachers at the university.

When a section for chemists was established in 1893 within the Polytechnic Society (see below) in Christiania, Norway was still a class society. The nobility, however, was abolished in the constitution approved in 1814. The King and the court were in Stockholm. The working class was growing, but so was the upper middle class. The number of people with an academic education was small, but growing. To study at the university required a high school diploma ("examen artium"), but you could become an engineer or a pharmacist without the high school diploma.[2]

Norway has always been an exporter of raw materials: in the nineteenth century in particular timber, fish and metals like silver, copper and iron. After the development of hydroelectric power early in the twentieth century, there was a growth in the export of fertilizers, aluminium and magnesium. It was always a challenge to the engineers and scientists in Norway to develop economic processes for transforming these raw materials to finished products for export and not only for the home market. The time period to be discussed in this paper was the pioneering years when cheap hydroelectric power became available, making it possible to manufacture products that could compete on the world market using electrochemistry. The societies to be discussed in this paper were important meeting places for educated men to discuss what to do in a small country on the outskirts of Europe struggling for independence from 1814 to 1905, which, when it became independent in 1905, should be able to stand on its own feet.

10.2   Norwegian Institutions Teaching Chemistry

10.2.1   Technical Schools and Military Schools

The mountain school at Kongsberg was closed when the university was founded in 1811 and its mineral and book collection transferred to the Kongelig Frederiks Universitet (Royal Frederick's University or the University in Christiania).[3] Only 20 candidates were educated in the period in which the school was in operation. The education of candidates for the mining industry was continued at the university until 1910. But this early education of engineers was not expanded to other fields at the university. Instead separate technical schools were established in the German tradition. The majority at the university wanted to keep the applied sciences out so that the professors could concentrate

---

[1] Oslo, the capital of Norway, was from 1624 to 1924 called Christiania, and from about 1870 also spelled Kristiania.
[2] Gran (1911).
[3] In 1939 the university changed its name to Universitetet i Oslo (the University of Oslo).

on pure science, even though the professors in the Faculty of Mathematics and the Natural Sciences were in favour of educating engineers within the university.[4]

The first technical school was established in Horten in 1855.[5] Later schools were established in Trondheim (1870), Christiania (1873) and Bergen (1875). Some of the candidates later studied engineering at German universities and technical colleges. In 1910, Den tekniske høyskole (the Norwegian Technical College) was established in Trondheim. Entrance to the college was limited to students with a high-school diploma or graduates from one of the technical schools. The number of engineers educated at the technical schools in Norway in the period from 1873 to 1892 with a degree in chemistry was 31 – about 7% of the total number of graduates. In the next 20 years the number of engineers educated in chemistry was 205 – 10% of the total number of graduates. The first seven certified engineers in chemistry graduated from the Norwegian Technical College in 1914, while an unknown number were educated abroad.

10.2.2 The University in Christiania

Before 1850, all students learned some chemistry as part of their first year of study (for "examen philosophicum" giving right to the title cand.philos.) at the Royal Frederick's University. In 1850, the university started a teaching programme in mathematics and the natural sciences for future high-school teachers (cand.real.). In 1890, 22 had graduated, but fewer than 10 graduated in the period from 1895 to 1908, after which the number of graduates slowly started to increase again.

The university-educated high-school teachers obtained a broad background in science before 1905. Afterwards the students could do a major in chemistry only and write a thesis based on their own research. Before 1914, ten such candidates graduated. The first Ph.D. (dr.philos.) in chemistry at the university was granted in 1904 to Haavard Martinsen.[6] The next Ph.D. in chemistry was granted to Claus N. Riiber. He was educated at the university without taking any final exam before he got his Ph.D. degree. He became professor of organic chemistry at the Norwegian Technical College. A few Norwegians went to Germany and took a Ph.D. there. From 1869, students without a classical background were fully accepted as students. Also the students at Den Kongelige Norske Krigsskole (the Royal Norwegian Military Academy) in Christiania were taught chemistry by the university professors.[7]

---

[4] Gran (1911).
[5] The school originally had the name Den tekniske skole (the Technical School) i Carjohansvern. Carljohansvern was at that time the main naval station in Norway established in 1818.
[6] Martinsen had taken a master's degree (diplomprüfung) in chemistry in Dresden. He was a promising scholar, but his scientific career ended in 1918 when he took over Bjølsen Valsemølle, a family company.
[7] Gran (1911).

10.2.3   Professors of Chemistry

When the teaching began at the university in 1814 there was only one chair in both chemistry and physics. It was held by the Norwegian Jac Keyser, who had been trained as a teacher of physics and mathematics from Det Pædagogiske Seminarium (College of Education) in Copenhagen.[8] He was granted permission to concentrate on physics from 1838, and Julius Thaulow became the first professor of chemistry. When Thaulow died, the next professors were the German Adolf Strecker, Peter Waage and Heinrich Goldschmidt. In 1872, a second chair in chemistry was established and Thorstein H. Hiordahl was appointed. He retired in 1918. The professors and their assistants taught students of pharmacy, medicine and science. The pharmacy "students" were not academic citizens and their final examination was not taken at the university until after 1932 when Farmasøytisk institutt (Institute of Pharmacy) was established within the University of Oslo. Hiortdahl served for many years as the member of commission arranging the examinations.

10.3   The Roots of the Norwegian Chemical Society

10.3.1   The Learned Societies in Norway

Det Kongelig Norske Videnskabers Selskab (The Royal Norwegian Society of Sciences and Letters) was founded in 1760 in Trondheim, and it is consequently the oldest learned society in Norway. Both Keyser and Thaulow were members. A separate society was, however, founded in the capital in 1857, entitled then *Videnskabsselskabet in Christiania*, which was later altered to *Det Norske Videnskaps-akademi* (the Norwegian Academy of Sciences and Letters). The academy had a limited number of members from all scientific disciplines. In the beginning all professors at the university became members.[9] However, due to lack of resources the academy has never played an important role in the country as similar academies in some other countries.

Several scientific/technical societies were founded both before and after 1850, but they did not survive. For instance, five students of chemistry founded *Den fysiske-kjemiske forening* (the Physics-Chemistry society) in 1858, but the society only lasted to 1861.[10] Later several of the students became active members of the academy.

---

[8] Pedersen (2004).

[9] Amundsen (1957–1960).

[10] Another prominent member of this student society was Hans Henrik Hvoslef, who was educated in pharmacy and attained a Ph.D. degree in Göttingen in 1856 with Friedrich Wöhler as supervisor. Hvoslef was assistant professor at the university for some years before he established his own retail pharmacy in Christiania. Another member was Frantz Pecker Möller, who also had a degree in pharmacy and who earned his Ph.D. degree in Heidelberg under the supervision of Robert W. Bunsen in 1861. Möller later improved his father's process for producing cod liver oil. A third member was Henrik Mohn, who became Norway's first professor of meteorology. On the history of the society see Hiortdahl (1917). Hiortdahl was a student at university while the society existed and he had known all the members.

10.3.2 The Technical Society

Norsk Kjemisk Selskap (the Norwegian Chemical Society) grew out of *Polyteknisk forening* (PF) (The Polytechnic Society) in Christiania. PF was founded in 1852, primarily on the initiative of Anton Rosing.[11] The roots of PF were in *Den Tekniske Forening* (the Technical Society) that was founded by Julius Thaulow in 1847.[12] Thaulow and Rosing (see below) both worked for the enlightenment of the people, and Rosing was very much engaged in improving conditions for workers. This was a very sensitive issue around 1850 in Norway when a number of radical leaders were arrested and served long prison sentences.[13]

The main target group for the Technical Society was artists and craftsmen. Thaulow wanted to stimulate the intellectual and practical development of craftsmen and artists in Christiania by exposing them to the natural sciences. He also lectured on the use of chemistry in agriculture and published a book on the subject, inspired by Liebig.[14] The society quickly became popular and attracted 300 members (in a city with a population at that time of 29 000).

10.3.3 The Polytechnical Society

During a meeting in 1852 in the Technical Society, Rosing discussed the idea of creating a Polytechnical Society with some of the other members. The idea was to create a society with a broader scope that should enhance the position of the

---

[11] Hans Anton Rosing was born in 1827 into a supportive family. Rosing went to Copenhagen in 1843 where he studied at Den Polytekniske Læreanstalt (The Polytechnic College) founded in 1829; see Chapter 3 in this book. He was an active student of science, technology and art, and he even took part in politics. Rosing was one of the central figures in the founding of *Den polytekniske forening* (The Polytechnic Society) in Copenhagen in 1845; see Fransen and Harnow (1996). Rosing left for Norway in 1849 without having finished his studies at the Polytechnic College. Rosing worked at the city gas works for a while and used his spare time to talk important members of the Technical Society into founding the Polytechnical Society. An opening for Rosing arose when the government wanted to establish an agricultural college, and Rosing received a scholarship to prepare for work as the chemistry teacher. He studied for a year with Julius Adolf Stöckhardt in Tharandt, after which he went to Paris, visiting Robert Bunsen in Heidelberg on his way. In Paris, he was accepted as a student by Jean-Baptiste Dumas. The Russian Leon Shishkov (1830–1908) became the second student and his good friend. Both played central parts in establishing the French Chemical Society (*Société chimique de Paris* [later France]); see Chapter 5 in this book. Back in Norway in 1858, Rosing taught at the Agricultural College in Aas, the last six years from his sick bed, until his death in 1867. One year before his death, Rosing married his cousin Hedevig Rosing, see Gleditsch (1954).

[12] Julius Thaulow founded the Technical Society when he was 34 years old, and he had taught chemistry at the university in Christiania since he was 27. The university board selected Thaulow to be trained as the first professor of chemistry with Jacob Berzelius as his mentor. Thaulow studied with Christoph Heinrich Pfaff in Kiel, Jac Keyser in Christiania, Heinrich Rose in Berlin, Justus von Liebig in Giessen and Jules Pelouze in Paris. He carried out independent research and published seven papers in Liebig's journal during his stay abroad. He worked at the university after his return, from 1844 as professor of chemistry. Unfortunately he died an untimely death in 1850, aged only 38. Bjørn Pedersen "Training of a young Norwegian chemist in the 1830s in Europe", Science in Europe/Europe in Science 1500–2000, paper presented in Maastricht 4th–6th November 2004.

[13] Zachariassen (1977).

[14] Thaulow (1841).

natural sciences in all parts of society. The Technical Society and the Poly-technical Society had, however, common administration and physical premises until 1871, when the Technical Society was merged with Haandverksforeningen (the Craft Union) founded in 1838. The Polytechnical Society continued as an independent society for the elite, functioning as a forum for cooperation between professors in the natural sciences at the university, leaders in industry and politicians.[15] Two of the most influential members from the 1860s were professor of applied mathematics Cato M. Guldberg and professor of physics Hartvig C. Christie.

It was from this society that a *Chemikergruppe* (Group of Chemists) emerged in 1893. A similar group had been established some weeks before by the electroengineers. The Group of Chemists in the Polytechnical Society devel-oped into an independent chemical society for the whole country with an increasing number of regional branches from 1926. The last remaining ties to the Polytechnical Society were finally formally broken in 1980. In the period covered in this paper, from 1893 to 1914, however, the chemists were organized as a group within the Polytechnical Society.

10.4   The Norwegian Chemical Society

10.4.1   A Meeting Place for People Interested in Chemistry

At the end of the nineteenth century, the chemical industry in Norway was almost non-existent. The increasing number of chemists in the Polytechnical Society wanted a forum of their own to enhance the importance of chemistry, to discuss questions of a chemical-technological nature and to stimulate the progress of chemical industry in Norway. Consequently, they formed their own section within the society as mentioned above in 1893. 54 persons were invited to the first meeting and 40 became members in the first year. As mentioned, only 31 chemical engineers had graduated in Norway before 1894 so a large proportion of the qualified "chemists" in the town became members in the first year. Among the 40 members were the chemists and mineralogists at the university, the technical school, the Agricultural college, the chemical control station, some brewery chemists, pharmacists, a couple of officers, chemists at the city gas works and a few industry owners. They were all living in, or close to, Christiania.

The number did not increase until after 1906 as shown in Table 10.1. That was when hydroelectric power plants started to be built in Norway in larger numbers. The power was cheap and had to be used locally. That stimulated the construction of chemical plants for producing calcium nitrate (Norwegian saltpetre) using a new process developed by the professor of physics Kristian Birkeland and the construction engineer and industrialist Sam Eyde.[16] Also other processes were developed later depending on cheap electric power and requiring chemical expertise.

---

[15] Fasting (1952).
[16] Egeland and Burke (2005).

Table 10.1 Members of the Norwegian Chemical Society, 1893–1914.

Year	Number of members
1893	30
1894	43
1895	45
1896	42
1897	46
1898	40
1899	42
1900	36
1901	45
1902	45
1903	49
1904	47
1905	50
1906	49
1907	52
1908	60
1909	67
1910	82
1911	94
1912	110
1913	125
1914	128

Only experts in chemistry and related sciences like medicine and pharmacy could become members of the Group of Chemists. Later this was further restricted to Norwegian citizens with an academic or equivalent theoretical education or practical experience in chemistry. In addition to these requirements, one had to be a member of the Polytechnical Society. The meetings were held in the premises of the mother society. The main part of a meeting was a talk followed by dinner. The first talk was given by Eyvind Bødtker[17] on ptomaines – foul-smelling decay products of proteins – not too inspiring a subject before dinner!

10.4.2 Chairmen

The executive committee had four members and the election period was one year. Waage was elected the first chairman from 1893 until 1898, when Thorup took over and served to 1904 as shown in Table 10.2. With few exceptions the chairman was a university professor. One exception was Nicolaysen who was headmaster (overlærer) at the technical school in Christiania. Halvorsen and Farup both became professors at the Norwegian Technical College in

---

[17] He was then only 26 years old and educated as a pharmacist. He had gained his Ph.D. in Leipzig under Johannes Wislicenus (1835–1902) the year before and had just started to work at the university in Christiania as an assistant to Hiortdahl. He became professor of organic chemistry in Christiania in 1918 when Hiortdahl retired. Dedichen (1932).

Table 10.2   Chairmen in the Group of Chemists in the Polytechnical Society,
1893–1916.

Chairman	Period
Professor Peter Waage	1893–1898
Professor Søren Thorup	1898–1904
Headmaster C. Nicolaysen	1904–1906
Professor Thorstein Hiortdahl	1906–1912
Professor Heinrich Goldschmidt	1912–1916

Trondheim, but at the time they were chairmen, they worked as independent consultants in Christiania.

10.4.3   The Activity within the Group of Chemists

During the first 25 years about 150 papers were presented. About one-third of the papers focused on a purely scientific subject while one-third presented an industrial subject. The rest were on analysis and control and on education and administration. The speakers also discussed science policy, for instance the establishment of an industrial laboratory in Norway as early as 1893, an institution that was not realized until the post-Second World War period. Another frequent topic was the discussion of the establishment of a technical college in Norway and where it should be located. Finally in 1900, Stortinget (the Norwegian Parliament) decided to locate the college in Trondheim. The Norwegian Technical College opened in 1911.[18]

The majority of the members of the society had a background in engineering (ingeniørkjemikere – engineering chemists). However, the dominant group was the chemists at the university in Christiania even though they were few. Professor Hiortdahl later became the first honorary member of the group of chemists.

10.4.4   Women as Members?

One woman was invited to the day of foundation in 1893: Caroline Steen, who was an assistant of Torup. Caroline Steen was Danish by birth. She was denied a higher education by her father, but after his death she pursued a career as a physician in Norway, graduating in 1896 at 42 years of age.[19] According to the

---

[18] Ingebrechtsen (1943); Koren (1918); Terjesen (1993).
[19] Steen was a brilliant teacher in domestic science both in Norway and Denmark and ended her career as an associated professor (docent) of nutrition science in Denmark. She was not allowed to practice as a physician in Denmark. Caroline Steen in *Dansk Kvindebiografisk leksikon* (2000), ed. Jytte Larsen. Steen was a co-author with Henriette Schønberg Erken of a Norwegian cookery book for schools that went through 16 editions. Schønberg Erken is famous in Norway for her *Stor Kokebok for større og mindre husholdninger* (*Large Cook Book for Greater and Smaller Households*) that was used in almost all families in Norway from 1895–1950. The book she wrote with Caroline Steen was called *Kokebok for skole og hjem* (*Cookbook for School and Home*).

regulations of the Polytechnical Society only men could be members; so after many discussions it was first decided that although women could not be members of the society, Steen still could be a member of a section. Later "men" in the regulation was interpreted as "human beings" so both men and women could be admitted.

10.4.5   Initiators

The real founding fathers of the Group of Chemists were Axel Krefting and Einar Simonsen. Both became students in 1885 and both worked afterwards as assistants at the chemical laboratory at the university. Krefting was educated in engineering before he became a university student. Krefting had studied chemical engineering and building engineering at Trondhjems tekniske Læreanstalt (The technical school in Trondheim) and mechanical engineering at the Königlich Technischen Hochschule in Hannover. From 1887 to 1893 he worked as a teacher in chemistry at the technical school in Trondheim. Later he became secretary for the Union for Craft and Industry and became editor for the journal they published.[20] Einar Simonsen took "examen philosophicum" and studied chemistry with Waage from 1886 to 1888. In 1890 he was hired as a teacher of chemistry at the technical school in Christiania. He acquired a travel grant to 11 European countries to study industry laboratories, material testing and forgery of foodstuff.[21]

Both were practical men. Amongst other things, Krefting developed a popular method for removing rust from archaeological items. He later became a social politician, contributed to the founding of the Norges arbeidsgiverforening (Employers' Federation) in 1900 and became the head of the administration there. Einar Simonsen developed a method for producing ethanol from sawdust and became a teacher in chemistry, technology and consumer goods at Oslo Handelsgymnasium (the Oslo School of Business and Economics) from 1900. Towards the end of his life he became the head inspector of the production of distilled spirits and malt in Christiania. Both he and Krefting became elected presidents of PF, Krefting in the period 1904–1908 and Simonsen 1909–1915.

These two young founding fathers were supported by two more-established chemists in the capital: the city chemist Ludvig Schmelck and the professor of physiology Sophus Torup. Schmelck was a student from 1874 and took Examen philosophicum the following year. He then held several positions, amongst others as assistant at the university. In 1881, he established the first private chemical laboratory in Christiania where he analysed a wide variety of samples from foodstuff to forensic analysis.[22] He was also a member of the Royal Norwegian Society of Sciences. Torup was interested in the carbon

---

[20] Krefting, Axel, in *Studentene fra 1885 til 25-årsjubileet* (Kristiania); Johannessen (2002).
[21] Einar Simonsen, in *Studentene fra 1885 til 25-årsjubileet* (Kristiania); Koren (1917); Halvorsen (1918).
[22] Ludvig Schmelck, in *Studenterne fra 1874 - 25 aars jubileumsbok* (Kristiania); Hiortdahl (1916).

dioxide transport in blood as well as in public health and diets. He gave valuable advice to Frithjof Nansen's and Roald Amundsen's polar expeditions, which enabled them to avoid becoming inflicted with scurvy.[23]

10.4.6   Change of Name in 1916

In 1916, when Goldschmidt was chairman, the name of the section was changed to *Norsk Kemisk Selskap* (the Norwegian Chemical Society, NKS). The main reason for the change of name was the connection to societies in other countries. The new name had to be approved by the Polytechnical Society, and the coupling to the mother association did not change. From 1911, the Group of Chemists was a member of the *International Association of Chemical Societies*, and from 1919 NKS became a member of the *International Union of Pure and Applied Chemistry*. NKS has since represented the Norwegian chemists in the appropriate international forums.

10.4.7   Social Tasks

The Group of Chemists was also used by the government and other state institutions to give professional advice on chemical questions. This was discussed at the regular meetings, and the society usually established a committee to prepare an answer. It was an active time both before and during the First World War. Before the war many new industries were established and during the war the challenge was to find substitutes for products that became limited in supply (Norway was neutral during the war). NKS also acted as a pressure group to improve the chemistry education in schools and at the university.

10.4.8   Norwegian Journals for Chemistry

Rosing founded not only three societies, but also two journals in the 1850s. In Norway, *Polyteknisk Tidsskrift* (the Polytechnic Journal), now called *Teknisk ukeblad* (the Technical weekly journal), is now the oldest journal of its kind in the world. In France he started a journal that in the 1860s became the *Bulletin de la Société chimique de France*, see Chapter 5 by Fell and Rocke.

Teknisk ukeblad contained from the start in 1854 many articles that can be listed as chemical, and many of them were written by Rosing. That continued after the Group of Chemists in the Polytechnical Society was founded in 1893. A special journal for chemical issues was discussed, and first realized in 1904 when Eivind Koren founded *Pharmacia* (Pharmacy), a journal for pharmacy and chemistry. The name of the journal changed many times. In the coming years the journal published reports from meetings in NKS and articles written by members of NKS. When Koren died in 1920 the responsibility for publishing the journal was transferred to the Polytechnical Society and the name was

---

[23] Getz (1975).

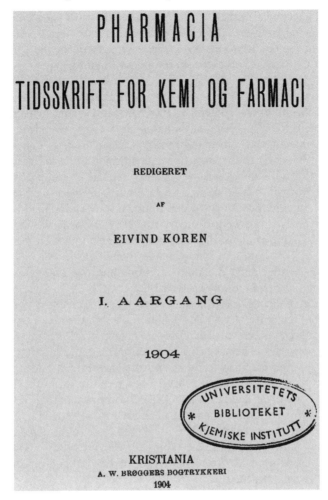

Figure 10.1   Title page of the first volume of *Pharmacia*, the first chemistry journal published in Norway. It is still published today, although the name and profile have been modified through the years. Today the title of the journal is *Kjemi*.

changed to *Tidsskrift for Kemi og Bergvesen* (Journal for Chemistry and Mining). Today the journal is called simply *Kjemi* (Chemistry), and the Norwegian chemists publish their scientific papers elsewhere (Figure 10.1).

10.5   Conclusion

The Norwegian Chemical Society first acquired its name in 1916 at the end of the time period to be covered in this book. Before that it was a Group of Chemists within the Polytechnical Society. The section was founded in 1893, while the Polytechnical Society dates from 1852. It was located in Christiania (now called Oslo). Not until the 1920s were similar local societies established in other parts of the country, and NKS became a society for the whole country representing the Norwegian chemists in international forums.

The main reason for the slow start was simply that there were few chemists, and hardly any chemical industry in Norway in the nineteenth century. From about 1870 the number of chemists increased and chemical engineers were educated from local polytechnic schools and from technical schools in Germany. Some of the university-trained science teachers concentrated on chemistry, particularly those few who attained a job at the university.

Another reason for the slow start was the political situation of the country. Before 1814 Norway was in union with Denmark and was ruled from Copenhagen. The king regarded Norway as a source of raw material and most of the civil servants in Norway were Danes. After 1814, Norway was in union with Sweden and in a more independent position, but the population was only about one million in 1814, most of whom made their living as small farmers and fishermen. Only after 1850 did the conditions improve and Norway had its cultural golden age from about 1860 to 1920.

In the group of chemists within the Polytechnical Society the two professors of chemistry at the university had the leading role, but the majority of members were chemical engineers, pharmacists or people working in the chemical industry or breweries. The society was an important meeting place for politicians, civil servants, engineers and scientists. Therefore it had an important influence on the development of Norway before and during the First World War when Norway was neutral, but lacked certain raw materials that had to be substituted. Before the war hydroelectric power was developed in Norway and because the energy was cheap and had to be used close to the place where it was produced new industries could be established. The most well-known is Norsk Hydro, established in 1905, producing fertilizers (in particular calcium nitrate, so-called Norwegian saltpetre).

References

Amundsen, Leiv (1957–1960), *Det Norske Videnskaps-akademi i Oslo*, Oslo.

Dedichen, Georg (1932), Professor Dr. Eyvind Bødker, *Tidsskrift for kjemi og bergvesen* **11**, 249–250.

Egeland, Alv. and William J. Burke (2005), *Kristian Birkeland. The First Space Scientist*, Springer.

Fasting, Kåre (1952), *Teknikk og samfunn, Den polytekniske forening, 1852–1952*, Oslo.

Fransen, Peter and Henrik Harnow (1996), *Fra teknisk selskab til studenterpolitisk interesseorganisation, Polyteknisk Forening 1846–1996*, Lyngby.

Getz, Bernard (1975), Torup, Sophus, *Norsk biografisk leksikon* **17**, 19–23.

Gleditsch, Ellen (1954), Hans Anton Rosing (1827–1867), *Fra fysikkens verden* **16**, 1–12.

Gran, Gerhard (1911), *Det Kongelige Fredriks Universitet 1811–1911*, Kristiania.

Halvorsen, B. F. (1918), Mindetale over kjemiker Einar Simonsen, *Tidsskrift for kemi, farmaci og terapi* **15**, 58.

Hiortdahl, Th. (1916), Ludvig Schmelck, *Tidsskrift for kemi, farmaci og terapi* **6**, 84.

Hiortdahl, Th. (1917), Den fysiske-kemiske forening av 1858, *Tidskrift for kemi, farmaci og terapi* **14**, 240–246.

Ingebrechtsen, Kristian (1943), Norsk kjemisk selskap 50 år, *Tidskrift for kjemi, bergvesen og metallurgi* **22**, 325–332.

Johannessen, Finn Erhard (2002), Krefting, Axel, *Norsk biografisk leksikon* **5**, 363–364.

Koren, Eivind (1917), Kjemiker Einar Simonsen 50 år, *Tidsskrift for kemi, farmaci og terapi* **14**, 237–240.

Koren, Eivind (1918), Norsk kemisk selskap. Gruppe av P.F. 2. mai 1893–2. mai 1918, *Tidsskrift for kemi* **15**, 141–152.

Pedersen, Bjørn (2004), Stereobilder og professor Keyser, *Fra fysikkens verden* **66**, 50–53.

Terjesen, Sven G. (1993), *Norsk Kjemisk Selskap 1893–1993*, Oslo.

Thaulow, M. C. Julius (1841), *Chemiens anvendelse i Agerdyrkningen*, Christiania.

Zachariassen, Aksel (1977), *Fra Marcus Thrane til Martin Tranmæl, Det Norske arbeiderparti fram til 1945*, Oslo.

CHAPTER 11

POLAND: Chemists in a Divided Country. The Long-lasting Genesis and Early History of the Polish Chemical Society, 1767–1923

HALINA LICHOCKA

11.1 Introduction

The continuity of the historical development of the Polish state was broken off in 1795, when the neighbouring powers Austria, Prussia and Russia finally divided the territory of Poland between themselves. Between 1795 and 1918, Poland did not exist as a sovereign state and this historical circumstance essentially affected its entire social and cultural life. Foreign governments made considerable efforts to perpetuate this with their policies, aiming to deprive Poles of their national identity and assimilate former Polish citizens into their own societies.[1] In the second half of the nineteenth century, the only Polish higher education establishments still in existence were those in the territory of the Austrian partition (Cracow, Lemberg). In other Polish territories, Polish schools were closed and partially replaced by Russian and German schools. For this reason, young Poles often studied at foreign universities. Many Poles studied chemistry in Riga, Dorpat, Warsaw, St. Petersburg, Moscow and Kiev as well as at German and Swiss universities. Concern for the development and dissemination of science in the mother tongue was part of the struggle by Poles to survive as a nation. The foundation of scientific societies was a good means towards this end, and such institutions were already

---

[1] See Davies (1981).

Creating Networks in Chemistry: The Founding and Early History of Chemical Societies in Europe
Edited by Anita Kildebæk Nielsen and Soňa Štrbáňová
© The Royal Society of Chemistry 2008

relatively abundant at the turn of the nineteenth century.[2] Societies were formed thanks to the financial support of individuals and in defiance of the policies of the partitioners. Accordingly, their activity was not always overt or legal. Such illegal groups were particularly numerous in the zone ruled by Russia, where freedom of association was most oppressively restrained.

Chemistry teachers and managers of chemical processing plants were considered professional chemists. In most cases, they were graduates of medical, pharmaceutical, natural sciences and technical studies; however, Polish chemists – working both in the territory of the former Poland and in exile – were perceived in accordance with their formal citizenship as Austrian, German and Russian scholars.

It was not until 1918 that Poland again found its place on Europe's map and no later than two years after that the country was forced into a defensive war with the Bolshevik Russia. After Poland had succeeded to repel the Bolshevik attack, the Poles could eventually start restoring their completely destroyed country and establish their own scientific and educational institutions.

The oldest scientific society in Poland that included the word "chemical" in its name was formed in 1767. *Warszawskie Towarzystwo Fizyczno-Chemiczne* (The Warsaw Physico-Chemical Society) set itself the goal of encouraging practical application of the achievements of science for the benefit of industry and agriculture. The society promoted the search for the exploitation of domestic natural resources and underlined the need for new factories. It publicized the benefits of applications of chemistry in crafts and pharmacy. The society's official journal was published in Polish and German under the title *Różne Uwagi Fizyczno-Chemicznego Warszawskiego Towarzystwa na rozszerzenie praktycznej umiejętności w fizyce, ekonomii, manufakturach i fabrykach, osobliwie względem Polski; Vermischte Abhandlungen der Physisch-Chemischen Warschauer Gesellschaft zur Beförderung der praktischen Kenntnisse in der Naturkunde, Ökonomie, Manufakturen und Fabryken, besonders in Absicht auf Polen* (Miscellaneous remarks of the Physico-Chemical Society of Warsaw towards the development of practical skill in physics, economics, manufacture and factories, with special regard to Poland). The Polish version was published simultaneously with the German one.

11.2   Involvement of Chemists in General Science Societies

In November 1800 representatives of scientific communities from all Polish lands founded *Towarzystwo Warszawskie Przyjaciół Nauk*[3] (The Warsaw

---

[2] Słownik (1978–1994), a four-volume collective publication of the Polish Academy of Sciences Library.

[3] Manuscript sources of information on the history of the Society may be found in the Archiwum Akt Dawnych (Central Archives of Historical Records) in Warsaw, in the collection: Towarzystwo Przyjaciół Nauk 1800–1832, Signature No. TPN 2–105. The activities of the Society have been the subject of many historical papers.

Society of Friends of Sciences). A Chemistry Section was formed in 1826 within the *Dział Umiejętności* (Division of Skills)[4] with several active members (among them Aleksander Chodkiewicz[5] and Jędrzej Śniadecki[6]). Foreign corresponding members of the Section included L. J. Gay-Lussac, H. Davy and L. J. Thénard. Emphasis was placed on practical application of the achievements of chemistry, analysis of mineral waters and brines, fermentation, tanning techniques, dyestuffs and gas engineering. The results of members' research and reports of developments in European science were presented during gatherings of the section and general meetings of the society, and were later published in scientific journals.

The Warsaw Society of Friends of Sciences received its funding from membership fees and donations, the latter sometimes being very generous.[7] The society owned an impressive library collection, acquired mainly by way of donations and available to the wider public.[8] The collections, together with the other property belonging to the society, were confiscated by the Russian authorities in 1832 and the society was then disbanded.

Formed a little later than its Warsaw counterpart, *Towarzystwo Naukowe Krakowskie* (1815–1872, The Cracow Scientific Society) was founded by 40 members.[9] In 1872 the Cracow Scientific Society was transformed into *Polska Akademia Umiejętności* (the Polish Academy of Skills), which is still active today. Chemists belonged to the department of natural and mathematical sciences, as well as the *Komisja Balneologiczna* (Balneological Committee, formed in 1858) and the *Komisja Fizjograficzna* (Physiographic Committee, established in 1865).

---

[4] Pharmacists formed a separate group in the Warsaw Society of Friends of Sciences within the Section of Medicine, Surgery and Pharmacy.

[5] Aleksander Count Chodkiewicz (1776–1838), a Polish aristocrat, writer, chemist and technologist. Member of numerous academic societies and institutions, including the Agricultural and Galvanic Societies of Paris and the Mineralogic Society in Jena. He assigned large sums of money to support financially disadvantaged scientists and artists. His palace in Warsaw housed a private chemical laboratory where he conducted research. The laboratory and library were open to talented enthusiastic chemists from Warsaw University. The most important publication by Chodkiewicz was *Chemia* (Chemistry) in seven volumes, presenting the entirety of current knowledge about inorganic, organic and analytic chemistry.

[6] Jędrzej Śniadecki (1768–1838). Following completion of his studies at the University in Cracow, Śniadecki went to Pavia, where he obtained a doctorate in medicine and philosophy. He further developed his medical knowledge in Vienna, Switzerland and the Netherlands and spent two years studying chemistry under Joseph Black in Edinburgh. Śniadecki worked as professor of chemistry at the University in Vilna. He was the author of Polish chemical nomenclature. His best-known work is *Teoria jestestw organicznych* (Theory of Organic Entities), published in the years 1804–1811, and later translated into German and French.

[7] Among the most generous benefactors was Stanisław Staszic (1755–1826), a priest educated in Poland and abroad, philosopher and naturalist, president of the Warsaw Society of Friends of Sciences for many years. On his initiative, and mostly from his private funds, an impressive edifice was erected for the Society in Warsaw, later called the Palace of Staszic.

[8] The most important donor of books for the library was Prince Aleksander Sapieha (1773–1812), a bibliophile, naturalist and traveller. He took interest in chemistry and had a private laboratory. He donated his collection of books to the Warsaw Society of Friends of Sciences and established a fund for the acquisition of new items.

[9] Archives of Cracow University (Archiwum Uniwersytetu Krakowskiego), collection: *Towarzystwo Naukowe Krakowskie*.

Throughout its lifetime, the Cracow Scientific Society[10] was the publisher of as many as eight serial journals: some purely academic and some aimed at disseminating scientific knowledge among society. The most important of these was *Rocznik Towarzystwa Naukowego Krakowskiego* (Annals of the Cracow Scientific Society), with 44 volumes published until 1872. Many dissertations in chemistry were published there, but chemists could also use the pages of the society's two other serials: *Przegląd Lekarski* (Medical Review) and *Sprawozdania Komisji Fizjograficznej* (Reports of the Physiographic Committee).

Beyond publishing in the above-mentioned journals, some members of the society printed their own works. In the short period of 1834–1836, the most important publication channel for Cracow chemists and pharmacists was *Pamiętnik Farmaceutyczny Krakowski* (The Cracow Pharmaceutical Diary), founded and run by Florian Sawiczewski, a professor of chemistry.[11] Chemistry-related topics were the explored in more than 25% of all articles published in this periodical.[12]

The *Towarzystwo Naukowe Warszawskie* (Warsaw Scientific Society) was formed in Warsaw in 1907 and it still exists today. It originally incorporated three departments: I – Linguistics and History of Literature; II – Historical, Social and Philosophical Sciences; III – Mathematical and Physical Sciences. The activity of Department III was focused on the establishment of laboratories and research institutions. By 1914, the society possessed a number of laboratories, including a Laboratory of Physiological and Pathological Chemistry, a Soil Science Laboratory and a Mineralogy Laboratory, where chemists conducted research, offered training to younger colleagues and carried out numerous chemical analyses on commission, which was a way of obtaining funds for the society. However, the main sources of funding for the society were membership fees, the sale of its own publications and donations from the Polish society.[13] The results of research studies and papers read at sessions were published by Department III in the series *Prace Towarzystwa Naukowego Warszawskiego. Wydział III Nauk Matematycznych i Przyrodniczych* (Papers of the Warsaw Scientific Society, Department III of Mathematical and Natural Sciences), which was first published in 1908.[14]

11.3   Chemists in Societies of Naturalists

A relatively large number of chemists were members of the *Polskie Towarzystwo Przyrodników im. Kopernika* (Copernicus Memorial Polish Society of

---

[10] Jabłoński (1967); Poradzisz (1984).

[11] Florian Sawiczewski (1797–1876), physician and pharmacist, professor of chemistry at Cracow University between 1829 and 1851, member of the Cracow Scientific Society and the Physiographic Committee. On completion of his studies at Cracow, he undertook a study trip to Paris, Vienna and Erfurt. The author of numerous papers in chemistry, pharmacy and pharmacognosy, published in Polish and German.

[12] Lichocka. (1986).

[13] The donations were often quite generous. In 1911 Józef Potocki, a Polish aristocrat, bestowed on the Society a tenement house at Śniadeckich Street in Warsaw, where numerous research laboratories were to be located.

[14] A total of 35 volumes were published in Warsaw between 1908 and 1933.

Naturalists), founded in Lemberg in 1875. The founding fathers of the society were eight professors of Lemberg universities, including three chemists. By the end of the same year, membership of the society had risen to 133. The number kept growing in subsequent years and a branch of the society was formed in Cracow in 1890. Just before the First World War the society numbered 392 members, of whom chemists accounted for approximately 25%.

The statutory objectives of the society[15] were to study Poland's natural world and foster the development of natural sciences, establish laboratories and research stations, augment the society's library and publish books and journals. Members met regularly at weekly scientific meetings; annual general meetings usually took place on 15th February, Copernicus' birthday. National conferences were also organized in collaboration with the medical community alternately in Cracow and Lemberg. During these conferences, chemists worked in a dedicated section on chemistry. Beginning in 1876, the society published a journal called *Kosmos*, with Bronisław Radziszewski,[16] an organic chemist, as the founder and editor-in-chief for over 30 years.[17] *Kosmos* is still in publication as a journal today.

11.4   Societies of Applied Science

11.4.1   Pharmaceutical Associations

Polish chemistry owes a lot, especially in terms of the quality control of raw materials and improving analytical methods, to the oldest Polish scientific society of pharmacists, founded in 1819 as the Pharmaceutical Division of the Vilna Medical Society. Apart from domestic members (ordinary and corresponding), there were foreign members, including Ferdinand Giese (professor of chemistry and head of the University of Dorpat) and Johann Wolfgang Döbereiner (professor of chemistry at Jena). The official publication of the Pharmaceutical Division was the *Pamiętnik Farmaceutyczny Wileński*[18] (The Vilna Pharmaceutical Diary), appearing in the years 1820–1822. It was the first scientific pharmaceutical journal published in Polish, and one of the first serials in Europe dedicated to the science and practice of pharmacy. The earliest periodicals of this kind were published in France and Germany at the turn of the nineteenth century. The *Pamiętnik Farmaceutyczny Wileński* was their

---

[15] Ustawy (1900); see also: 1875–1899 (1900).
[16] Bronisław Radziszewski (1838–1914) is known as the discoverer of the conversion of nitriles into carboxylic acid amides. The Radziszewski reaction is an accepted method in organic synthesis. Radziszewski studied in Warsaw and Moscow. Actively involved in an unsuccessful anti-Russian popular uprising in 1863, he sought shelter in Belgium, where he worked at Friedrich August Kekulé's laboratory in Ghent and was conferred a doctorate in 1867. He followed this with three years spent at the laboratory of Prof. L. Henry in Louvain. In 1870, Radziszewski became a professor at the Technical Institute in Cracow. In 1872 he assumed the post of professor of chemistry at Lemberg University.
[17] Siemion (1987) and (1999).
[18] Lichocka (1981) and (1990).

direct successor. It was not until some time later that similar journals started to be published in Italy, Switzerland, Spain and other European countries.[19]

Of all the pharmaceutical societies in Polish lands in the second half of the nineteenth century, chemists benefited the most from collaboration with *Galicyjskie Towarzystwo Aptekarskie* (Pharmacist Society of Galicia) in Lemberg, founded in 1868[20] and operating until 1939. The collaboration was particularly intense in education and dissemination of science, holding "best scientific paper" competitions, and editorial work, with textbooks written and published in analytical chemistry, materials science and chemical nomenclature. Through the efforts of the society a three-year Pharmaceutical School was opened in Lemberg in 1874, followed by a Chemical Laboratory, specializing in toxicology and analysis of food. The society's journal, from 1871, was the biweekly *Czasopismo Towarzystwa Aptekarskiego* (*Journal of the Pharmacist Society*), with 45 annual volumes between 1871 and 1914.

Other associations of pharmacists in the second half of the nineteenth century were more professionally oriented and chemists were less involved in their work. However, both professions shared an obvious interest in the chemical drug industry, which started to develop quite early on Polish land.[21]

A similar pattern in the involvement of chemists could be seen in associations concerned with agricultural sciences and farming-related occupations. There were few chemists involved, either as teachers in agricultural schools or as engineers and technologists at enterprises processing agricultural and forest products.

11.4.2 Museum of Industry and Agriculture

In 1875, on the initiative of landed gentry and factory owners, a social and scientific corporation was founded in Warsaw under the name *Muzeum Przemysłu i Rolnictwa* (The Museum of Industry and Agriculture). Initially, there were not many members (51 in 1886), but the number grew to 373 by 1900. At that time the *Muzeum* had a library of about 2000 volumes and four specialized museums (Agriculture, Industry, Technology and Ethnography) were established.

Chemists represented a significant percentage of the *Muzeum's* membership.[22] Most of the members were university graduates employed in industrial facilities in the Russian zone of partition. Through the efforts of these highly qualified experts, a chemical laboratory became the first to be created at the museum, as early as 1876. The laboratory carried out analyses on commission for industry and agricultural technology and improved methodology for quality assessment of industrially processed food and controlling the composition of fertilizers. To disseminate the knowledge, the facility organized lectures and exhibitions for the

---

[19] Colloque (1990).
[20] Dewechy (1928).
[21] Kikta (1972).
[22] Kabzińska (1990).

general public, and offered courses in pure and applied chemistry for students of *Klasy Rzemieślniczo-Przemysłowe* (Crafts and Industry Classes), a vocational school affiliated with the *Muzeum*. However, the most important role of the chemical laboratory was that of a workshop for many a young chemist engaged in their own research. It was in this laboratory that Maria Curie-Skłodowska, the future Nobel Prize Laureate, acquired professional skills as chemist.

11.5   Technological Societies

With the territory of Poland divided by frontiers, initiatives to form a scientific, technical or social institution often had to be implemented as the simultaneous formation of several societies under similar names, profiles and statutes, functioning as surrogate branches of an organization that would otherwise have extended over the entire area of former Poland. A good example is found in the attempt, in the 1870s, to unite Polish communities of engineers and technologists. Towards this goal, three technical societies were formed in 1877: *Towarzystwo Politechniczne* (The Polytechnical Society) in Lemberg, *Krakowskie Towarzystwo Techniczne* (The Cracow Technological Society) and *Towarzystwo Techniczne in Poznań* (The Poznan Technological Society). The Russian partition was outside the geographical reach of these institutions and similar societies were only founded there more than 20 years later because creation of any Polish organization was at this time treated as a political offence. *Stowarzyszenie Techników* (Association of Technologists) in Warsaw was formed in 1898, and its namesake organization in Vilna came into being in 1903.

The oldest journal published by the Lemberg Polytechnical Society was the monthly *Dźwignia* (Lever, 1877–1883), followed by *Czasopismo Techniczne* (Technological Journal), published in collaboration with the Cracow Technological Society and featuring science- and technology-oriented articles as well as activity reports of the two societies.[23] Extensive collaboration between the individual communities of engineers was instrumental in organizing all-Poland congresses of technologists, which took place in Lemberg, Cracow and Warsaw (a total of seven congresses before the First World War).[24]

As in its Lemberg counterpart, membership of the Cracow Technological Society was open to graduates of technical universities or equivalent university-level schools and to graduates of higher schools of industry, but four years work experience was required.

Warszawskie Koło Chemików (Warsaw Club of Chemists) was formed in February 1909. The Club of Chemists within the Association of Technologists in Warsaw continued the traditions of an earlier chemical society and accepted the entire programme and aims of the former corporation, which was called

---

[23] Piłatowicz (1990).
[24] Kalabiński (1964).

Sekcja Chemiczna (Chemistry Section)[25] and was part of *Oddział Warszawski* (the Warsaw Branch) of the *Towarzystwo Popierania Rosyjskiego Przemysłu i Handlu* (Society for the Promotion of Russian Industry and Commerce).

11.6   The Warsaw Branch of the Society for the Promotion of Russian Industry and Commerce. The Chemistry Section

The *Towarzystwo Popierania Rosyjskiego Przemysłu i Handlu* (Society for the Promotion of Russian Industry and Commerce) was formed in St. Petersburg. The Warsaw branch of this society was established in 1884 by Ludwik Krasiński,[26] a Polish aristocrat, land owner and industrialist with a university background in chemistry. Its chemistry section held its first meeting in February 1887. Biweekly meetings were then organized regularly. The section was active for a little more than 20 years, with 50–60 active members. Its official organ, called *Chemik Polski* (Polish Chemist, Figure 11.1)), was published in the years 1901–1918, first as a weekly and later on a biweekly basis. The journal contained papers delivered during meetings of the chemistry section and specialized articles exploring various areas of pure and applied chemistry, submitted by authors in Poland and in emigration.

The chemistry section had its own library, initially with books and journals from private collections that grew in time thanks to additional donations, purchases and subscriptions. The collection included complete sets of issues of 15 foreign journals spanning a period of over 20 years.

A major achievement of the chemistry section was the standardization of the Polish chemical nomenclature of inorganic compounds. Consultations on a draft nomenclature were conducted by correspondence and also occupied a few consecutive meetings of the section. Published reports[27] state that there were 60 people who contributed to the discussion, including 16 chemists with a PhD. degree, 6 professors of chemistry, 18 holders of the MSc. degree in chemistry who worked at university-level schools, 12 chemical engineers employed by industrial enterprises, 3 factory owners or directors, 3 pharmacists and 2 physicians. Following a great deal of discussion, a final proposal was presented by the chair of the Chemistry Section Bronisław Znatowicz,[28] during the ninth Congress of Polish Physicians and Naturalists in Cracow in 1900.

---

[25] Szperl (1917).

[26] Ludwik Krasiński (1833–1895) studied in Paris, and owned mines in Poland and abroad (*e.g.* a pyrite mine in Spain) and landed estates, where he built chemical processing facilities: sugar refineries, distillery, yeast and starch factories and an asphalt plant. He was a founding member and a regular benefactor of the Museum of Industry and Agriculture in Warsaw.

[27] Grabowski (1900).

[28] Bronisław Znatowicz (1851–1917), chemist, professor at Warsaw University, co-founder and first editor-in-chief of the journal *Chemik Polski*, author and translator of chemical textbooks and monographs. He investigated the reduction and nitration of aromatic compounds and the production of dyes from tar, and studied the phenomenon of phosphorescence and electrolysis of organic compounds.

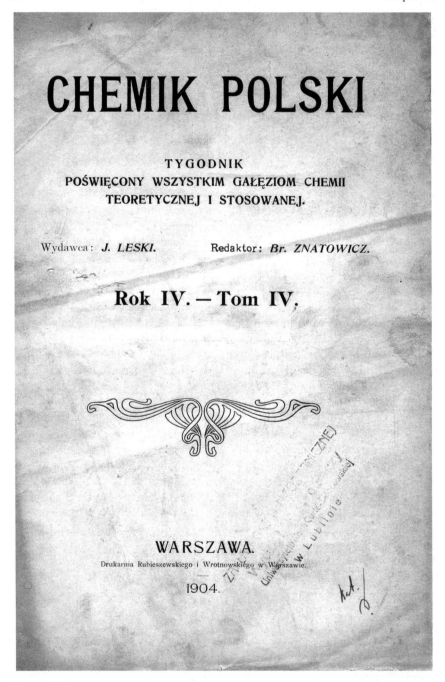

Figure 11.1   Title page of *Chemik Polski* published in Warsaw 1901–1918 by *Sekcja Chemiczna w Warszawskim Oddziale Towarzystwa Popierania Rosyjskiego Przemysłu i Handlu* (Chemistry Section of the Warsaw Branch of the Russian Society for the Promotion of Industry and Commerce).

The proposal was accepted and the implementation of a standardized chemical nomenclature began all across former Polish territory.

When the opportunity arose, in the wake of the revolution of 1905, for the chemistry section to detach itself from the Russian Society for the Promotion of Industry and Commerce, with hopes for more freedom of activity, two options were considered. One idea was to form an all-Polish association of chemists, but it proved politically impossible at that time. The other, more feasible, option was to join (transferring all property and publications) the Association of Technologists in Warsaw as an autonomous Club of Chemists within that corporation. It was this second option that was finally chosen in 1909.

Even though Poland was not an independent state and all Polish universities in the Russian partition had been disbanded, the number of Polish chemists was growing. Shortly before the First World War there were about 2000 graduate chemists of Polish descent working in Poland and abroad. The faculty of chemistry at the Polytechnic University in Riga had during its lifetime more than 200 Polish chemical engineers among its graduates. Their scientific output was equally significant. For example, more than 1000 papers written by Poles appeared in the German journal *Berichte der Deutschen Chemischen Gesellschaft* in the years 1869–1919.[29] Bibliographic studies by J. Zawidzki[30] demonstrate that in 1891–1900 Poles wrote 997 articles in various areas of chemistry published in European scientific journals.[31]

11.7   Consolidation of Polish Communities of Chemists 1919–1924

When, in 1919, Poland reappeared on the political map of Europe, it was finally possible to re-establish the institutions of an independent state. In the wake of these processes, the decision to reactivate the *Towarzystwo Fizyczno-Chemiczne* (Physico-Chemical Society), originally founded 150 years earlier, seemed viable. On 1st March 1919 the building of the *Centralne Towarzystwo Rolnicze* (Central Agricultural Society) in Warsaw hosted a meeting of professors of chemistry from the universities and polytechnic institutes of Warsaw, Cracow and Lemberg, who set up an Organizing Committee of fourteen university professors and commenced work on the society's statute. The statute was drafted by Jan Zawidzki. The Committee began preparatory work by consulting with Clubs of Chemists in Cracow and Lemberg, the Chemical Section at the *Stowarzyszenie Techników* (Association of Technologists) in Warsaw and

---

[29] J. Zawidzki (1959), 39. Speech delivered at the inaugural meeting of the Polish Chemical Society on 1st November 1919.

[30] Jan Wiktor Zawidzki (1866–1928), physico-chemist, graduate of the Riga Polytechnical University, wrote his doctorate under Prof. W. Ostwald in Leipzig. Co-founder of the Polish Chemical Society and the first editor-in-chief of the Society's journal *Roczniki Chemii* (Annals of Chemistry); author of numerous papers on chemical kinetics and autocatalysis as well as history of chemistry.

[31] *Royal Society of London, Catalogue of Scientific Papers 1800–1900*, London, 1867–1925, Vol. I–XIX.

the nascent Physical Society in Warsaw. These consultations led to the decision
to form two independent institutions: the *Polskie Towarzystwo Chemiczne*
(Polish Chemical Society) and the *Polskie Towarzystwo Fizyczne* (Polish
Physical Society).

11.7.1   The Polish Chemical Society Comes into Being

The Organizing Committee's next move was to turn to the most prominent
Polish chemists residing in Poland or abroad, inviting them to contribute to the
establishment of the Polish Chemical Society. The first meeting of the society
took place as early as 29th June 1919 at the Warsaw Polytechnic Institute. The
event, attended by 37 chemists, was mainly concerned with organizational issues,
including election of the board of the society and adoption of the statute. The
first meeting of the board took place on the same day. An inception ceremony
was scheduled for 1st November 1919 in Warsaw. Among the participants were
representatives of the state authorities, universities and academic societies.

The chief premise underlying the formation of the *Polskie Towarzystwo
Chemiczne* (Polish Chemical Society – PTCh) was the need for integration of
the groups and communities of Polish chemists that existed in the former three
partitions, now liberated and again forming one country. This goal was
reflected in the statute (par. 2): "The seat of the society shall be the capital
city of Warsaw. Local branches of the society may be formed in other cities and
towns of the State of Poland and shall respect the appropriate local laws of
association." A statutory goal of the society was "the encouragement of
progress of chemical science and propagation thereof among the public, as
well as representation of the professional interests of chemists, both researchers
and those industrially employed".[32]

In cities where Clubs of Chemists had previously been active, branches of the
society were formed first. The earliest was the Lemberg Branch, established on
1st December 1919, followed by the branch in Lodz (27th March 1920) and, in
June 1920, by two more branches in Cracow and Poznan. In Vilna, where the
former Russian authorities had disbanded all university-level schools, academic
life had to be restored from scratch, and a local branch of the society was only
established there in 1924.

The statute listed the following sources of funding for the society: "1) annual
membership fees; 2) profits generated by enterprises; 3) voluntary contributions
and subsidies; 4) donations and bequests to the society; 5) interest on capital
owned by the society and, generally, profits from its property." In practice, with
vast resources invested in publishing work, the society faced financial problems
from the outset. Financial reports for the years 1919–1924 indicate that the
society's publications were partly funded by regular grants from the Ministry of
Religious Denominations and Public Education and from industrial enterprises,

---

[32] Statut Polskiego Towarzystwa Chemicznego (Statute of the Polish Chemical Society), archives of
the Maria Skłodowska-Curie Museum in Warsaw.

which were the society's sponsoring members. Significant funding also came from the Trade Union of Large-Scale Chemical Industry Workers.

11.7.2 Membership

There were 118 founding members of the PTCh and their number increased quickly, especially owing to the establishment of local branches. For example, the Lodz branch comprised 40 members when it was formed, and the branch in Cracow 31. Accurate data are missing for the other branches. The first Congress of the Polish Chemical Society, held in Warsaw in 1923, brought together 784 participants, but it should not be supposed that all of them were members. In 1929, the society had 640 ordinary members. Assuming that the number of chemists with a university or engineering degree working in Poland at that time was nearing 2000, members of the society accounted for little more than 30% of all university graduates in chemistry (Figure 11.2).

Apart from the founding members, the Polish Chemical Society recognized the following categories of members: extraordinary, ordinary, sponsoring and honorary. Extraordinary membership was awarded to those "of unblemished reputation, working in a chemistry-related occupation or interested in the progress of chemical science".[33] The decision to award this kind of membership was taken by the board in response to a written motion signed by three ordinary members. Following a one-year trial membership and on payment of the requisite fees, an extraordinary member was entitled to a regular membership (with the exception of students, who could only become extraordinary members). Sponsoring members were persons or institutions who supported the society with sums of money higher than the regular membership fees. An honorary membership was conferred on those who particularly contributed to the development of chemistry. In 1924 the society included seven honorary members, namely Henri Le Chatelier, Maria Skłodowska-Curie, Albin Haller, Georges Urbain and Charles Moureu from Paris; Paul Sabatier from Toulouse; and Emil Godlewski from Cracow.

The members included very few representatives of natural sciences other than chemistry (pharmacists, physicians, agricultural engineers) as they could join their respective scientific-professional societies.

Ordinary members of the PTCh were mostly graduates of faculties of chemistry at domestic and foreign universities and polytechnic schools. The official report on the first Congress of the society (1923) states that 29.7% of the participants occupied research positions at universities/polytechnics, 42.2% were engineers employed by industry and 28.1% were students. Of the 92 papers delivered during the Congress, 60 were presented by chemists employed at universities/polytechnics and the remaining 32 by industrial chemists.[34] This shows that while the majority of members of PTCh were associated with various branches of the chemical industry in Poland, research was mainly undertaken by university professors, the only exception being the branch in

---

[33] *Ibid.*
[34] Sprawozdanie (1923).

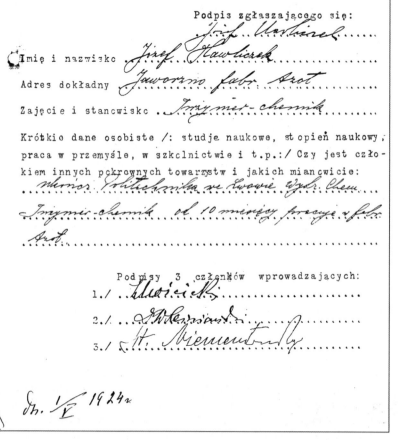

Figure 11.2   Membership application to the board of the Polish Chemical Society, May 1924. (Possession of the Museum of M. Sklodowska-Curie in Warsaw.)

Lodz, which was the most industrialized Polish city and at the same time lacked university-level institutions. The Lodz branch had been actively involved in research from its early days.

The board of the Polish Chemical Society elected in 1919 comprised 14 people, including 11 professors of universities and polytechnic schools: Leon

Marchlewski[35] (president), Stefan Niementowski[36] (vice-president), Jan Zawidzki (vice-president), Stanisław Bądzyński,[37] Karol Dziewoński,[38] Kazimierz Jabłczyński,[39] Bolesław Miklaszewski,[40] Tadeusz Miłobędzki,[41] Ignacy Mościcki,[42] Józef Pietruszyński,[43] Ludwik Szperl,[44] Józef Strassburger [vel Strasburger],[45] Wojciech Świętosławski[46] and Józef Zawadzki.[47]

11.7.3 Everyday Routines

The primary voice of the PTCh was the general meeting of its members, which took place once a year, usually in the first quarter. During these meetings

---

[35] Leon Marchlewski (1869–1946), organic chemist, professor of chemistry at Jagiellonian University in Cracow, president (rector) of the University from 1926 to 1928, member of the Polish Academy of Skills and the Warsaw Scientific Society. Discovered similarities in the chemical structure of chlorophyll and haemoglobin in 1912. Author of numerous publications, including the academic textbook *Chemia Organiczna* (Organic Chemistry).

[36] Stefan Niementowski (1866–1925), professor of general and analytical chemistry at the Lemberg Polytechnic Institute, twice elected president of the Polytechnic Institute, member of the Polish Academy of Skills, discovered methods of synthesis of γ-hydroxyquinolines and 4-oxoquinazolines, later called *Niementowski reactions*.

[37] Stanisław Bądzyński (1862–1929), professor of hygiene and physiological chemistry at Lemberg University and Warsaw University (from 1919). Member of the Polish Academy of Skills, co-editor of the German journal *Jahresbericht über die Fortschritte der Tierchemie* between 1898 and 1914.

[38] Karol Dziewoński (1876–1943), engineering technologist, graduate of the Lemberg Polytechnic Institute, professor of organic chemistry at Cracow University from 1911, member of the Polish Academy of Skills.

[39] Kazimierz Jabłczyński (1869–1944), graduate of the faculty of chemistry at Zurich Polytechnic Institute, carried out his PhD. research at Heidelberg, was awarded the doctor's degree at the University of Fribourg (Switzerland). He went on to work at Basle University and, from 1914, in Warsaw. Professor of inorganic chemistry at Warsaw University.

[40] Bolesław Miklaszewski (1871–1941), chemist and economist, professor of chemistry at Warsaw Polytechnic Institute, Minister of Religious Denominations and Public Education between 1923 and 1924.

[41] Tadeusz Miłobędzki (1873–1959), chemist, professor of inorganic and analytical chemistry at Cracow University, Warsaw Polytechnic Institute and Poznan University. President of the Polish Chemical Society in 1928, editor-in-chief of the Annals of Chemistry in the years 1935–1939.

[42] Ignacy Mościcki (1867–1946), inventor, engineering chemist, graduate of Riga Polytechnic Institute, owner of several tens of patents. Between 1892 and 1912 Moscicki lived in Fribourg in Switzerland, where he developed and introduced a method for gross production of nitric acid from air and water in an electric arch. Professor of chemical technology at the Lemberg Polytechnic Institute from 1912. He received *honoris causa* doctorates from many European universities. From 1926 to 1939 Mościcki was president of the Republic of Poland.

[43] Józef Pietruszyński (?), owner and director of a sugar-works near Warsaw.

[44] Ludwik Szperl (1879–1944), graduate of Warsaw University, research assistant at Zurich Polytechnic Institute between 1904 and 1909, professor of general and inorganic chemistry at Warsaw Polytechnic Institute from 1919.

[45] Józef Strasburger (1860–1924), engineering chemist, graduate of Riga Polytechnic Institute and University in Würzburg.

[46] Świętosławski Wojciech A. (1881–1968), physico-chemist, professor of Warsaw Polytechnic Institute and Warsaw University, vice-president of the *International Union of Pure and Applied Chemistry* from 1934 to 1940, and Minister of Religious Denominations and Public Education 1935–1939. Member of the Polish Academy of Skills.

[47] Józef Zawadzki (1886–1951) professor of physical chemistry and chemical technology at Warsaw Polytechnic Institute, president of Warsaw Polytechnic Institute in the years 1936–1939, Member of the Polish Academy of Skills, president and vice-president of PTCh (re-elected several times).

reports were heard, new authorities selected and papers presented. Day-to-day activity of the society was carried out by its branches. The work of the Warsaw branch was organized by Zarząd Główny (the main board). Scientific meetings of the Warsaw branch, accompanied by the delivery of lectures, took place, on average, 15 times a year. Meetings of other branches were organized at a similar frequency. Lectures delivered during meetings were subsequently published, in abstracted form, in official reports. A review of the reports shows, for example, that a total of 157 lectures were delivered in 1923. Lectures and papers encompassed the following areas: issues in chemical theory; chemistry in association with other sciences (mostly geology and biology); chemical industry applications of the results of research; latest developments in chemistry in Poland and abroad; biographies of eminent chemists; the importance of the chemical industry for the development of the national economy; and the application of chemistry to the munitions industry, medical treatment, pharmaceutical industry and agriculture. Most of these reports were prepared by members who were professors of universities and polytechnic institutes, while some were written by industrial chemists.

An early project completed by the Polish Chemical Society was the establishment of a *Komisja Terminologiczna* (Terminology Committee), whose task was to ensure homogeneity of Polish chemical terminology, especially with regards to names of organic compounds, which, at that time, were not harmonized and were subject to regional variation due to German and Russian influences. The Terminology Committee brought together chemists, linguists and representatives of various natural sciences. Discussions during the first Congress of the PTCh led to the formation of two sections within the Committee, namely *Sekcja Pedagogiczna* (Pedagogic Section) and *Sekcja Przemysłowa* (Industrial Section), both commencing their activity in 1923. The Pedagogic Section collaborated with *Ministerstwo Wyznań Religijnych i Oświecenia Publicznego* (Ministry of Religious Denominations and Public Education) to prepare chemistry syllabuses for secondary schools and organized lectures and conferences for teachers focused on teaching methodology. The Industrial Section, through meetings, papers and expert reports, supported activities of the state authorities geared towards intensifying the development of the chemical industry in Poland.

The PTCh collaborated with other scientific societies and public organizations in the country. Particularly lively contacts were maintained with the *Towarzystwo Naukowe Warszawskie* (Warsaw Scientific Society), *Polskie Towarzystwo Fizyczne* (Polish Physical Society), *Związek Zawodowy Wielkiego Przemysłu Chemicznego* (Trade Union of Large-Scale Chemical Industry Workers) and *Polskie Towarzystwo Farmaceutyczne* (Polish Pharmaceutical Society). Collaboration was most often in the form of open meetings, during which lectures were delivered, and guest participation in sessions, congresses and conferences of the other organization.

Foreign contacts were also thriving. Representatives of the PTCh attended *International Union of Pure and Applied Chemistry* conferences, including the first meeting in Rome in 1920. Collaboration with European chemical societies

Table 11.1   Polish Chemical Society – exchange of foreign periodicals.[a]

Great Britain	1. Journal of the Chemical Society
	2. Journal of the Society of Chemical Industry
Austria	3. Monatshefte für Chemie
Belgium	4. Bulletin de la Société Chimique de Belgue
Czechoslovakia	5. Rozpravy České akademie věd a umění
France	6. Bulletin de la Société Chimique de France
	7. Chimie et Industrie
	8. Revue de Produits Chimiques
Spain	9. Annales de la Sociedad Espagnola de Fisika y Quimica
Japan	10. Japanese Journal of Chemistry
Canada	11. Canadian Chemistry and Metalurgy
Romania	12. Buletinul Societei de Chimie din Romania
United States of America	13. Botany
	14. Journal of the American Chemical Society
	15. Journal of the Industrial and Engineering Chemistry
	16. Physics
Switzerland	17. Helvetica Chimica Acta

[a] Data based on A. Hajkiewicz-Mochniej (1992).

was through the exchange of journals (see Table 11.1). In 1924, the PTCh commenced collaboration with the *Czechoslovak Ceramics and Glass Society*, watching over the production of cement, brick and construction ceramics.

According to the 1920 statute, the Polish Chemical Society achieved its goals by: "organizing regular scientific sessions; organizing lectures open to the public; organizing congresses of chemists; participation in the work of and projects undertaken by other societies, both in Poland and abroad, with the aim of encouraging the development of chemical science; publication of specialist journals, devoted to the science of chemistry and the chemical profession, as well as of separate scientific publications; founding of libraries, collections and chemical laboratories; conferring awards and research grants in chemistry and applications thereof; providing specialist and professional information to state authorities and public organizations." The society was committed to fulfilling its statutory tasks and any impediments were usually due to financial problems. Library collections and chemical laboratories were provided by former Clubs of Chemists, now termed sections of the society.

11.7.4   Publications

In the period of consolidation of the Polish Chemical Society, its publishing activity was limited to two periodicals: *Roczniki Chemii* (Annals of Chemistry), the successor of the *Chemik Polski* (The Polish Chemist, established 1901), and *Przemysł Chemiczny* (Chemical Industry), successor of *Metan* (Methane, published in Lemberg from 1916). The *Chemik* published original contributions and review papers, annual activity reports of the PTCh, biographies and jubilee papers, as well as reviews and notes on new publications in chemistry. The authors of these texts were employees of universities, research institutes, private chemical laboratories or industrial plants or were secondary school teachers. Most contributions came from authors working in universities or polytechnic institutes, with some sent by foreign contributors. Articles were printed in Polish with French summaries. Later (from 1926 onwards) additional summaries in English and German were also included.[48] The other PTCh periodical, *Przemysł Chemiczny*, was focused on technical chemistry and new technologies. Both periodicals maintained high academic standards. Their target readers were those with an academic or professional interest in various fields of pure and applied chemistry. Copies were also sent abroad under library exchange agreements.

11.8   Demarcation

As chemistry developed and the number of people working in this field grew, the need arose for a community of specialists that would provide a platform for academic debate and mutual criticism of works. Poland had a rich tradition of organized communities in science, the first of which dated back to the Period of Enlightenment in the late eighteenth century (see Table 11.2). Moreover, in the special circumstances of political dependence, all forms of association, including associations of chemists, played a role in the struggle to maintain Polish national identity and develop and disseminate science in the mother tongue. As a result, aspirations of the individual communities or professions were not regarded as matters of primary importance. This attitude was reflected in the history of formation of chemical communities in the form of specialized clubs or sections within larger associations with more outreach. Basic change came with the advent of the Polish Chemical Society. Some separatist tendencies had already been seen during the pre-establishment organizational work, strong enough to thwart the plan to reactivate the organization in the form of a physico-chemical society. In the later period, the PTCh, while collaborating with other scientific societies, industry and various public organizations, set considerable store by maintaining its identity and prestige and by defending the interests of its members. One example of this was the internal regulations for the editorial staff of the Annals of Chemistry, which stipulated that only works by members of the society were eligible for publication. However, it quickly became clear that taking an elitist attitude too far was economically unfeasible and the society later looked more favourably at individuals and institutions associated with various branches of chemistry-based manufacturing and commerce, who would join the society as sponsoring members.

11.9   Conclusions

The history of the Polish Chemical Society (*Polskie Towarzystwo Chemiczne*) is not a typical one, just as neither was Poland's fate in the last two centuries, compared with the rest of Europe, especially due to the political and social

---

[48] Hajkiewicz-Mochniej (1992).

Table 11.2 Polish scientific societies in which chemists were involved, 1767–1919.

Name of the society	Place	Years of existence
Towarzystwo Warszawskie Fizyczno-Chemiczne (Warsaw Physico-Chemical Society)	Warsaw	1767–1769
Towarzystwo Warszawskie Przyjaciół Nauk (Warsaw Society of Friends of Sciences)	Warsaw	1800–1832
Towarzystwo Lekarskie Wileńskie (Vilna Medical Society)	Vilna	1805–1939
Towarzystwo Naukowe Krakowskie (Cracow Scientific Society)	Cracow	1816–1872
Towarzystwo Lekarskie Warszawskie (Warsaw Medical Society)	Warsaw	1821–1951
Galicyjskie Towarzystwo Aptekarskie (Pharmacist Society of Galicia in Lemberg)	Lemberg	1869–1939
Zjazdy Lekarzy i Przyrodników Polskich (Congresses of Polish Physicians and Naturalists)	Cracow	1869–1939
Polskie Towarzystwo Przyrodników im. Kopernika (Copernicus Memorial Polish Society of Naturalists)	Lemberg, now Warsaw	1874–present
Muzeum Przemysłu i Rolnictwa (Museum of Industry and Agriculture)	Warsaw	1875–1939
Towarzystwo Politechniczne (Lemberg Polytechnical Society)	Lemberg	1877–1939
Towarzystwo Techniczne Krakowskie (Cracow Technical Society)	Cracow	1877–1939
Towarzystwo Popierania Rosyjskiego Przemysłu i Handlu. Oddział Warszawski (Warsaw Branch of the Russian Society for the Promotion of Industry and Commerce)	Warsaw	1884–1914
Stowarzyszenie Techników w Warszawie (Association of Technologists)	Warsaw	1898–1935
Stowarzyszenie Techników Polskich w Wilnie (Association of Technologists)	Vilna	1903–1939
Centralne Towarzystwo Rolnicze w Warszawie (Central Agricultural Society)	Warsaw	1907–1929
Towarzystwo Naukowe Warszawskie (Warsaw Scientific Society)	Warsaw	1907–present
Chemiczny Instytut Badawczy (Chemical Research Institute)	Lemberg, Warsaw	1916–1939
Polskie Towarzystwo Farmaceutyczne (Polish Pharmaceutical Society)	Warsaw	1919–present
Polskie Towarzystwo Chemiczne (Polish Chemical Society)	Warsaw	1919–present
Polskie Towarzystwo Fizyczne (Polish Physical Society)	Warsaw	1919–present

Table 11.3   The chronicle.

1767	*Warszawskie Towarzystwo Fizyczno-Chemiczne* (The Warsaw Physico-Chemical Society). Warsaw
1800	*Towarzystwo Warszawskie Przyjaciół Nauk* (The Warsaw Society of Friends of Sciences). Warsaw
1815	*Towarzystwo Naukowe Krakowskie* (The Cracow Scientific Society). Cracow
1871	*Polska Akademia Umiejętności* (The Polish Academy of Skills). Cracow
1875	*Towarzystwo Przyrodników Polskich im. Mikołaja Kopernika* (The Copernicus Memorial Polish Society of Naturalists). Lemberg
1877	*Towarzystwo Politechniczne* (The Polytechnical Society). Lemberg
	Krakowskie Towarzystwo Techniczne (The Cracow Technological Society). Cracow
	Towarzystwo Techniczne w Poznaniu (The Poznan Technological Society). Poznan
1881	*Towarzystwo Popierania Rosyjskiego Przemysłu i Handlu* (The Society for the Promotion of Russian Industry and Commerce). St. Petersburg
1884	*Oddział Warszawski* (The Warsaw Branch) of the Society for the Promotion of Russian Industry and Commerce. Warsaw
1887	*Sekcja Chemiczna* (The Chemistry Section) of the Warsaw Branch of the Society for the Promotion... *etc.* Warsaw
1909	The Chemistry Section joins *Stowarzyszenie Techników w Warszawie* (The Association of Technologists in Warsaw) as an autonomous Koło Chemików (Club of Chemists) within that corporation
1919	06. 29. The first meeting of *Polskie Towarzystwo Chemiczne* (PTCh) (The Polish Chemical Society). Warsaw
1919	*Oddział Lwowski* (The Lemberg Branch) of PTCh. Lemberg
1920	*Oddział Łódzki* (The Lodz Branch) of PTCh. Lodz
	Oddział Krakowski (The Cracow Branch) of PTCh. Cracow
	Oddział Poznański (The Poznan Branch) of PTCh. Poznan
1923	The first Congress of the Polish Chemical Society. Warsaw
1924	*Oddział Wileński* (The Vilna Branch) of PTCh

circumstances in a country divided between Austria, Prussia and Russia. The first Polish scientific society, having in its name the adverb "chemical", was founded in Warsaw in 1767 (see Table 11.3). At the time, when the Polish population lived divided by the border lines of the annexing powers, that is in the time span covered by this book, scientific societies were being established in each part of the former Poland, trying to embrace the whole of it with their activities. In these societies, sections were created grouping representatives of different branches of sciences, including chemistry. As chemical faculties at universities and higher education establishments of chemical profiles were being established, the number of scientific societies' members who had completed higher chemical education was gradually rising. While the process of uniting the Polish territories was going on after the First World War, chemical associations and sections, which had been active at universities and industrial centres, started to integrate. Chemical engineers and professors of chemistry, who had previously worked at various European universities, started coming back. It was only then that it became possible to establish one Polish Chemical Society, covering the whole of Poland. The Polish Chemical Society, founded in 1919, grouped chemists from higher education establishments and scientific institutes as well as those working for industry. The statutory aims of the society were realized by its six territorial branches. Once a year, the general meeting of its members was held, responsible for choosing the management.

The branches organized regular scientific sessions and lectures for the wider public, and collected books and specialist papers. The management coordinated all the society's activities, acted as publisher and conducted international cooperation. The Polish Chemical Society was an institution that reflected the development of all fields of chemistry.

References

Colloque International. La presse pharmaceutique dans le monde de sa naissance a 1840 (1990), Société d'Histoire de la Pharmacie, Paris.

Davis, N. (1981), *God's Playground. A History of Poland.* Volume II: *1795 to the Present*, Oxford University Press, Oxford.

Davis, N. (1982), *God's Playground. A History of Poland.* Volume I: *The Origins to 1795*, Columbia University Press, New York.

Dewechy, F. (1928), Z przeszłości Towarzystwa Aptekarskiego we Lwowie, *Wiadomości Farmaceutyczne* **12**, 149–151.

Flis, M., Wójcik, W. and Kotarbina, H. (1972), *Polskie towarzystwa naukowe od XV wieku*, Wykaz, Warsaw.

Gerber, R. (1977), *Studenci Uniwersytetu Warszawskiego 1808–1831. Słownik biograficzny*, Ossolineum, Warsaw.

Grabowski, A. (1900), *Polskie słownictwo chemiczne. Rzecz przedstawiona w imieniu chemików warszawskich pod obrady IX Zjazdu Lekarzy i Przyrodników w Krakowie*, Rubieszewski and Wrotnowski Press, Warsaw.

Hajkiewicz-Mochniej, A. (1992), Roczniki Chemii w latach 1921–1939, *Analecta* **2**, 123–184.

Hubicki, W. (1977), Chemia, in *Historia nauki polskiej*, Ossolineum, Wrocław, Vol. 3, 468–477.

Jabłoński, Z. (1967), *Zarys dziejów Towarzystwa Naukowego Krakowskiego (1815–1872)*. PWN, Krakow.

Kabzińska, K. (1990), Organizacje chemików polskich na przełomie XIX i XX wieku i ich rola w rozwoju chemii w Polsce, *Kwartalnik Historii Nauki i Techniki* **4**, 561–583.

Kalabiński, B. (1964), Zjazdy techników polskich w latach 1882–1917, *Studia i Materiały z Dziejów Nauki Polskiej D* **4**, 3–47.

Kikta, T. (1972), *Przemysł farmaceutyczny w Polsce (1823–1939)*, PZWL, Warsaw.

Konopka, S. (1966), Polskie towarzystwa lekarskie w XIX wieku, *Służba Zdrowia* **18**, No. 10, 1–5; No. 11, 1–6.

Kulecka, A., Osiecka, M. and Zamojska, D. (2000), *Którzy nauki, cnotę, Ojczyznę kochają. Znani i nieznani członkowie Towarzystwa Królewskiego Warszawskiego Przyjaciół Nauk. W dwusetną rocznicę powstania Towarzystwa*, Archives PAS, Warsaw.

Leppert, W. (1917), *Rys rozwoju chemii w Polsce do roku 1830*, Warsaw.

Lichocka, H. (1981), *Pamiętnik Farmaceutyczny Wileński 1820–1822, Bibliografia analityczna zawartości*, Polish Academy of Sciences, Warsaw.

Lichocka, H. (1986), *Pamiętnik Farmaceutyczny Krakowski 1834–1836, Bibliografia analityczna zawartości*, Polish Academy of Sciences, Warsaw.

Lichocka, H. (1990), The Vilna Pharmaceutical Diary 1820–1822, The Journal Character, in *The Pharmaceutical World's Press from its Beginning to 1840*, Société d'Histoire de la Pharmacie, Paris.

Michalski, J. (1977), Warunki rozwoju nauki polskiej, in *Historia nauki polskiej*, Vol. 3, Ossolineum, Wrocław, 1–351.

Michałowicz, H. and Gutry, C. (1967), *Roczniki Towarzystwa Warszawskiego Przyjaciół Nauk 1802–1830, Bibliografia zawartości*, Ossolineum, Wrocław.

Offmański, M. (1907), *Dzieje Warszawskiego Towarzystwa Przyjaciół Nauk 1800–1832*, Warsaw.

Piłatowicz, J. (1990), Polskie czasopisma techniczne przed i w okresie I wojny światowej, *Kwartalnik Historii Prasy Polskiej* **29**, 1/4, 7–9.

Poradzisz, J. (1984), Komisja Balneologiczna w Krakowie 1858–1877, *Studia i Materiały z Dziejów Nauki Polskiej* **B**, 81–125.

Rodnyj, N. I. (1977), Problemy nauki i jej rozwoju u chemików XIX stulecia, *Człowiek i Światopogląd* 60–78.

Siemion, I. (1987), *Reakcje imienne chemików Polaków*, PWN, Warsaw.

Siemion, I. (1999), *Bronisław Radziszewski i lwowska szkoła chemii organicznej*, Wrocław University Press, Wrocław.

Słownik polskich towarzystw naukowych (1978–1994), 4 vols., Łoś, J. and Sordylowa, B., ed., copyright Zakład Narodowy im., Ossolińskich, Wrocław.

Sprawozdanie z I Zjazdu Chemików (1923), *Roczniki Chemii* **3**, 41.

Szperl, L. (1917), O działalności Sekcji Chemicznej i Koła Chemików w latach 1887–1917, *Chemik Polski* **15**, 132–137.

Ustawy polskiego Towarzystwa Przyrodników im. Kopernika we Lwowie z dnia 22 października 1874 r. (1900), *Kosmos* **25**, 5/6, 266–268.

Zawidzki, J. (1959), O rozwoju chemii w Polsce, in *Szkice biograficzne*, PWN, Warsaw, 29–40.

1875–1899, Dwudziestopięciolecie Polskiego Towarzystwa Przyrodników im. Kopernika (1900), *Kosmos* **25**, 5/6, 268–365.

CHAPTER 12

PORTUGAL: Tackling a Complex Chemical Equation: The Portuguese Society of Chemistry, 1911–1926

VANDA LEITÃO, ANA CARNEIRO AND ANA SIMÕES

The *Sociedade Chimica Portugueza* (Portuguese Society of Chemistry – SCP) emerged in the early twentieth century, following the creation of the *Laboratorio Municipal de Chimica do Porto* (Municipal Laboratory of Chemistry of Oporto – LMCP) and a scientific journal. The process leading to its foundation followed a path opposed to the one leading to the creation of various chemical societies in mid-nineteenth-century Europe, in which the society came first, and the journal ensued afterwards.[1] Two other differences oppose the national and the international cases. The SCP emerged outside the capital city, Lisbon, in Oporto, a dynamic northern city with a strong tradition in trade, agriculture and industry, in which Port wine has played a central role. Furthermore, after the first 15 years of its existence, the society lost its autonomous status, and incorporated physicists among its members and changed its name accordingly. Despite the time span covered by most contributions to this volume, this chapter focuses on the period extending from the creation of the SCP, in 1911, right after the establishment of the I Republic (1911–1926), until its transformation into the *Sociedade Portuguesa de Química e de Física* (Portuguese Society of Chemistry and Physics – SPQF) in 1926. Given the particular circumstances which surrounded its creation, in close association with the

---

[1] For the British societies of chemistry, and their later amalgamation, as well as the French *Société chimique*, see Brock (1992), 440–446; Fox and Weisz (1980), 269–272; Carneiro (1992), 230–239.

Creating Networks in Chemistry: The Founding and Early History of Chemical Societies in Europe
Edited by Anita Kildebæk Nielsen and Soňa Štrbáňová
© The Royal Society of Chemistry 2008

LMCP and a pre-existing journal, special attention is given to these genetic factors. They are therefore part and parcel of the process of the creation of the SCP, and for this reason they will be given special emphasis.

12.1 Background: The Municipal Laboratory of Chemistry

In nineteenth-century Portugal, the most relevant laboratories of chemistry were located at the University of Coimbra, created following the reform of higher education of 1772, and at the Polytechnic School of Lisbon and the Polytechnic Academy of Oporto, both technical schools for higher education founded in 1837, in the context of the educational reforms of Portuguese Liberalism (in both institutions the construction of laboratory facilities took quite some time).[2] During the nineteenth century, chemical research at Coimbra was negligible. At Oporto the lack of appropriate laboratory facilities very much limited scientific production, and only the laboratory of the Polytechnic School of Lisbon was productive both in fundamental research on organic chemistry and applied chemistry, especially between the 1860s and the 1890s. The leading chemists of this institution were Agostinho Vicente Lourenço, who had been trained at the research laboratory of Adolphe Wurtz in Paris and held extensive contacts abroad, and António Augusto Aguiar, who was trained by Lourenço.[3] Both produced original research on fundamental organic chemistry which was acknowledged internationally. In the field of application, the chemists José Júlio Bettencourt Rodrigues and Júlio Máximo de Oliveira Pimentel also produced relevant work.[4]

Having a very different character, the LMCP was created in 1882,[5] modelled on the Municipal Chemical Laboratory of Paris. Falling under the jurisdiction of the Oporto city council, its purpose was "to protect the consumer against fraud". The chemist Ferreira da Silva, director of the Polytechnic Academy of Oporto, was appointed its first director.

Ferreira da Silva enrolled at the University of Coimbra in 1872, graduating in natural philosophy four years later. He was staff member of the Polytechnic Academy of Oporto since 1877, first as a substitute professor of chemistry, and then as professor (after 1880). There he took upon himself the task of reforming chemical teaching. Formed as a chemist in Portuguese institutions he travelled often abroad to attend meetings, visit exhibitions and establish contacts with foreign chemists with whom he maintained often close relationships.[6] Ferreira da Silva strove to shape this laboratory into a research centre devoted to applied chemistry, very much in the spirit and orientations of Liberalism, which

---

[2] This reform took place during the government of Passos Manuel.
[3] Herold and Carneiro (2005).
[4] Leitão (1998).
[5] The LMCP was officially created in 1882, Ferreira da Silva was appointed its director in 1883 and the LMCP began functioning in 1884 when its regulations were published.
[6] Aguiar, A. (1924a) (1924b); Carvalho (1953); Simões (1931).

extended to the early twentieth century, and in particular of the policies implemented during the so-called "Regeneration" period (1850–1877), a Portuguese version of capitalism, which favoured the applications of science as a means of achieving economic and social progress.

Despite policies promoting applied sciences, and particularly engineering,[7] chemistry was not among the main priorities of the "Regeneration" rule as well as in the period that ensued. During this period, governmental efforts materialized in state-led initiatives oriented towards the control over territory, namely civil construction, roads and railways, the latter becoming symbolic of the technical novelties taking place in Portugal during the second half of the nineteenth century. Private initiative was scarce, and Portuguese chemical industry merely manufactured soaps, candles, sulfuric and hydrochloric acids and fertilizers (nitrates), a group of industrial ventures typical of eighteenth-century chemical industry.[8] Synthetic dyestuffs, the chief driving force behind European chemical industry in the second half of the nineteenth century, had no expression in the country.[9]

With the creation of the LMCP, and the prospect of proper financial and human means, Ferreira da Silva aimed at reaching standards of chemical practice absent from the understaffed and under-funded laboratories existing in Portuguese schools for higher education. During the 25 years of its life, the LMCP provided a public service to the region, by carrying out analysis of milk, vegetable oils, wines, water and mineral waters, *etc.*, and became closely linked to the local pharmaceutical and medical schools, in matters associated with hygiene and toxicology. The outcome of some of the services Ferreira da Silva provided gave him great notoriety, nationally and abroad, at times followed by public controversy amply publicized in national newspapers. One of these cases involved the accusation of a fraud in the manufacture of Portuguese wines exported to Brazil, which Ferreira da Silva settled proving that the salicylic acid it contained was a natural by-product of reactions involved in wine manufacture, not a fraudulent addition. The other case concerned the death of a child, one of the heirs to a great family fortune, whom Ferreira da Silva showed had been poisoned by a relative, who used a chemical compound for which supposedly there were no means for its identification in Portugal.[10]

The LMCP was recognized nationally and abroad, and became a centre of "scientific propaganda", that is, a centre publicizing the uses of chemistry. Also considered a "true school of chemistry",[11] since December 1903, the Oporto City Council, upon request of the *Centro Farmacêutico Português* (Portuguese Pharmaceutical Centre), authorized the laboratory to provide technical instruction to pharmacists. These teaching activities were soon extended to encompass the teaching of general chemistry, toxicology, food chemistry and pharmaceutical chemistry. In addition, its library was open to the public. The training provided

---

[7] Diogo (1994).
[8] Homburg, Travis and Harm (1998); Bradbury and Dutton (1972).
[9] Meyer-Thurow, G. (1982) and Homburg (1992).
[10] Aguiar, A. (1924a), (1924b).
[11] Ferreira da Silva (1909), 221–222.

was meant to be an alternative to the recourse to foreign chemists who were contracted by various official institutions in order to carry out chemical analysis in the fields of food chemistry, hygiene and industrial chemistry.

In 1907, the city council closed down the LMCP, allegedly for economic reasons. The council considered its budget too high in relation to the work produced,[12] but according to a Brazilian chemist, the Laboratory spent 82 000 *reis*[13] during 15 years while the administrative bureau of the council spent almost double that amount (143 000 *reis*).[14] In fact, it is plausible to conjecture that economic interests might have conspired in order to eliminate an organization, which certainly had the potential to help economic activities linked to agriculture and industry, but could also undermine established privileges and interests in matters pertaining to quality control of food and drinks. On the other hand, political and personal factors might have also played a determining role. In fact, the opposition to the LMCP came especially from a Republican council representative who strove to assert himself through the press by writing pamphlet-like articles criticizing the importance given to the chemist and professor Ferreira da Silva.[15] Criticisms haunted Ferreira da Silva's career even in his capacity of head of the SCP, which he founded in 1911.[16]

Following the closure of the LMCP, various associations of wine, milk and beer producers expressed their support to the laboratory praising its services,[17] but they were unable to join forces and fund this laboratory or create another of the same kind on a private basis, a move which was not part of the Portuguese entrepreneurial culture, in contrast with various private initiatives of this kind taking place in other European countries. The contributions, impact and reasons behind the end of the LMCP still deserve further investigation, but the impact of its closure prompted Ferreira da Silva to launch a new strategy which was to materialize in the creation of the SCP.

12.2   The *Revista de Chimica Pura e Applicada* (Journal of Pure and Applied Chemistry)

Ferreira da Silva was also behind the foundation, in 1905, of the *Revista de Chimica Pura e Applicada* (Journal of Pure and Applied Chemistry – *RCPA*, Figure 12.1).[18] This journal, the first Portuguese specialized chemical journal, succeeded to the *Revista Chimico-Pharmaceutica* (Chemical-Pharmaceutical Journal), created by a group of students of the Pharmacy School of

---

[12] Ferreira da Silva (1909a), IX–XV.
[13] Portuguese currency of the time. It is not possible to give the corresponding sum in euros.
[14] N/A (1924b), 227.
[15] Leite (1907), reprinted in Ferreira da Silva (1909a), 133–137.
[16] N/A (1917a), 400.
[17] Many associations, mainly pertaining to wine production and trade and pharmacy, reacted against the closure of the LMCP. The Academy of Sciences of Lisbon also joined this movement as well as the directors of the chemical laboratories of the University of Coimbra, and the Polytechnic Schools of Oporto and Lisbon.
[18] In 1958 the journal changed its name to *Revista Portuguesa de Química* (Portuguese Journal of Chemistry).

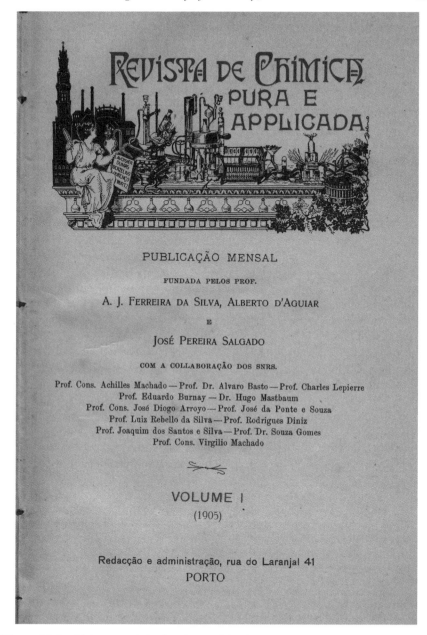

Figure 12.1 Title page of the first issue of the first volume of *Revista*, 1905.

Oporto between 1903 and 1904, an initiative soon taken over by some university lecturers.

Other participants in the project were Alberto de Aguiar, who taught at the Medical School of Oporto, and José Pereira Salgado, who taught at the chemical laboratory of the Polytechnic Academy of Oporto, whose creation owed a great deal to Ferreira da Silva.[19] The journal became in this way a local joint enterprise, based in Oporto, and congregating a group of professionals of different backgrounds, in particular medicine,[20] whose fields of expertise were mainly the applications of chemistry.

In the journal's first issue it was argued that the restructuring of higher education and in particular of pharmaceutical teaching, which took place in 1902, was expected to lead to an increase in chemical research. With this reform, chemistry provided pharmacy and medicine with a scientific status as chairs of pharmaceutical chemistry, food chemistry and toxicology were created in the schools of pharmacy, together with courses on sanitary chemistry at the medical schools of Oporto, Lisbon and Coimbra, whose laboratories underwent more or less extensive reorganization. Simultaneously, this reform enabled pharmacists to emancipate themselves from medical doctors, who had been in charge of teaching pharmaceutical courses in pharmacy schools.[21] In this process pharmacists tightened their ties to chemists in order to affirm their profession, a move which was facilitated in Oporto by the fact that the LMCP, directed by Ferreira da Silva, delivered lectures on toxicology and food chemistry to pharmaceutical students. In this way, chemistry found a clientele among pharmacists, who became a strong presence and were to play an important role in the SCP.

Ferreira da Silva was optimistic that the few existing agricultural stations, chemical industries and technical laboratories, together with those devoted to hygiene and medical clinics, would provide enough new research material to support a chemical publication. But the journal had, from the outset, the dual purpose of compiling research work carried out by Portuguese chemists in national laboratories, and simultaneously providing an outlet for popularizing both fundamental and applied chemistry, among students and people interested in keeping up with recent developments in the realm of chemistry.[22]

A close inspection of the ordering of areas within the journal reveals commonalities and differences between the local chemical context and the

---

[19] This laboratory became in 1991 part of the Faculty of Sciences of Oporto, following the Republican reform of the teaching system.

[20] Pereira Salgado graduated in medicine and became a doctor in physics and chemistry and in industrial chemical engineering, and taught at the laboratory of chemistry of the Polytechnic Academy of Oporto, and later the Faculty of Sciences; Alberto de Aguiar was also a graduate in medicine and taught hygiene, bacteriology and parasitology from 1911 to 1917, when he became professor of biological chemistry and general pathology.

[21] Ferreira da Silva (1914), 114–115. Concerning the reform of pharmaceutical teaching by the Minister Hintze Ribeiro, Ferreira da Silva argued that the above reform "entailed the development of the teaching of chemistry in our country, through the creation in new schools of the course of pharmaceutical chemistry, a discipline which should have been taught among us for a long time, as well as the course of legal chemistry applied to hygiene, a course very useful now that medical and legal services were remodelled and inspection of food and drink has been strengthened." See Ferreira da Silva (1914), 136.

[22] N/A (1905).

international one.[23] As was typical in most European countries, bacteriology was included in chemistry. However, the journal still ascribed the first place to inorganic chemistry as compared to organic chemistry, which came only second, an ordering which is striking given the impact of the latter in the second half of the nineteenth century and the early-twentieth century.

Quite surprisingly, the fourth position was occupied by mineral and geological chemistry, as if one were suddenly taken back to the seventeenth-century roots of eighteenth-century mineral chemistry. Mineral chemistry took in this way precedence over medical and pharmaceutical chemistry, probably testifying to the still-prevailing reluctance of Portuguese medicine and pharmacy to go beyond clinical practice in the adoption of scientific methods based on research. Internationally, the level of analysis had already made the transition from the organ to the cell and from the cell to the molecule, paving the way for the emergence of biochemistry in the early-twentieth century, through the intermediate stage of biological (or physiological) chemistry.[24]

Despite the first three positions ascribed to the main mid-nineteenth-century divisions of chemistry, overall the local emphasis on the applications of chemistry is paramount. This fact still echoed and was consistent with the rhetoric of progress of the "Regeneration", by emphasizing scientific development that could contribute to the blossoming of local industries and the exploration of national mineral and agricultural resources. Moreover, the emphasis on applications enlisted potentially public support, given its immediate impact on daily life.

Ferreira da Silva was keen on publicizing the creation of the journal abroad, in order to gather international support for it. The journal was welcomed by the influential *patron* of French chemistry, Marcelin Berthelot, as well as by various similar societies across Europe, even extending to the American Chemical Society.[25]

Despite the scarcity of effective contributors to the journal, 12 on the whole,[26] during the first four years the journal released 11–12 issues, annually. Chemical topics addressed in the papers did not correspond to their order of importance as stated in the journal's aims. Effective priority was given to sanitary chemistry, technical and industrial chemistry, general chemistry and analytical chemistry. In the first years and until the closure of the LMCP, the journal's topics mirrored its activities to such an extent that one can claim that the journal was a privileged outlet for the scientific production of the

---

[23] The areas were ordered as follows: Inorganic Chemistry; Organic Chemistry; Analytical Chemistry; Mineral and Geological Chemistry; Agricultural Chemistry; Sanitary and Food Chemistry; Bacteriology and Hygiene; Technical and Industrial Chemistry; Medical Chemistry (Biological and Pathological); Pharmaceutical Chemistry; Toxicological Chemistry; Medical Hydrology; Bibliography; Review of Journals; The Portuguese Chemical Movement; Miscellanea and Correspondence; Literature and History of Chemistry.

[24] Carneiro (2006); Amaral (2001).

[25] Amorim da Costa (1997).

[26] The list of regular contributors included Aquiles Machado, Álvaro Basto, Charles Lepierre, Eduardo Burnay, Hugo Mastbaum, José Arroio, Ponte e Sousa, L. Rebelo da Silva, Rodrigues Dinis, Santos e Silva, Sousa Gomes and Virgílio Machado.

laboratory. Between 1884 and 1907, the main areas of work of the LMCP focused on what was then called sanitary chemistry, in particular chemical analysis of the water supplied to the city of Oporto, wine, beer, olive oil, milk and dairy products, preserves, meat and salt, among others. Toxicology, notably in matters pertaining to reactions of alkaloids and chemical examination of blood stains, together with commercial chemistry, in particular analyses carried out on goods dispatched from Oporto customs and subject to complaints, also had a prominent role. But the laboratory also engaged in hydrological chemistry, by studying mineral waters with medical applications, as well as agricultural, pharmaceutical and clinical chemistry, serving in this way physicians and hospitals.

Besides strictly chemical articles, contributions under the heading Scientific Literature played a dominant role, in quantitative terms, although one should bear in mind that often they were short and conveyed superficial assessments. They included news, reviews of papers and books, papers on the history of chemistry and mainly obituaries. Also papers devoted to the teaching of chemistry, including lecture notes and course syllabuses (Chemical Education) played an important part in the journal's output. Therefore, it comes as no surprise that the main contributors to the journal were the chemists operating at the LMCP. More than one-third of the articles published were authored by Ferreira da Silva and his collaborators as well as by authors working in chemical laboratories, two performing functions of directors – Hugo Mastbaum and Achilles Machado.

Following the closure of the LMCP, Ferreira da Silva's contributions were dramatically reduced, and the author took refuge in the writing of biographical notes, eulogies and scientific notes.

In the foreword of the journal published in 1909, Ferreira da Silva complained about the consequences of this decision.[27] Portuguese Port wine, prior to these unfortunate events, had been analysed in France in the laboratories run by Berthelot and the famous agricultural chemist Jean-Baptiste Boussingault, in Germany by the analytical chemist Carl Fresenius, a former research student of Justus von Liebig, and in Denmark by one of the leading "zymotechnologists" Alfred Jørgensen, founder of the Laboratory of Fermentology and of the journal *Zymotechnisk Tidende*.[28] Following the closure of the Laboratory the same happened again. A similar situation occurred with the study of Portuguese minerals and salts, whose composition was analysed in France. In Ferreira da Silva's opinion, this state of affairs was a consequence of the lack of appropriate laboratory facilities in the country.

In this context, the editorial board decided to put emphasis on the dissemination of analytical methods and instruments to be used in laboratories devoted to applied chemistry,[29] as well as giving more space to sanitary and agricultural chemistry, fields more easily perceived as useful. In addition, it

---

[27] Ferreira da Silva (1909b).
[28] Bud (1992) and (1994).
[29] Jørgensen (1909a) and (1909b).

began publishing notes on chemistry for secondary school students, such as those by Sophus Mads Jørgensen, a Danish inorganic chemist who contributed an article for pupils of secondary schools.[30]

Two years later, in 1911, Ferreira da Silva once again bitterly regretted the fate of the laboratory, and complained about the lack of support and encouragement.[31] "Years of struggle and effort led to dismay and solitude rather than to work and initiative! However, the *Revista de Chimica Pura e Applicada* has been published regularly, with the purpose of fostering the good cause of the science of chemistry, and the promotion of scientific culture, which is immensely delayed among us! By an unfortunate coincidence it was during this period that the sledge hammer hit the first chemical institution of Oporto and the work carried out there was abruptly interrupted, despite the palpable, evident and indisputable services provided!"

In the same article, Ferreira da Silva expressed his wish and hope of gathering support from his colleagues in order to create national institutions devoted to the promotion of chemistry.

12.3   The *Sociedade de Chimica Portugueza* (Portuguese Society of Chemistry)

By 1911, Portugal had gone through a major political change. In 1910, the monarchy was overthrown and the First Republic established. The priority of the government became the consolidation of the regime, which nevertheless was to be marked by political instability, accentuated by the First World War. The Republicans launched a major reform of the teaching system in 1911. The Polytechnic Academy of Oporto and the Polytechnic School of Lisbon, twin creations of Liberalism meant to oppose the traditional teaching associated with the University of Coimbra, were converted into Faculties of Sciences, together with the Faculty of Natural Philosophy at Coimbra. The medical and pharmaceutical schools were converted into faculties of medicine and of pharmacy in Oporto, Coimbra and Lisbon. With this reform a *licence* in physics and chemistry was created in the new faculties of science. Simultaneously the Technical Institute for Higher Education was founded in Lisbon.

Contrary to the "Regeneration", the leading intellectuals of the so-called 1911 generation favoured fundamental research rather than the applications of science, and advocated the creation of research schools and a concept of university modelled on the German one.[32] The Republican reforms of higher education were later to produce some significant results, especially on account of the efforts and initiatives of a group of distinguished members of the 1911

---

[30] Jørgensen, S. M. (1909b).
[31] N/A (1911).
[32] Amaral (2001), 159–177.

generation, despite the unfavourable context of the subsequent political cycle.[33] The Republican rule was short lived and, in 1926, Portugal entered a long period of obscurantism with António Salazar's dictatorship, which lasted until 1974.

12.3.1   Aims and Organization

From the mid nineteenth century onward Europe underwent a movement towards the creation of research schools and of specialized scientific societies, which reflected the growing scientific specialization taking place throughout the century.[34] Research schools were fundamental to the establishment of new scientific disciplines and traditions by training young scientists, and the emergence of societies played an essential role in the professionalization of practitioners of different fields of scientific enquiry, contributing decisively to their social recognition. No similar movement took place in nineteenth-century Portugal. The absence of research schools in chemistry or in any other scientific field, as well as of scientific societies, is notorious.[35]

The Portuguese Society of Chemistry (SCP) was basically the brainchild of Ferreira da Silva, who looked for another solution to promote chemical research in Portugal after the closure of the Municipal Laboratory of Chemistry of Oporto (LMCP). His awareness of the international network of chemical societies spreading all over Europe gave him a clue to the problem. With a considerable time gap when compared with other European countries, the SCP was founded in 1911, appropriating as its publication outlet the existing *Revista de Chimica Pura e Applicada* (*RCPA*, Figure 12.2).[36]

Guided by his willingness "to preach the Gospel of Chemistry" in Portugal, Ferreira da Silva strove to eradicate what he considered to be the unfortunate "barrier between pure and applied chemistry" and to demonstrate the importance of the alliance between chemistry and industry in solving people's basic needs.[37] Later, in 1917, another member of the society, the physicist Almeida Lima, drew attention to two important questions: while he supported the creation of the SCP, he expressed mixed feelings regarding its success; if the SCP failed, the consequences were in his view more damaging than the absence of such a society. He also warned that applied chemistry could not be consistently implemented without the support of fundamental chemistry, a domain which unfortunately had little expression in the country.[38]

---

[33] Among others, these initiatives materialized in the creation of national institutions oriented to the promotion and funding of scientific research, notably the Junta de Educação Nacional (National Board for Education, JEN), created in 1929, and Instituto de Alta Cultura (Institute of High Culture, IAC), founded in 1936, under whose auspices young Portuguese scientists were sent abroad to be trained in scientific research. Ramos do Ó (1999) and Amaral (2001), 171–177.

[34] Geison and Holmes (1993); Crosland (2003); Morrel (1972); Fruton (1988).

[35] Herold and Carneiro (2005).

[36] For example, the Chemical Society of London was founded in 1841, and the Société Chimique de France in 1857.

[37] Ferreira da Silva (1914), 41.

[38] N/A (1917c), 403; (1917f), 413.

Figure 12.2   Title page of the eighth volume of *Revista*, 1911, documenting that the journal was the organ of the *Sociedade Chimica Portugueza*.

On 28 November 1911, Ferreira da Silva requested various personalities to attend a preliminary meeting. On 28th December, the founding members of the SCP met at the laboratory of mineral chemistry of the Lisbon Polytechnic School.[39] In addition to the promoter, the meeting was attended by Álvaro Basto, José Pereira Salgado, Carl von Bonhorst (teacher at the Escola Marquês de Pombal, a secondary technical school), César Lima Alves (professor at the Agricultural Institute for Higher Education), Armando Artur de Seabra (serving in the 18th Agricultural Section in Lisbon), Artur Cardoso Pereira (professor at the Faculty of Medicine) and Hugo Mastbaum. They were all entrusted with the task of launching the society.[40]

The purposes of the SCP were clear: to promote regular and extraordinary scientific meetings; to publish a scientific journal to channel research on chemistry and allied sciences; to establish a chemical library open to the public; and to exchange information, publications and know-how with scientific societies in Portugal and abroad. With the purpose of becoming part of an international network of similar institutions, the creation, aims and scope of the new society were publicized among its foreign counterparts, notably the *Chemical Society of London*, the *Deutsche chemische Gesellschaft* and the *Société de chimie physique de Paris*.

The statutes of the SCP were approved on 28th December 1911, and published in 1912.[41] Ferreira da Silva was elected president, Mastbaum secretary and A. Seabra treasurer. This first executive committee decided to establish the headquarters of the SCP in Lisbon, and to form an administrative committee composed of three chemists, two from Lisbon and one from the University of Coimbra, which had not participated in any prior initiatives leading to the society's foundation. Coimbra's absence was probably the outcome of the status it conferred to chemistry and chemical research, subsidiary in many respects to medicine. Throughout the nineteenth century, chemical courses at the University of Coimbra, the only university in the country, were attended mainly by medical students, generally oriented to pursue careers as physicians.

On 26th January 1912, a meeting to elect the administrative committee of the SCP was held, the statutes were read and some minor changes were introduced.[42] The administrative committee was composed of a president, two vice-presidents, two secretaries, one of them holding also the position of librarian, a treasurer and three voting members. The secretaries, the treasurer and one of the presidents should reside in Lisbon. Mandates were annual, but could be followed by a second one. According to the statutes, the society was supported by the fees paid by members, donations and the income from the selling of publications.[43]

---

[39] N/A (1912a).
[40] Alberto de Aguiar, Achilles Machado, Severiano Monteiro and Charles Lepierre supported unconditionally the idea but could not attend the meeting.
[41] N/A (1912b).
[42] The president was Ferreira da Silva; Achilles Machado and Álvaro Basto became vice-presidents; Mastbaum was first secretary; Cardoso Pereira was second secretary; Armando Seabra became the treasurer; Carlos von Bonhorst, César Lima Alves and Pereira Salgado were voting members.
[43] N/A (1912b), 5.

Meetings were held monthly and decisions made and papers delivered were to be published in the society's journal. While monthly meetings were held in Lisbon, extraordinary meetings were to be held at Oporto and Coimbra, whenever the executive committee thought it convenient, a decision which shows the society's intent to decentralize its activities. Concerning the practical implementation of the routines outlined in the statutes of the SCP little can be said, because no archival material on which to base an evaluation has survived.

In 1916, the SCP underwent reorganization, splitting into three sections, Oporto, Coimbra and Lisbon.[44] The Oporto section included since its inception a physics group composed of three elements,[45] a fact that is quite revealing of the reduced opportunities for physicists to exercise their profession and to intervene in Portuguese society. Within a political and cultural framework which favoured application, the realms in which physicists could have played a role were dominated by engineers. Such was the case of electricity and of its uses.[46] The weaknesses of the Portuguese scientific community facilitated associations with scientific practitioners of different backgrounds as a means of counterbalancing adversity. Therefore, disciplinary autonomy was not considered part of a demarcation strategy to such an extent that in 1926 the SCP gave way to the SPQF.[47] According to Aguiar, who was appointed president of its executive committee, this state of affairs showed "the very deep crisis which affects the experimental sciences, in particular chemistry and physics due to a lack of a scientific culture and of laboratory facilities". In his view, this situation led to "the desertion of the most dedicated and interested chemists, who were led to look for more profitable jobs".[48] The board of directors of the former society recognized that "the situation has not improved from the material and scientific point of view (...). The situation of our (chemical) laboratories have increasingly deteriorated".[49]

Only in 1974 was the SCP divided into two separate institutions, the *Sociedade Portuguesa de Química* (Portuguese Society of Chemistry – SPQ) and the *Sociedade Portuguesa de Física* (Portuguese Society of Physics – SPF). Three years later, the two societies were finally able to find an appropriate building to house them, which has been kept to this day. The aims of the SPQ were enumerated and were basically the same as those formulated back in 1911. Rather than a forum for specialized scientific debate and an instrument for

---

[44] The scientific activity of the three sections cannot be fully assessed, because the minutes of their meetings were not regularly published, in a clear violation of the statutes.
[45] They were Francisco Paula Azeredo, Alexandre Sousa Pinto and Álvaro Machado. Francisco Paula Azeredo (1859–?) was a military engineer and graduated from the University of Coimbra in natural philosophy and mathematics and was given the title of doctor in physics and chemistry. Alexandre Sousa Pinto (1880–?) also graduated from the University of Coimbra in natural philosophy and mathematics and was given the title of doctor in physics and chemistry. He taught physics at the Polytechnic Academy of Oporto and held political and administrative posts. Álvaro Machado (1879–1946) had a background in natural philosophy and medicine and became a professor of physics of various secondary schools and later of the Faculty of Sciences of Oporto.
[46] Matos *et al.* (2005), 41–50.
[47] Gieryn (1983).
[48] Aguiar (1928).
[49] Aguiar (1928).

professional recognition and affirmation of chemistry and of its subdisciplines, the SPQ was meant to be merely a locus of open debate, aimed at promoting mainly the progress of chemical practice and the teaching of chemistry.[50]

12.3.2 Membership

According to the statutes of 1912, the SCP established various categories of members: honorary, permanent, correspondents and associates, with no limit to the number of members.[51] The honorary category included individuals whose scientific contributions to the progress of chemistry were considered highly relevant. They were selected by the society and their number should not exceed ten. The permanent members, in addition to the founding members, could be proposed by the board of directors, or any member resident in Portugal, their election depending on their scientific work and proof of their scientific contributions to the development of pure and applied chemistry. The correspondents were foreigners elected upon presentation of their chemical credentials. Finally, the associates were Portuguese and foreigners who simply showed an interest in pure or applied chemistry. Thus there were no professional restrictions, and virtually any pharmacist, physician or engineer could become a member. By making membership solely dependent on a personal interest in chemistry, the constitution of the SCP reflected the undefined status of chemistry within Portuguese science and society, the lack of recognition of the professional status of chemists and the vulnerabilities of the small Portuguese chemical community in which practitioners of fundamental chemistry were scarce.

Admission to the SCP was made upon request of the candidate on a special form, subscribed by two permanent members. The election should take place in the first meeting following candidacy and was successful if there was a majority of votes. Permanent members paid an annual fee amounting to 5000 *reis* (then the Portuguese currency) and both the correspondents and associates should pay 2500 *reis*. All members were entitled to receive the society's journal, attend meetings, submit papers and participate in scientific discussions. Only permanent and honorary members could be part of the board of directors.

In 1912, the SCP possessed 76 members, but the first complete list was only published in 1915, the total number of members having increased to 146.[52] Despite the fact that the society had incorporated physicists since 1926, their number is inexpressive. The year of 1928 was selected because a list of members was published again in this year, the total number of members being 214.

---

[50] Ferreira da Silva (1914), 39–42.
[51] N/A (1912a).
[52] These numbers should be contrasted with other societies. For example, in 1914 the *American Chemical Society* had 6091 members, the *Deutsche Chemische Gesselschaft* 3356 members, the *Société chimique de France* 1857 members, and The *Chemical Society of London* 3202 members. N/A (1914), 7.

Table 12.1   Number of members of *Sociedade de Chimica Portugueza* by professional occupation in 1915 and 1928, excluding corresponding and honorary members.

Occupations	1915 (127 members)	1928 (1926) (182 members)
Analyst or chemical researcher	30 (23.6%)	15 (8.2%)
University lecturer of chemistry and/or physics	4 (3.1%)	7 (3.8%)
Teacher of physics and chemistry (secondary education)	3 (2.7%)	8 (4.4%)
Pharmacist and/or university lecturer of pharmacy	15 (11.8%)	27 (14.8%)
Physician and/or university lecturer	7 (5.5%)	23 (12.6%)
Lecturer in technical schools (higher education)	8 (6.3%)	11 (6.0%)
University lecturer (sciences other than chemistry and physics, and humanities)	11 (8.7%)	6 (3.3%)
Engineer and university lecturer or teacher	0	28 (15.4%)
Agriculturalists and/or University Lecturer	12 (9.4%)	10 (5.5%)
Military officer	0	5 (2.7%)
Industrialist	6 (4.7%)	2 (1.1%)
Others[a]	13 (10.2%)	6 (3.3%)

[a] The category "others" included judges, students, geologists, *etc.*

Table 12.1 shows the number of members by professional occupation in 1915 and 1928, as listed in the society's journal.[53] The corresponding members which amounted to 15 in 1915 and 32 in 1928 were excluded, together with the honorary members, 4 in 1915 but not stated in 1928.[54] Corresponding members were primarily from Brazil, Spain and Italy, certainly due to cultural affinities.

It becomes apparent from Table 12.1 that analysts and chemical researchers amounted to 23.6% of the members in 1915, but decreased to 8.2% in 1928 (excluding corresponding members, who were foreigners living and working abroad, and honorary members), though research proper was scarce. University lecturers held positions primarily in the Faculties of Sciences of Lisbon, Oporto and Coimbra, the Technical Institute, the Agricultural Institute, the Faculty of Medicine and the Pharmacy School (in which chemical teaching was administered). Other institutions included the Navy School, the Agricultural School of Coimbra, the Technical Faculty of Oporto and the Army School. The accumulation of various positions is striking, especially in education. Thus physicians teaching in higher education accumulated with clinical practice. Teaching in higher education was not separated from teaching in secondary

---

[53] Occupations of some members could be classified differently, in view of the assessment of their careers. For example, Charles Lepierre was classified as a university lecturer, despite his substantial work on analytical chemistry, notably mineral waters.

[54] They were Armand Gautier, honorary professor of the Faculty of Medicine in Paris; Dioscoride Vitali, honorary professor of the Royal University of Bologna; Icilio Guareschi, professor of the Royal University of Turin; and José Rodriguez Carracido, professor of the Faculty of Pharmacy of the Central University of Madrid.

education and it was common for a person to hold a position both at a *lycée* and a faculty. Chemistry had no autonomy and specialization was negligible both in teaching and in laboratory positions. It was currently associated with physics in secondary education, and in higher education with pharmacy, medicine, physics and agriculture; not until 1911 was a university *licence* in physics and chemistry created, which partially explains the situation. Members working in laboratories had various backgrounds ranging from pharmacy, medicine, agriculture, and physics and chemistry, and carried out primarily routine analyses associated with toxicology, food chemistry, sanitary chemistry, hygiene and pharmaceutical chemistry. It is worth noting that by comparing the list of members of 1915 with that of 1928, the number of pharmacists using the title *Farmacêutico-Químico* (Pharmacist-Chemist) instead of simply Pharmacist, increased considerably. Research on fundamental chemistry nationwide was negligible. Chemists working in industry and industrialists were scarce, a result of the vulnerabilities of the Portuguese chemical industry.

Only the list of members of 1915 provides the date of admission, which makes it impossible to ascertain the extent of regular admissions from then onwards. However, it is worth noting that the first women were admitted into the SCP after 1916.[55] In succeeding years, the number of high-school teachers and university lecturers grew considerably, as did the number of pharmacists. The number of medical doctors and chemists declined, and it seems that in this group a subtle opposition separated medical doctors using chemistry and chemists with a medical background.[56] Finally, the number of engineers grew, probably as an outcome of the reform of technical teaching in 1911.

Table 12.1 also shows a significant overall increase in members working in schools for higher education, together with secondary school teachers. In 1928, the number of members belonging to the Oporto delegation of the SCP dominated those of Lisbon, unlike in 1915. Among foreign members, Brazilians increased dramatically from 9 in 1915, to 21 in 1928.

12.3.3   The Appropriation of an Old Journal by the New Society

The SCP adopted as its official periodical the *Revista de Chimica Pura e Applicada* (*RCPA*) (Figure 12.3), which began being published in Lisbon. According to its guidelines it should publish the minutes of sessions as well as scientific articles presented in the regular monthly meetings; scientific papers published in other journals (with the required permission); and generally all scientific contributions which could promote the development of chemistry.

Figure 12.4 represents main areas included in the journal as the official publication of the SCP. These areas were grouped in view of our reading of the contents of the articles. Sanitary chemistry emerges as the main field, followed

---

[55] They were Bertha Peixoto, an industrial chemist, and Maria Emilia Salvador, graduate in physics and chemistry, both members since 1916; Etelvina Pereira dos Santos and Maria José Rodrigues, both graduates in physics and chemistry, became members in 1917. See N/A (1917b), 417 and (1917g), 430.
[56] N/A (1917d), 426.

Figure 12.3 Frontispieces of the Journal of Pure and Applied Chemistry (*Revista de Chimica Pura e Applicada*) before and after the creation of the Portuguese Society of Chemistry (1905 and 1911).

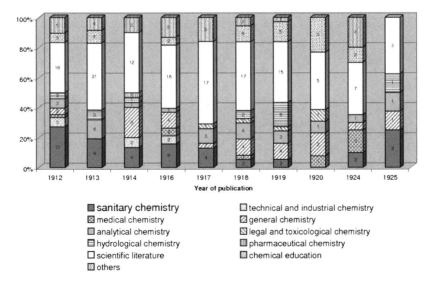

Figure 12.4 Main areas included in the *Revista de Chimica Pura e Applicada* (Journal of Pure and Applied Chemistry), from 1912 to 1925.

by technical and industrial chemistry, and general chemistry, but the number of articles published in each of these fields was much reduced. Articles under the headings Scientific Literature and Chemical Education continued to be dominant in the journal's contents.

Major contributors to the journal were still Ferreira da Silva and Hugo Mastbaum, but the former now changed his focus from chemistry to historical notes and obituaries. Achilles Machado and Alberto Aguiar continued to be contributors, but not Pereira Salgado; Cardoso Pereira, who had international experience from attending foreign laboratories in France and Germany, emerged now as a major author in this second phase of the journal's life.

Following the journal's orientation, established prior to the creation of the SCP, and the peculiarities of the Portuguese chemical context, the articles presented at meetings focused mainly on chemical analysis of butter, mineral waters, urine, *etc.* By 1914, the executive committee was expressing in the Annual Report its dissatisfaction with the workings of the SCP, which failed to comply with the statutes in what concerned the organization of meetings (only seven meetings were held), the participation of members and the number of original papers published in the journal. In 1914, the number of issues was reduced to two per year.[57]

Ferreira da Silva's explanation for this state of affairs was rather rhetorical by masking the debilities of Portuguese chemistry, and blaming the excessive modesty of Portuguese chemists:[58] "(. . .) this situation is probably explained by the excessive modesty and shininess of Portuguese chemists. How many interesting observations, which certainly deserve to be published, are left at the corner of a drawer? Generally people only wish to publish what is perfect and final." In fact, the levels of original scientific production in the country were generally very low.

In 1915, the publication of *RCPA* was discontinued and the following year a second series appeared with the same title. The journal kept its former orientation, but printed basically translations of papers published in foreign journals. In 1917, following the creation of the physics section in Oporto, the journal began accepting articles on physics,[59] and survived until 1920 owing to the commitment of Ferreira da Silva and the editor Pereira Salgado. The emphasis on chemical analysis and lecture notes was striking: Achilles Machado focused on the application of electricity to chemical analysis,[60] and the Italian Giovanni Costanzo, who taught radioactivity and dioptrics at the Technical Institute of Lisbon, supplied his lecture notes.[61] The *RCPA* was again discontinued in 1921, 1922 and 1923. Ferreira da Silva died in 1923 and

---

[57] Number of issues published between 1905 and 1925: 1905–12; 1906–12; 1907–11; 1908–11; 1909–9; 1910–9; 1911, 1912, 1913–12; 1914–2; 1915–0; 1916–4; 1917, 1918–5; 1919–3; 1920–2; 1921, 1922, 1923–0; 1924–3; 1925–1.

[58] Ferreira da Silva, Mastbaum and Cardoso Pereira (1914), 3.

[59] Machado (1918). This article written by a physics professor dealt with the functioning of the nonius, a mathematical instrument invented by the sixteenth-century Portuguese mathematician Pedro Nunes.

[60] Machado (1920).

[61] Costanzo (1920) and (1919 and 1920), 220–255.

Aguiar took over as chief editor in 1924. Three issues were published in 1924 followed by one in 1925. In the 1924 volume, articles devoted to Ferreira da Silva are paramount (*c.* 30%).[62] In 1925, two scientific papers, one authored by Egas Pinto Basto, on the role of phosphoric acids in soils and olive oils,[63] and the other by Couceiro da Costa, on thermochemical studies of distillation and crystallization, were published.[64] These were the first significant papers authored by chemists working at the University of Coimbra.[65]

The vicissitudes that affected the journal mirrored the fragile situation of chemical research and practice in Portugal. In this context, it is not surprising that lecture notes and materials became an increasingly important bulk of the journal's publications. The journal assumed progressively its major function as an outlet for the dissemination of laboratory chemical techniques and analytical methods, chemical teaching and the popularization of chemistry through obituaries, news, *etc.*

12.4 Concluding Remarks

The history of the creation of the SCP and of its first years of existence is particularly revealing. When contrasted with the patterns of development of similar institutions in scientific centres throughout Europe, many differences emerge. The journal preceded the society; the society did not emerge in the capital city; the number of chemists actually working in laboratories decreased with time. Despite being a small number of individuals mainly devoted to the teaching of physics and often of physics and chemistry, in 1926 the SCP was reorganized after more than a decade of existence, its new name highlighting this dual disciplinary identity. It was then realized that the practice of experimental sciences in general was underdeveloped, and only an alliance between chemists and physicists would make them stronger. It becomes evident that generalizations put forward on the basis of well-known cases dealing with centres of scientific production give way to incorrect conclusions when applied to institutions of the so-called European scientific peripheries such as Portugal.

The SCP is among the first specialized scientific societies created in Portugal, despite the difficulties surrounding professionalization and specialization in Portuguese science. In this context, the question of the society's delayed emergence, when compared with foreign similar institutions, becomes a minor detail in an otherwise complex and fascinating story. The peculiarities of the political, social and scientific context enable one to understand the problems it faced, and the role played by specific individuals becomes an important factor in a difficult equation. The crucial role played by the chemist Ferreira da Silva, who was previously behind the success of the LMCP, the contours of the

---

[62] N/A (1924a–1925), (1924b).
[63] Pinto Basto (1925).
[64] Couceiro da Costa (1925).
[65] Amorim da Costa (1997), 340.

interactions he established with the international community and the communication channels available to him, should be scrutinized in much more detail.

The society's development shows how much it fulfilled functions contrasting with those one has in mind when extrapolating from case studies stemming from similar institutions born in scientific centres. The journal of the SCP increasingly played a hybrid role of a popularization journal, and an outlet for dissemination of analytical techniques and lecture notes, rather than including scientific research. Its contents show how much the functions played by science in a so-called peripheral country differed from those it plays in the so-called centres. The lack of a critical mass of fundamental chemical research which could support the development of areas within applied chemistry made it particularly difficult to sustain a journal on the basis of original research. Within the journal, chemistry assumed its major function first through its applications to basic needs, and afterwards as a discipline to be taught in secondary schools and institutions for higher education. The importance of obituaries and historical notes within the journal shows how important it was for a group of chemists to establish and consolidate by all means a national historical tradition in Portuguese chemistry.

The non-correlation between the emergence of the SCP and a process of consolidation of chemistry as an autonomous discipline carving an identity of its own within the scientific brotherhood shows not only how careful one should be when referring to the notion of chemical community in the local context, but also how much more fruitful it becomes to articulate first the contours of a scientific community in general terms, depicting the unstable balance of power within it, and then to specify the process of its differentiation in scientific areas. One then understands how a chemical society could host a minority group of physicists and become a joint society for a period of almost 50 years. This question is closely dependent on the perceived utility of scientific disciplines in changing political contexts. It becomes evident that in a country which barely underwent the first stages of industrialization, the political context of Liberalism ("Regeneration"), with its rhetoric of progress and immediate applications, entailed a hierarchy of the sciences, in which engineering occupied the first place, medicine/pharmacy and chemistry a distant second, and no role was ascribed to physics. Only in the mid-1950s and 1960s, with the applications of atomic and nuclear physics to the uses of atomic energy, did Portuguese politicians turn to physics and physicists for help.

Thus, the SCP was an association in which practising chemists were a minority with rather eclectic backgrounds, ranging from pharmacy, medicine, natural philosophy, and physics and chemistry. Furthermore, they were far from being specialists, engaging simultaneously in various fields of chemistry, and performed multiple tasks ranging from teaching to administrative and other official functions. The process of emancipation of chemistry from pharmacy, natural history and medicine, which had occurred from the late-eighteenth to the early-nineteenth centuries throughout Europe, began in Portugal one century later. For these reasons, instead of affirming and consolidating an existing and flourishing profession, the society was created within a strategy seeking to gather support among all those who could help in disseminating the science of chemistry, irrespective of their professions and backgrounds. In practice, the society propagated methods and techniques that could be applied to practical purposes associated with the needs of everyday life, and tried to make chemistry sufficiently attractive, popular and potentially useful in order to be perceived as a profession worth pursuing.

Acknowledgements

A word of gratitude is due to Prof. António Manuel Nunes dos Santos for kindly supplying us with relevant documents. The authors wish also to thank the Secretary General Portuguese Society of Chemistry (SPQ), Prof. Fernando Pina, for kindly allowing us to carry out research in the society's library, and Mrs. Cristina Campos for her kind assistance.

Abbreviations

LMCP	*Laboratorio Municipal de Chimica do Porto* (Municipal Laboratory of Chemistry of Oporto)
RCPA	*Revista de Chimica Pura e Applicada* (Journal of Pure and Applied Chemistry)
SCP	*Sociedade Chimica Portugueza* (Portuguese Society of Chemistry)
SPF	*Sociedade Portuguesa de Física* (Portuguese Society of Physics)
SPQ	*Sociedade Portuguesa de Química* (Portuguese Society of Chemistry)
SPQF	*Sociedade Portuguesa de Química e de Física* (Portuguese Society of Chemistry and Physics)

References

Aguiar, A. (1899), O Prof. Ferreira da Silva. Notas Biographicas, *Jornal da Pharmacia Magalhães* **5**, 5–38.

Aguiar, A. (1923), *Discurso pronunciado no Laboratório da Faculdade de Ciências em homenagem ao eminente químico e professor Dr. António Joaquim Ferreira da Silva, em 9 de Dezembro de 1922*, Edição da Revista de Semiótica Laboratorial, Porto.

Aguiar, A. (1924a), Dr. António Joaquim Ferreira da Silva, *Revista de Chimica Pura e Applicada* **16**, 7–11.

Aguiar, A. (1924b), Notas biográficas do Dr. António J. Ferreira da Silva, *Revista de Chimica Pura e Applicada* **16**, 11–33.

Aguiar, A. (1928), Untitled Foreword, *Revista de Chimica Pura e Applicada* **18**, 1–2.

Amaral, I. (2001), *As Escolas de Investigação de Marck Athias e de Kurt Jacobsohn e a Emergência da Bioquímica em Portugal*, Unpublished Ph.D. dissertation, New University of Lisbon, Lisbon.

Amorim da Costa (1997), The Mirror of the Portuguese Chemical Laboratories in the First Decades of the Twentieth Century, *Centaurus* **39**, 332–342.

Bradbury, F. R. and Dutton, B. G. (1972), *Chemical Industry. Social and Economic Aspects*, Butterworths, London.

Bud, R. (1992), The Zymotechnic Roots of Biotechnology, *British Journal for the History of Science* 25, 127–144.

Brock, W. H. (1992), *The Fontana History of Chemistry*, the Fontana Press, London.

Bud, R. (1994), *The Uses of Life. A History of Biotechnology*, Cambridge University Press, Cambridge.

Carneiro, A. (2006), After Mateu Orfila: Adolphe Wurtz and the status of Medical, Organic and Biological Chemistry at the Faculty of Medicine, Paris (1853–1884), in Bertomeu Sanchez, J. R., Belmar, A. G. and Nieto-Galan, A., ed., *Chemistry, Medicine and Crime: Mateu Orfila (1787–1853) and his Times*, Science History Publications, New York.

Carneiro, A. (1992), *The Research School of Chemistry of Adolph Wurtz, Paris, 1853–1884*, Unpublished Ph.D. dissertation, University of Kent, Canterbury.

Carvalho, R. (1953), *Ferreira da Silva, homem de ciência e de pensamento (1853–1923)*, Tipografia da Livraria Simões Lopes, Porto.

Costanzo, G. (1920), Fórmulas para os dióptricos e para os reflectores de revolução, *Revista de Chimica Pura e Applicada* 5, 117–128.

Costanzo, G. (1919 and 1920), Notas da lições de radioactividade dadas no Instituto Superior Técnico de Lisboa, *Revista de Chimica Pura e Applicada* 4, 206–229, 284–310 and 5, 15–35, 220–255.

Couceiro da Costa (1925), Aplicação da Termodinâmica ao estudo da distilação e cristalização fraccionadas, *Revista de Chimica Pura e Applicada* 17, 28–37.

Crosland, M. P. (2003), Research Schools of Chemistry from Lavoisier to Wurtz, *British Journal for the History of Chemistry* 36, 333–336.

Diogo, M. P. (1994), *A Construção de uma Identidade Profissional. A Associação dos Engenheiros Civis Portuguezes, 1869–1937*, Unpublished Ph.D. Dissertation, New University of Lisbon, Lisbon.

Ferreira da Silva, A. J. (1909a), ed., *A Questão do Laboratorio de Chimica Municipal do Porto (1907–1908)*, Imprensa Portugueza, Porto.

Ferreira da Silva, A. J. (1909b), Untitled Foreword, *Revista de Chimica Pura e Applicada* 5, 1–3.

Ferreira da Silva, A. J. (1914), *Ciência e Crenças*, Cruz and Livreiros, ed., Braga.

Ferreira da Silva, A. J., Mastbaum, H. and Cardoso Pereira (1914), Relatorio relativo ao ano de 1913, *Revista de Chimica Pura e Applicada* 10, 1–4.

Fox, R. and Weisz, G., ed. (1980), *The Organization of Science and Technology in France 1808–1914*, Cambridge University Press, Cambridge.

Fruton, J. (1988), The Liebig Research Group – a Reappraisal, *Proceedings of the American Philosophical Society* 132, 1–66.

Geison, G. L. and Holmes, F. L., ed. (1993), Research Schools. Historical Reappraisals, *Osiris* 8, 3–248.

Gieryn, T. F. (1983), Boundary-work and demarcation of science from non-science: strains and interests in professional ideologies of scientists, *American Sociological Review* 48, 781–795.

Herold, B. J. and Carneiro, A. (2005), Portuguese organic chemists in the 19th century. The failure to develop a school in Portugal in spite of international links, in É. Vámos, ed., *Proceedings of 4th International Conference on History of Chemistry, Communication in Chemistry in Europe across Borders and across Generations*, Budapest, vol.1, 25–48.

Homburg, E. (1992), The emergence of research laboratories in the dyestuffs industry, 1870–1900, *The British Journal for the History of Science* **25**, 91–111.

Homburg, E., Travis, A. S. and Harm, G., ed. (1998), *The Chemical Industry in Europe, 1850–1914, Industrial Growth, Pollution and Professionalization*, Kluwer Academic Publishers, Dordrecht.

Jørgensen, G. (1909a), Sur la détermination de quelques acides végétaux organiques, *Revista de Chimica Pura e Applicada* **5**, 193–208, 225–233.

Jørgensen, S. M. (1909b), Páginas para os estudantes dos cursos secundários, *Revista de Chimica Pura e Applicada* **5**, 183–185, 212–217, 268–275.

Leitão, V. (1998), *A Química Inorgânica e Analítica na Escola Politécnica de Lisboa e na Academia Politécnica do Porto no século XIX (ensino e investigação)*, Unpublished M.Sc. dissertation, New University of Lisbon, Lisbon.

Machado, A. R. (1918), Estudo geral do nónio: Sua origem. Teoria. Construção e Usos práticos, *Revista de Chimica Pura e Applicada* **13**, 40–60.

Machado, A. (1920), Aplicação da ponte de Wheatstone ao estudo da marcha da hidrolise da ureia pela urease da soja híspida, *Revista de Chimica Pura e Applicada* **5**, 50–56.

Matos, A. C. *et al.* (2005), *A Electricidade em Portugal. Dos Primórdios à 2ª Guerra Mundial*, Museu de Electricidade/EDP, Lisboa.

Meyer-Thurow, G. (1982), The Industrialization of Invention: a Case Study from the German Chemical Industry, *Isis* **73**, 363–381.

Morrel, J. B. (1972), The chemist breeders: the research schools of Liebig and Thomas Thomson, *Ambix* **19**, 1–46.

N/A (1905), O nosso programa, *Revista de Chimica Pura e Applicada* **1**, 1–3.

N/A (1911), *Revista de Chimica Pura e Applicada* **1**, 1–2.

N/A (1912a), *Revista de Chimica Pura e Applicada* **1**, 1–3.

N/A (1912b), *Estatutos da Sociedade Chimica Portugueza, aprovados em assembleia geral dos sócios fundadores em 28 de Dezembro de 1911*, Typographia a vapor da "Enciclopédia Portugueza", Porto.

N/A (1914), Sociedade Chimica Portuguesa. Relatório da Direcção ao anno de 1913, *Revista de Chimica Pura e Applicada* **10**, 1–25.

N/A (1917a), 2ª Acta. Sessão Administrativa e Scientifica de 30 de Abril de 1915, *Revista de Chimica Pura e Applicada* **12**, 398–400.

N/A (1917b), 2ª Acta. Sessão Scientifica de 18 de Janeiro de 1916, *Revista de Chimica Pura e Applicada* **12**, 416–418.

N/A (1917c), 4ª Acta. Sessão Scientifica de 20 de Abril de 1917, *Revista de Chimica Pura e Applicada* **12**, 403–408.

N/A (1917d), 7ª Acta. Sessão de 8 de Agosto de 1916, *Revista de Chimica Pura e Applicada* **12**, 424–426.

N/A (1917f), 7ª Acta. Sessão Scientifica de 23 de Novembro de 1917, *Revista de Chimica Pura e Applicada* **12**, 412–414.

N/A (1917g), 9ª Acta. Sessão de 19 de Maio de 1917, *Revista de Chimica Pura e Applicada* **12**, 429–430.

N/A (1924–1925), *Revista de Chimica Pura e Applicada* **16**, 7–82, 124–132.

N/A (1924b), Dr. António Joaquim Ferreira da Silva. I- Homenagem da Sociedade Brasileira de Química ao sábio químico português Ferreira da Silva, *Revista de Chimica Pura e Applicada* **16**, 222–229.

Pinto Basto, E. (1925) Retogradação do ácido fosfórico nos adubos compostos, *Revista de Chimica Pura e Applicada*, **17**, 1–15.

Ramos do Ó (1999), *Os Anos de Ferro – o dispositivo cultural durante a política do espírito, 1933–1949*, Editorial Estampa, Lisboa.

Simões, O. (1931), *Elogio histórico do sócio efectivo da Academia das Ciências de Lisboa, Doutor António Joaquim Ferreira da Silva*, Imprensa da Universidade, Coimbra.

CHAPTER 13

RUSSIA: The Formation of the Russian Chemical Society and Its History until 1914

NATHAN M. BROOKS, MASANORI KAJI AND
ELENA ZAITSEVA

The Russian Chemical Society (*Russkoe Khimicheskoe Obshchestvo*, RKhO) was first organized in 1868 by a group of young research-oriented chemists in St. Petersburg. Some of the organizers of the society had been advocating its establishment for nearly ten years and a number of them had attended meetings of foreign chemical societies, upon which they modelled their new society. However, as with many aspects of Russian society and culture, the adoption of a foreign model was moulded to fit the Russian conditions. Even though it appears that at least some of the organizers of the society had an expansive vision of what the society could undertake as its mission, the Russian political reality necessitated a narrow, purely scientific, role for the society for most of its existence. Yet even this circumscribed role was new and unusual for Russia at this time. Although the society was organized by St. Petersburg chemists, the organizers clearly had the goal of a nationwide organization that could knit together the largely isolated community of chemists in Russia through regular meetings, a yearly congress and a journal. The society would continue to function as the only nationwide organization of chemists in Russia until after the 1917 Bolshevik Revolution. This chapter will examine the formation of the Russian Chemical Society in 1868 and very briefly some of its history in the following years.

Creating Networks in Chemistry: The Founding and Early History of Chemical Societies in Europe
Edited by Anita Kildebæk Nielsen and Soňa Štrbáňová
© The Royal Society of Chemistry 2008

13.1   Scientific Institutions in Russia during the Eighteenth and Nineteenth Centuries

Western science came late to Russia. The first major scientific institution in Russia was the Imperial Academy of Sciences, founded in 1725 by Peter the Great, who envisioned the Academy as the wellspring of an entire educational and scientific system.[1] This was not to be, however. The Academy remained the only important scientific institution in Russia for the rest of the century and never adequately fulfilled its instructional intentions as envisioned by Peter the Great. Moreover, while it was necessary that the Academy at first be staffed with foreigners, it did very little to prepare Russians to become academicians. By the end of the century, the Academy had acquired great prestige and several of its members achieved great scientific renown, but it was the merest speck in the great sea of Russia and with no prospects of growing larger.

The situation would gradually begin to change in the nineteenth century with the rise of the universities. While Moscow University had been founded in 1755, it trained few students until well after the turn of the nineteenth century. Influenced by the Enlightenment, Tsar Alexander I (r. 1801–1825) during the first years of the nineteenth century created a complete educational system from primary to university education, supervised by the Ministry of National Enlightenment.[2] One of the main goals of Alexander and other Russian officials was to educate a sufficient number of Russians to fill the rapidly expanding Russian bureaucracy, which included the professors at the higher educational institutions. However, this process would take time. Like the situation with the Academy of Sciences in the early eighteenth century, the universities were forced to hire many foreigners at first. While many of these foreigners were forced out of their positions for various reasons or decided to leave Russia, very few of the Russians who assumed the teaching positions at the universities and other higher educational institutions were competent teachers or scholars. One of the major problems involved with staffing the universities was that, in general, the landowning nobility – along with the clergy, the social group with the best educational preparation – did not aspire to positions at the higher educational institutions. This fact also influenced the prestige of the universities in a society that put a high value on rank, status, awards and other indicators of prestige.

During the second quarter of the nineteenth century, Tsar Nicholas I (r. 1825–1855) and the Russian state pursued a rather paradoxical policy in regards to higher education. On the one hand, Nicholas distrusted the political and social views of students and professors, and he sought to limit the enrolment in higher education to groups he considered most politically reliable (the landowning nobility), while at the same time imposing a rather strict censorship of all publications, both domestic and imported. At the same time,

---

[1] For information about chemistry during the eighteenth century in Russia, see Brooks (1989), chapters 1–2; and the relevant sections of Solov'ev (1985).

[2] The most complete account of the university system during the first half of the nineteenth century can be found in the multi-volume work by Petrov (2002–2003).

however, Nicholas and his educational officials (in particular, S. S. Uvarov, Minister of National Enlightenment from 1833–1849) considerably strengthened the higher educational system by improving the physical facilities and upgrading the training of the professoriate, including sending some students to western European countries for advanced training. In addition, Nicholas placed considerable emphasis on technical education, expanding the existing technical institutions while founding new ones. All of these measures strengthened the education of Russian chemists and helped prepare the groundwork for the development of a scientific community. Nicholas's reign ended particularly badly for education, however. Frightened by the events of 1848, Nicholas acted to curb any political ferment in Russia by prohibiting study abroad and severely limiting the enrolment of students in higher educational institutions, among other measures. Finally, the staggering defeats in the Crimean War (1853–1856) sent a shock wave throughout Russian society. With Nicholas's death towards the end of the war, the possibility for substantial reforms throughout Russian society was open.

13.2 Chemistry in Russia during the First Half of the Nineteenth Century

During the first half of the nineteenth century, chemistry professors in Russia had a "local" orientation.[3] That is, they sought to obtain status and recognition through activities that would be valued by their university colleagues and administration, as well as by notables in the local community. These activities often included serving on university and governmental committees, participating in local agricultural or other societies, performing consultations on various topics for the government, and so forth. Most of the chemistry professors at the universities during the first half of the nineteenth century did conduct at least some research during their careers, usually as part of their advanced degrees. However, the focus of their attention was not on the production of original chemistry research that would be reported in national or international scientific journals. Some chemists did continue to pursue original chemistry research during most of their academic careers, such as F. F. Reuss, G. I. Gess, N. N. Zinin and K. K. Klaus. Reuss taught at Moscow University from 1804 and throughout the 1820s, while Gess held the chair of chemistry at the Academy of Sciences from 1830–1850 and taught at various institutions in St. Petersburg during these years.[4] Both Zinin and Klaus taught at Kazan University in the late 1830s–1840s and pursued their chemistry research in private laboratories.

Chemists in Russia would begin to move away from this local orientation towards a more professionalized attitude beginning in the late 1850s, in part as a result of the impact of the Crimean War. The defeat in the Crimean War shocked many educated Russians, convincing them that Russia was backward

---

[3] For a more detailed discussion, see Brooks (1998).
[4] For information about Reuss and Gess see Zaitseva (2001) and Zaitseva (2000).

and needed to modernize and adopt a wide range of western models. What is more important, however, is that this attitude also permeated the top levels of the government and was shared by Tsar Alexander II (r. 1855–1881). Education was one of the first of many areas to undergo reforms. The ban on foreign travel was lifted soon after the death of Nicholas I and within a few years a large number of young Russians were being sent abroad for advanced training in many different disciplines at state expense. At about the same time, military officials began to modernize military education by incorporating more academic subjects and making the education more rigorous as well. Chemistry received considerable attention in these military reforms, both in the curriculum and the facilities. The reforms for university education took slightly longer to complete and a new University Statute was finally unveiled in 1863 after years of debate. The new statute provided considerable autonomy for the universities, along with substantial increases in both teaching staff and facilities, such as chemistry laboratories.

The impact of these reforms was felt most strongly in St. Petersburg, where a small group of young chemists began to coalesce in the 1850s. This group included D. I. Mendeleev, A. P. Borodin, P. P. Alekseev, A. A. Verigo, N. N. Sokolov, A. N. Engel'gardt, and others. These young Russians had received their education at various higher educational institutions in St. Petersburg, a significant number of them at higher military schools. They shared several beliefs, including the view that the most important activity of a chemist was original laboratory research. Moreover, this research was to be evaluated according to the standards of the international chemical community, which they desired to join. The older chemists in St. Petersburg had a local orientation, whereas the young chemists of the late 1850s–1860s developed an international outlook. The young Russian chemists rejected the older chemists' local orientation and embraced the values of the international chemical community, which valued original chemical research and its publication in specialized chemistry journals.

Another factor that had great influence on these young chemists was the resumption of travel abroad for advanced study. The contact with research-oriented chemists in Western Europe had a profound impact on how they viewed their identity as chemists. They absorbed the belief that the main goal of a true chemist was to perform original chemistry research, which would be of interest to the international chemistry community, and publish the results in a specialized chemistry journal. Of course, Russian students had been sent abroad for advanced training in chemistry throughout the first half of the nineteenth century. However, the milieu the Russian chemistry students returned to following the Crimean War was very different from the earlier years. For one thing, far more students went abroad at this time and many of them returned to St. Petersburg without the prospect of a permanent teaching position. Also, many young chemists made or solidified friendships during their time abroad that they continued after their return to Russia.

Thus, we can see a critical mass of young chemists gathering in St. Petersburg in the late 1850s–early 1860s. In accordance with the young chemists' high value on original chemistry research, they naturally needed access to adequate laboratory facilities as well as an outlet in which to publish the results of their research. Laboratory facilities were still primitive at this time at most higher educational institutions in the city, while access was extremely limited even to graduates of a particular institution. Additionally, no specialized chemistry journal was being published in Russia at this time and Russians would have to look abroad to publish their work in a foreign-language chemistry journal.

Two young chemists, A. N. Engel'gardt (1832–1893) and N. N. Sokolov (1826–1877), tried to remedy this situation by founding a private chemistry laboratory (1857) and publishing a specialized chemistry journal in 1859 (*Khimicheskii zhurnal N. N. Sokolova i A. N. Engel'gardta*, the Chemistry Journal of N. N. Sokolova and A. N. Engel'gardt) in Russian.[5] After less than three years, both of these efforts failed, glaring indications of the weakness of chemistry in Russia at this time. Even though the chemists who used the laboratory paid a modest fee, the expenses of the laboratory proved to be difficult for Engel'gardt (the main source of funds for the laboratory) to keep funding the necessary purchases of equipment and supplies. The situation with the journal was different. As it turned out, the journal was not completely independent and was actually just a separate printing (with some minor changes) of the chemistry articles from another, government-related periodical, The *Gornyi zhurnal* (*Mining Journal*), whose editorial board Sokolov had recently joined. Sokolov massively increased the publication of chemistry articles for the Mining Journal, but mainly on organic chemistry topics which angered the other members of the editorial board.

13.3   Early Attempts to Form a Russian Chemical Society

At about the same time as the private chemistry laboratory closed and the Chemical Journal ended (1860), articles began to appear in the local St. Petersburg Press arguing for the formation of a chemical society in the city. Some of these explicitly pointed to the model of the Chemical Society of Paris and at least one of the authors sent his article from Paris, where he had been attending meetings of the Chemical Society. In addition, we have evidence from letters that various young Russian chemists in St. Petersburg, including L. N. Shishkov, were "bustling here about organizing a chemical society…".[6] Shishkov had recently returned from Paris where he had taken an active role in the affairs of the new Chemical Society there. Some of the calls for the founding of a chemical society in St. Petersburg explicitly argued for the society to have a laboratory and a journal. It seems clear that the young Russian chemists in St. Petersburg felt themselves hampered in the pursuit of their research by the lack of suitable laboratory facilities, since few of them had

---

[5] See the more detailed discussion in Brooks (1995) and Kaji (1993).
[6] Quoted in Mladentsev and Tishchenko (1938), 227–228.

obtained permanent positions at higher educational institutions. A call for a future chemical society to sponsor a laboratory seems only logical with the young chemists' belief in the importance of original chemistry research. Many of them had just returned to St. Petersburg from advanced study abroad and they found the contrasts stark between Russia and Western Europe. Similarly, the desire for a specialized chemistry journal to take the place of the defunct Chemical Journal also seems logical.

However, these efforts to found a specialized chemical society in St. Petersburg at this time were not successful. Why? One reason may be the difference in attitudes between the young research-oriented Russian chemists and their older generation of chemistry professors. These older professors generally held the local orientation described above and many of them held older theoretical ideas about chemistry. In an entry in his diary for 1861, Mendeleev described an attempt by him and Shishkov to persuade academician Iu. F. Fritsche to become the head of a future chemical society: "Shishkov and I expressed our opinions, saying that no one else besides Fritsche could initiate this society ... [because] he alone stands above petty interests ... [Fritsche replied] ... What do you want from me? I do not have the strength to catch up with you ... I fear that if I bring you together in a society I will compromise my work in the future. I respect ... your views, your direction. Your theoretical frame of mind sometimes is incomprehensible to me. I sympathize with the society you desire. I am wholly ready to assist its realization, but I will not become its head – because I am not able to guide it in this direction. I fear the direction for the fact that I am not sure that I will not be compromised."[7] At this time, Fritsche was the only chemistry academician. Evidently, Mendeleev and Shishkov felt that Fritsche's high status was essential in establishing a chemical society. In fact, when the Russian Chemical Society eventually was founded in 1868, its first president was N. N. Zinin, who had been elected academician in chemical technology in 1865. Similarly, Mendeleev and Shishkov believed that Fritsche could draw together all of the chemists in St. Petersburg. Fritsche's refusal indicates a generational gap between older chemists and the emerging cadre of research-oriented chemists. While Fritsche did pursue original research, he was to a great extent cut off from current theoretical ideas in chemistry.

In the years immediately after these events, chemists in St. Petersburg continued to meet informally to discuss their research, but no concrete steps to establish a formal society were taken. There is evidence that some of the young chemists in St. Petersburg preferred informal meetings to a formal organization.[8] In addition, there were distinct theoretical differences in the views of many of the leading young chemists, as well as between the younger and older chemists.

Perhaps the largest impediment for organizing a formal society was the attitude of the state officials. Organizing a voluntary society in Russia was far more difficult at this time than in most countries of Western Europe or the United States. State approval was necessary for any formal society, even if the proposed

---

[7] *Nauchnoe nasledstvo*, **2** (1951), Moscow: Izd-vo Akademii Nauk SSSR, 164.
[8] See Kozlov (1958), 21.

society received no funding from the government. Moreover, since state authorities were extremely concerned about the possibility of political activity in any formal society, they tended to permit only a very few such societies to be formed. Only 25 voluntary societies were formed during the reign of Nicholas I. While the conditions were eased somewhat early in the reign of Alexander II, as part of the general relaxation of controls involved in the Great Reform Era, the restrictions imposed on the organizers of a society were still significant. However, the University Statutes of 1863 contained a provision that allowed the universities to establish general or specific scientific societies, with the right to organize laboratories and museums, to send research teams to differing parts of the empire, to seek private donations for scientific works and to publish scholarly studies.[9]

In spite of these new regulations, serious obstacles remained to confront the organizers of a voluntary society. The University Statutes allowed the universities to establish societies, but this implied that the organizations thus formed would have a local character. Leading government officials were especially concerned about nationwide organizations that could serve as ready springboards for political activity. Local organizations did not arouse such great distrust as nationwide ones. The new regulations divided the responsibilities for voluntary associations and societies among the different ministries, with natural science societies being approved by the Ministry of National Enlightenment. Thus, the attitude of the Minister of National Enlightenment was a crucial factor in forming a society such as a chemical society.

The proponents of a formal chemical society were faced with opposition from the Ministry of National Enlightenment. Even in the years after 1863, the relatively liberal Minister of National Enlightenment Golovnin refused to allow the organization of a nationwide general natural science society that would hold yearly congresses.[10] Additionally, the contents of an anonymous letter to the ministry petitioning for the formation of an official chemical society – and the response of the ministry – suggest the firm opposition by the ministry to a chemical society. The letter argues that since the ministry "dismisses the initiative of local, influential chemists" to form a chemical society, then the ministry should draw up the statutes for such a society.[11] The letter, written in German and signed "a young chemist", contains a petition entitled "Ueber den Zustand der Chemie in St. Petersburg", consciously modelled on Liebig's famous polemic. The letter specifically equates chemistry with laboratory research, but does not call for the chemical society to sponsor a laboratory, a change from previous proposals that perhaps indicates the improving conditions for laboratory research in the city. The author of the letter specifically uses the London and Paris chemical societies as models for the chemical society in St. Petersburg, with no mention of a nationwide organization. The author envisages an expansive role for the chemical society, with the publication of a journal as its most important function. This role would include the society's

[9] Rudakov, Ucheniia uchrezhdeniia, in section Rossiia in *Entsiklopedicheskii slovar' [Brokgauz and Efron]* (St. Petersburg, 1909), 414–416.
[10] For a summary of Golovin's views see Sinel (1973), 29–73.
[11] Central State Historical Archive, St. Petersburg, f. 733, op. 142, d. 237, ll. 1–3 op.

supervision of the teaching of chemistry in secondary schools and control over the publishing of chemistry textbooks. It was also proposed that the society would take an active role in local affairs, including the supervision of public sanitation and hygiene. Finally, the letter sharply criticizes the older generation of chemists for not wanting to organize a chemical society, as well as for their "outdated ideas." The ministry did not respond to this letter, continuing its apparent opposition to the formation of a formal chemical society.

13.4   The Formation of the Russian Chemical Society

During these years, chemists in St. Petersburg continued to meet informally, as no official society was in the offing. The climate for organizing a scientific society soon changed, however, following D. A. Tolstoi's appointment as Minister of National Enlightenment in 1866, a step that was intended to signal a growing conservatism in government policy following the failed assassination attempt against the tsar earlier in that year. While Tolstoi is normally regarded as a reactionary figure in Russian society, he actively supported the formation of a number of scientific organizations and societies that had been opposed by his predecessor. Soon after Tolstoi's appointment, he agreed to allow a nationwide congress of naturalists to meet in St. Petersburg. The arguments Tolstoi used to back his views are striking, as he firmly rejected the previous minister's desire to keep Russia's scientists isolated from each other in fear of political activity and instead called for the state to support the gathering of scientists as a way to promote the growth of science. In addition, he also believed that more financial resources should be given to science in order to allow Russian scientists to catch up with the west and avoid such debacles as the Crimean War.[12] It seems that Tolstoi believed Russian scientists would actually be *less* inclined to participate in political activities if they faced fewer restrictions on their scientific activities.

This congress – the First Congress of Russian Naturalists – took place in late December 1867–early January 1868 and involved hundreds of scientists from all over Russia. This congress was particularly important for chemists as Tolstoi told several of the chemists there that he would allow the formation of a chemical society.[13] This involvement of the minister needs to be emphasized, because later recollections of several chemists implied that the chemists at the congress suddenly realized "that there was sufficient scientific strength necessary for the formation of the chemical society".[14] The drafting of the statutes of the society began shortly after the congress and proceeded smoothly.[15] Mendeleev took the lead in this and was joined by other young chemists in St. Petersburg, while the older chemists (including Zinin, Fritsche,

---

[12] For example see Tolstoi's report to the tsar for 1866: Central State Historical Archive, f. 733, op. 117, d. 53, ll. 1–4 ob.

[13] See the letter from Markovnikov to Butlerov about the congress: Bykov, ed. (1961), 252.

[14] Menshutkin, *ZhRFKhO*, **25** (1894), prilozh., p. 27.

[15] This needs to be stressed because one of the main historians of the society implies that government officials hindered the formation of the society at this time; see Kozlov (1958), 25–26.

Voskresenskii and others) were conspicuously absent.[16] The statutes contained no unusual provisions and simply stated that the society's function would include holding monthly meetings in St. Petersburg, holding a yearly conference that would bring all society members together in one place, the publication of a specialist chemistry journal and the sponsoring of public lectures. An individual could become a member by having three current members support his membership and he could be anyone "who is occupied with the teaching of chemistry or who presents works on chemical topics".[17]

The initial membership in the society reflected its origins as an informal gathering of chemists in St. Petersburg. Almost all of the initial members were from St. Petersburg and only gradually over the course of the first decade did the proportion of members from St. Petersburg drop to about half. The prevalence of members from St. Petersburg reflected the relatively large number of educational institutions in the capital. In addition to the university, St. Petersburg had a technological institute, various institutes affiliated with different ministries (such as the Medical-Surgical Academy and the Forestry Institute), as well as a myriad of secondary schools that provided employment for chemists. It is interesting to note, however, that a considerable number of the initial membership of the society was from St. Petersburg University, but not involved directly with chemistry. This likely was a consequence of the society being established at St. Petersburg University under a provision of the 1863 University Statues. Thus, various professors of natural science at St. Petersburg University were initial members of the Russian Chemical Society, but they dropped their membership soon after the society gained relatively solid footing.[18]

Thus, the Russian Chemical Society was formed by a group of young Russian chemists who wanted to emulate the conditions for chemistry that they had experienced during their travels in Western Europe. An important aspect of this was the creation of a society that could organize regular meetings and sponsor the publication of a research journal. All of these were important components of the professionalization of chemistry in Russia that began to take place in the late 1850s and 1860s. There was a great diversity in theoretical views among the Russian chemists involved in the Russian Chemical Society. What seems to have been most important for them was the fundamental belief in the goal of chemistry as a laboratory-based research science.

13.4.1   Structure and Membership

The Russian Chemical Society (RKhO) remained the only nationwide chemical society in the Russian Empire from its organization in 1868 until after the 1917 Bolshevik Revolution. The society remained located in St. Petersburg at the

---

[16] Mendeleev (1954), *Sochineniia* **25**, 707.

[17] *ZhRKhO* (1869), **1**, pp. 2–3.

[18] These professors apparently wanted the Russian Chemical Society to be created, but once it was established, they did not feel it necessary to maintain their membership. What this implies is that the society was seen as a specialized scientific society for chemists, not a broader scientific society that would encompass various academic disciplines.

290 *Chapter 13*

university during all these years and chemistry professors at the university regularly served as important officers of the society. For example, Nikolai A. Menshutkin served as the editor of the Journal of the Russian Chemical Society (*Zhurnal russkogo khimicheskogo obshchestva*) for its first 31 years, from 1869 to 1900.[19] Chemists living outside of St. Petersburg typically became members of local natural history societies, many of which had separate chemistry sections, but the RKhO remained their disciplinary focus.

However, the structure of the RKhO changed in its early years to a form that was quite unusual among other European chemical societies. In 1872, F. F. Petrushevskii, a professor of physics at St. Petersburg University, and Dmitrii I. Mendeleev, professor of chemistry at St. Petersburg University, organized the Russian Physical Society. At this time Mendeleev was intensely involved with research and precise measurements of gases, which perhaps stimulated his interest in helping found the physical society. Like the RKhO, the Russian Physical Society (*Russkoe Fizicheskoe Obshestvo*; RFO) was based at St. Petersburg University and also drew most of its membership from the local St. Petersburg area for many years. In addition, the statutes of the two societies were quite similar. The RKhO assisted the new Physical Society at first by publishing physics articles in its journal and with other financial measures. In 1876, Mendeleev proposed the unification of these two societies into one society with two sections. Mendeleev's arguments for this unification are interesting. While he noted the overlapping scientific boundaries of chemistry and physics, Mendeleev believed that a unified society would provide institutional as well as scientific benefits by increasing its influence within Russian society. He emphasized that the RKhO and the RFO were two of the very few societies in Russia that did not receive subsidies from the state, but were weak and could benefit by joining together into one society. Also, Mendeleev believed there would be scientific gains from increasing contacts between chemists and physicists at the general meetings of the joint society.[20] The RKhO elected a commission to investigate the idea of combining the two societies, which then drew up a set of statutes for the combined society based on consultation with the RFO. The draft statutes were approved by the RKhO in September 1876, but their approval first by St. Petersburg University and then by the Ministry of National Enlightenment took longer, although no major changes in the statutes were required. The Russian Physical-Chemical Society (RFKhO), with individual chemistry and physics sections, became official in 1878 (Figure 13.1).[21]

The two sections operated independently for the most part, while the combined society alternated presidents from the two sections. The main joint activity of the society was a combined meeting of the two sections held every year. However, despite Mendeleev's hopes, there is little evidence that the combination of the two societies aided either their scientific or their social activities. Also, not all chemists were satisfied with the combined societies over

---

[19] Menshutkin (1908), *Zhizn' i deiatel'nost' Nikolaiaia Aleksandrovicha Menshutkina*, St. Petersburg.
[20] This manuscript is quoted in Kozlov (1958), 35–37.
[21] *ZhRFKhO, ch. kh* (1879), **11**, iv-v. For the years after 1878, we will use the abbreviation RKhO to denote the chemistry section of the RFKhO, unless otherwise indicated.

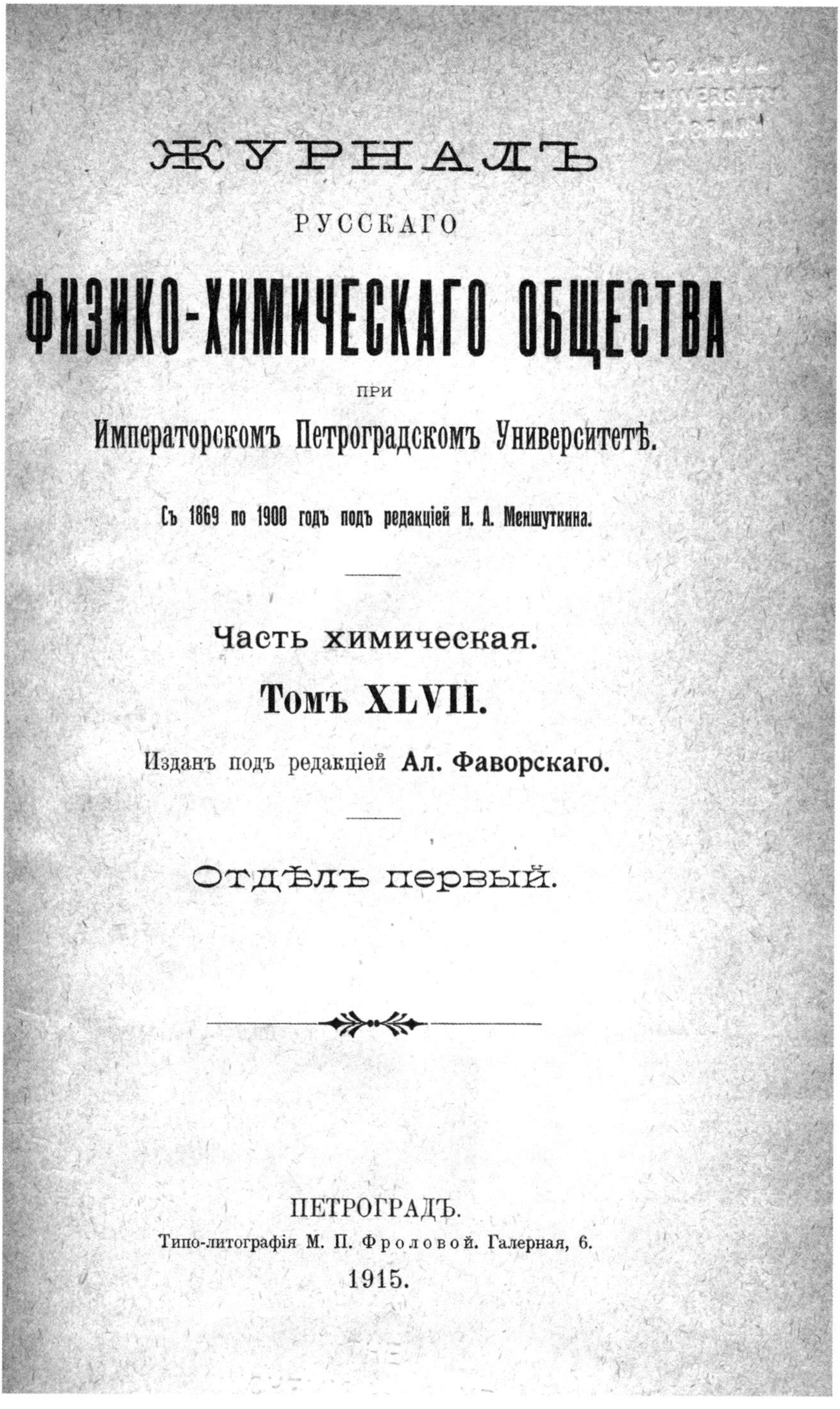

ЖУРНАЛЪ

РУССКАГО

ФИЗИКО-ХИМИЧЕСКАГО ОБЩЕСТВА

ПРИ

Императорскомъ Петроградскомъ Университетѣ.

Съ 1869 по 1900 годъ подъ редакціей Н. А. Меншуткина.

Часть химическая.
Томъ XLVII.

Изданъ подъ редакціей **Ал. Фаворскаго.**

ОТДѢЛЪ ПЕРВЫЙ.

ПЕТРОГРАДЪ.

Типо-литографія М. П. Фроловой. Галерная, 6.

1915.

Figure 13.1 Title page of the 47th volume, 1915 of *Russkoe fiziko-khimicheskoe obshchestvo* (Russian Physical-Chemical Society).

Table 13.1 Membership of the Russian Chemical Society by city.

Year	Total	St. P.	Mos.	Kazan.	Kiev	Kharkov	Warsaw	Others
1868	47[a]	39 (83%)	3	0	2	1	0	2
1869	60	42 (70%)	4	7	2	2	1	2
1872	80	48 (60%)	8	7	4	2	2	9
1877	119	56 (47%)	10	9	10	3	7	24
1893	244	86 (35%)	32	8	11	9	8	97
1912	387	161 (42%)	53	9	14	13	15	122

Source: calculated from information contained in the Journal of the Russian Chemical Society.
[a] V. V. Markovnikov believed that the stated membership for the society at its founding did not accurately account for all of the membership. He believed that the initial membership was about 60, instead of 48 as listed in the society's journal. If that was the case, however, this means that the society grew in membership very little during its first year. *Zhurnal Russkogo Fiziko-khimichesko-go Obshchestva, chast' khimicheskaia. XXV let RKhO (1868–1893)* **25** (1893), otd. 2, 64.

the years. In the early years of the twentieth century, there was even a commission established to discuss splitting the RFKhO into separate physics and chemistry societies, but this commission's proposal to split the society into two did not gain sufficient votes to pass at the general meeting of the society.[22]

The membership of the RKhO increased gradually and regularly from the society's founding until about 1900, while the membership fluctuated in the turbulent years between 1900 and 1914. At its founding (1868), the society had 35 members and the membership grew to 131 in 1880, to 232 in 1890 and 326 in 1900 (Table 13.1). During the first decade of the twentieth century, the membership varied, but usually was in the range of 350–380. This instability perhaps was a result of the disruptions caused by the 1905–1907 revolution, as well as the political uncertainty leading up to and after these years. After 1910, the membership resumed its steady increase, with a surprising jump from 409 members in 1913 to 512 members in 1914. No note was made about this sharp increase in membership in the society's minutes published in the Journal of the Russian Chemical Society for 1914 and it is difficult to ascertain the reason for this dramatic jump in membership. Since the membership records published in the journal relate to the membership on 1st January of each year, the 1914 figure would not have been swelled by dislocations related to the First World War.

Who became members of the RKhO? We can draw some tentative conclusion based on the membership lists printed in each volume of the society. However, only for some years do the lists contain information about the occupation of the members and where they lived, and even then not every member's occupation is identified. Still, these lists can provide us with valuable information about individuals that is often very difficult to find otherwise. In the original statutes (1868), individuals could become members of the RKhO if they were teachers of chemistry or if they had published a work on chemistry or had written a manuscript work on chemistry.[23] Apparently, these requirements

---

[22] *ZhRFKhO, ch. kh.* (1907), **39**, II.
[23] Ustav Russkogo Khimicheskogo Obshchestva, *ZhRKhO* (1869) **1**, 2.

were deemed too restrictive and in 1872 the statutes were changed to allow an individual to become a member of the society if three current members of the society supported the candidate and that the candidate was involved with chemistry in some way or was a teacher of chemistry or a related field. "These changes in the statutes give a greater possibility than previously for a wider circle of individuals to take part in the activities of the society."[24] There were no exclusions to membership if the prospective member met the rather broad requirements of the statutes. Thus, for example, pharmacists were able to join the society, although very few did join over the years of the society's existence up to 1917.

It is clear from the initial requirement for membership that the society's founders envisioned most of its members would be chemists teaching at academic institutions. This indeed turned out to be true, as approximately 80–90% of members for all of the society's history up to 1914 were faculty members or advanced students at higher educational institutions, including both universities and technical institutes. However, relatively few teachers at primary or secondary institutions became members. Perhaps most striking, only about 5–10% of the society's members were engaged in some type of chemical industry.[25] This is surprising, since other chemical societies, the German Chemical Society for example, had large percentages of members employed in the chemical industry. The reasons for the small number of RKhO members being employed in the chemical industry are difficult to ascertain. On the one hand, much of the chemical industry in Russia was dominated by foreign owners, who probably employed non-Russians in most technical positions. In general, these foreigners were not integrated into the social and cultural life of Russia; thus, they often did not join the RKhO. On the other hand, Russian-owned chemical industry concerns typically did not utilize very high technology in their operations. This was noted by V. V. Markovnikov, chemistry professor at Moscow University, in his remarks on the 25th anniversary of the founding of the RKhO.[26] Thus, one would not expect these simple chemical industry operations to employ very many skilled chemists who would then be expected to join the RKhO. However, several prominent industrialists and entrepreneurs who were interested in chemistry did become members of the society, including E. F. Bremme, G. A. Krestovnikov, I. A. Kumberg, A. V. Pel', S. I. Prokhorov, and others. Furthermore, it is likely that most industrialists preferred to join the societies specifically designed for industrialists and entrepreneurs that were founded during the second half of the nineteenth and early twentieth centuries (including the Society for the Encouragement of Industry and Trade, the Society for the Assistance in the Development of Chemical Industry and Trade and the St. Petersburg Society for the Assistance, Improvement, and Development of Factory Industry, among others).

---

[24] Otchet o deiatel'nosti Russkogo Khimicheskogo Obshchestva v 1872 godu (1873), *ZhRKhO* **5**, 3.
[25] Kozlov (1958), 38.
[26] Markovnikov (1955), 688.

In addition to chemists in academic institutions and the chemical industry, toward the turn of the century, several governmental organizations established chemistry laboratories. The chemists at these institutions were highly skilled and typically joined the RKhO. An example of these types of institutions was the Central Chemistry Laboratory of the Ministry of Finance, one of whose heads was N. D. Zelinskii, who joined this institution following his resignation from Moscow University in 1911 in protest over actions of the Ministry of National Enlightenment. Chemists at other institutions also joined the RKhO, including, for example from the 1912 membership list, chemists employed at the Vladimir Provincial Tax Administration, the Scientific-Technical Laboratory of the Naval Administration, the Chemistry Laboratory at the Port of Sevastopol and the Imperial Nikitinskii Garden in Yalta, among others.

One interesting aspect of membership in the RKhO is the number of women who became members. Although not large in number, women became members of the RKhO from the very earliest years of the society. In the 1870s, about three women were members in any given year, while this number increased to six in the 1880s and to eight to ten by the 1890s. At first, these women members had received their educations abroad but, with the rise of advanced separate courses for women (known as "Bestuzhev courses"), more women were able to obtain their education inside Russia as well as begin to teach in these settings. Several women played important roles in the RKhO by preparing abstracts of foreign chemistry articles for the journal of the RKhO, as well as by serving as correspondents of the society with foreign chemistry societies (including Iu. V. Lermontova and O. A. Davydova).

Several different types of memberships were established from the early years of the society. All members were categorized as either "local" (meaning that they lived in St. Petersburg) or "outside the city" (meaning that they lived outside of St. Petersburg), although each type of member paid the same 10 rubles in dues.[27] While we saw that chemists from St. Petersburg originally established the society, residents of this city remained dominant throughout the history of the society up to 1917. In fact, as the society grew, it established separate "councils" from each group to deal with matters of the society that came up between the yearly meetings, undoubtedly to give the members from outside of St. Petersburg more influence. The society also established a category of "permanent" (*postoiannye*) members. These could be members who had given the society at least 100 rubles in a lump sum or who had performed some special service for the society. For example, in 1871, A. N. Engel'gardt and P. A. Lachinov were made permanent members of the society after they donated a large collection of chemical preparations to the society.[28] Furthermore, the society designated a very few members as being "honoured" (or "honorary", *pochetnye*) members of the society for especially great scientific achievements. In the years before 1914, four chemists were honoured with this title: N. N. Zinin (in 1879), Mendeleev (in 1880),

---

[27] During these years, a laboratory assistant typically earned a salary of 600–800 rubles each year, while a university professor earned 2500–3500 rubles each year.

[28] *ZhRKhO* (1872) **4**, 3.

A. M. Butlerov (in 1882) and N. N. Beketov (in 1903). In addition to honoured members from Russia, a few prominent foreign chemists were given this title as well, such as Svante Arrhenius in 1909. Finally, the society continued to print both Butlerov's and Mendeleev's names in the membership lists after their deaths (1886 and 1907, respectively).

13.4.2 Finances

The finances of the RKhO remained in a precarious position during all the society's existence up to 1917. For the first 25 years of the society, the only funds for operating the RKhO came from dues, income from the sale of the journal and a few bequests. For example, the income from dues was 924 rubles in 1877, with a total income of 2189 rubles. In 1892, the amount from dues reached 1615 rubles and the society's total operating budget for the year was 5075 rubles, which included 1539 rubles left over from the previous year. The only additional income for the society came from subsidies given in a few emergency situations, such as in 1871 when the council (*Sovet*) of St. Petersburg gave the society 300 rubles to help cover its financial needs. In addition, St. Petersburg University also gave 500 rubles to the society in 1878 to cover the society's deficit and the university continued this subsidy every year after, increasing the amount to 1000 rubles in 1906. In addition to the subsidies given to the society by St. Petersburg University, several other institutions gave subsidies to the society after 1893. For example, in 1914, the Ministry of National Enlightenment gave the society 1000 rubles, Riga Polytechnical Institute gave 300 rubles, St. Petersburg Technological Institute gave 100 rubles, Mikhail Artillery Academy gave 98 rubles and Kazan University gave 44 rubles and 75 kopeks.[29] Numerous members donated money to the society and named the society as a beneficiary after their deaths. The largest sums were given by G. G. Gustavson, A. M. Zaitsev, I. F. Basilevskii, L. N. Shishkov, and others. Upon his death in 1907, for example, N. K. Iatsukovich, a professor of technical chemistry at Kharkov University, gave the RKhO "All of my financial capital ... for the publication of books on pure and applied physical and chemical subjects, as well as to assist young students from St. Petersburg University who are sent abroad for advanced study in these same subjects ...".[30]

However, even these subsidies and the income from bequests did not always cover the immediate needs of the society and members would need to take quick action to assist the society. For example, on 26th August 1898, at a dinner for chemists attending the tenth Congress of Russian Naturalists, in Kiev, the chemists discussed the financial difficulties of the RKhO. N. D. Zelinskii proposed taking up a collection for the society and 600 rubles were collected. S. N. Reformatskii sent a list of the donors to N. A. Menshutkin, the editor of the society's journal.[31]

---

[29] *ZhRFKhO, ch. kh.* (1914) **46**, v.
[30] Quoted in Kozlov (1958), 58.
[31] Kozlov (1958), 58.

These financial pressures are important to consider because they limited the range of action of the society. Even though the society ostensibly was an independent organization, in reality it could only function with the approval of St. Petersburg University, as well as that of the national government. Thus, for example, during the political upheavals of 1905–1907, when numerous other organizations became politically active, the RKhO did not join in overt political demonstrations, perhaps because of its precarious financial situation.

13.4.3 Publications

During the entire time of the society's existence from its founding in 1868 through to 1917, the main activities of the RKhO remained the publication of its journal and the holding of regular meetings at which the members would present and discuss their research activities. For those members living outside of St. Petersburg, the journal was the overriding reason for them to join the society, especially during the first years of its existence. For example, in 1869, V. V. Markovnikov, professor of chemistry at Kazan University at that time, wrote to his former mentor A. M. Butlerov, who had just moved to St. Petersburg University, about his (Markovnikov's) attempts to recruit members for the newly founded RKhO. "I have recruited six members for the Russian Chemical Society ... Zaitsev, Glinskii, and even Chugunov have promised to join ... But I have had to promise them that the society intends to begin the publication of the journal within the next year ..."[32] The RKhO did indeed start to publish the journal in 1869, but the first issues were rather thin. The first volume contained 38 research articles totalling 272 pages. From this modest beginning, the journal regularly expanded in the numbers of research articles, as well as in the number of pages in each volume (see Table 13.2). The numbers of articles published each year gradually increased during the first three decades of the society from an average of about 49 each year during 1870–1879 to an average of about 62 each year during 1890–1899. However, the number of papers published sharply increased during the following decade, rising to about 107 per year. The numbers of articles published during the troubled decade 1910–1919 remained high with an average of 119 published each year. Undoubtedly, the pressures of the First World War and the Russian Revolution of 1917 and the Civil War (1918–1920) complicated the publication of the journal in various ways, including the lack of paper for the journal. While the absolute number of articles increased over the society's existence from its founding until 1917, the average numbers of pages in each article expanded at an even steeper rate. During the society's first decade (1870–1879), the average number of pages in a research article was about 7.66. This figure rose to 9.47 during 1880–1889, but sharply expanded to 12.28 in 1890–1899 and to 12.64 in 1900–1909.

In addition to the original research articles, the journal in 1873 also began publishing in a second section abstracts of foreign chemistry publications and

---

[32] Quoted in Plate (1953), 78.

Table 13.2   Original research articles published in the journal of the RCS.

Volume	Year	No. of original research articles:total for 10 yrs.	Number of pages	
			Total p. for 10 yr.	Ave. pages/art.
1	1869	38	272	7.16
2–11	1870–1879	493	3776	7.66
12–21	1880–1889	616	5832	9.47
22–31	1890–1899	625	7675	12.28
32–41	1900–1909	1066	13469	12.64
42–51	1910–1919	1190	15510	13.03

Source: Kozlov (1958), 518.
Note: the totals for Vol. 1 refer only to that volume. The other listings for totals refer to the indicated decade.

other items of interest to Russian chemists, such as the proceedings of various scientific meetings. This second section expanded dramatically, even eclipsing in size the number of pages of the first section (original research articles). These abstracts of foreign chemistry publications were especially important to young Russian chemists, who often were not as fluent in foreign languages (particularly German) as the older generation of chemists. The second section of the journal also contained the protocols of chemistry papers given at other regional and national scientific societies in Russia, such as the Society of Admirers of Natural Science, Anthropology and Ethnography (*Obshchestvo liubitelei estestvoznaniia, antropologii i etnografii*). Finally, the second section occasionally published reports about significant chemistry work being done abroad, such as a description of Sabatier and Senderens' research on catalytic hydrogenation published in 1905 by N. N. Sokovkin.[33]

Original research articles in organic chemistry clearly dominated the pages of the journal for its entire existence up to 1917. During the first decade, articles in organic chemistry formed almost 80% of the journal's contents, while this percentage dropped to an average of about 55 in the three subsequent decades (see Table 13.3). The second most popular category for research articles was inorganic and analytical chemistry, which grew from about 9% in the early years of the journal to 21% during the decade 1900–1909, but slipped to 14% in the following decade. The number of publications in colloid and physical chemistry varied over these years, moving from 4% in the early years of the journal to 13% in the following decade, but slipping to 8% during 1880–1889. However, publications in this category grew strongly during the next two decades, comprising 16% of all research articles during the decade 1900–1909 and 27% in 1910–1919. Finally, it is interesting to note that only relatively few articles on subjects of applied chemistry appeared in the pages of the journal. For most of its existence, only about 7% of its articles related to applied chemistry topics. The exceptions to this were the two decades 1880–1889 and

---

[33] *ZhRFKhO, ch. kh.* (1905) **37**, otd. 2, 189–207.

Table 13.3 Subject distribution of articles published in the journal of the RCS.

Year(s)	Total #	Organic	Inorganic/anal.	Colloid/physical	Applied	Obit./other
1869	38	82%	10%	7%	2%	0%
1870–1879	493	77%	9%	4%	7%	3%
1880–1889	616	55%	14%	13%	15%	3%
1890–1899	625	58%	20%	8%	12%	2%
1900–1909	1066	54%	21%	16%	7%	2%
1910–1919	1190	50%	14%	27%	7%	2%

Total # = total number of articles published each year.
Organic = articles in organic chemistry; Inorganic/anal. = articles in inorganic and analytical chemistry; Colloid/physical = articles in colloid and physical chemistry; Applied = articles in applied chemistry; Obit./other = obituaries and articles on the history of chemistry. Percentages may not equal 100 due to rounding.
Source: Calculated from materials in Kozlov (1958), 521, 540, and *Zhurnal Russkogo Fiziko-Khimicheskogo Obshchestva*.

1890–1899, when 15% and 12% of all articles were on applied topics. It is likely that this increase in applied articles reflected the prolific publications of V. V. Markovnikov and his research school which undertook a sustained investigation of the chemical composition of petroleum from Russian deposits during these years.

13.4.4   Meetings and Other Functions

In addition to publishing original research articles in its journal, the other important activity of the RKhO was holding regular meetings. Typically, nine or ten meetings were held in St. Petersburg each year, with no meetings held during the summer months. The meetings were held in the chemistry auditorium at St. Petersburg University, again indicating the dominance of St. Petersburg University in the society's functions. If a larger meeting space was needed, the sessions were moved to the chemistry auditoriums at the Technological Institute or to the Forestry Institute, both also in St. Petersburg. For most meetings, however, about 35–50 members attended. The number of papers read and discussed at these meetings increased steadily from about 8 during the decade 1870–1879 to about 15 in 1900–1909. The disruptions of the First World War, the Russian Revolution of 1917 and the subsequent Civil War greatly impacted the activities of the RKhO. During that decade, only 79 meetings were held, averaging only about 8 papers presented at each session (See Table 13.4). The discussions of the papers would often last well past the end of the meeting and continue at a nearby restaurant, as recalled by N. A. Menshutkin in 1893.[34] Frequently the debates would be spread over the course of several meetings, despite the fact that the meetings typically were quite lengthy. Sometimes, the discussion could become very heated, especially when

---

[34] *ZhRFKhO, ch. kh. XXV let RKhO* (1893), **25**, otd. 2, 25.

Table 13.4   Number of meetings of the RCS and number of communications presented.

Year	Number of meetings	Number of communications
1869	9	84
1870–1879	90	736
1880–1889	88	804
1890–1899	95	1104
1900–1909	98	1459
1910–1919	79	614

Source: Kozlov (1958), 63.

members of the society held quite different theoretical views on an issue. In these cases, the RKhO would organize special sessions to debate and test the differing viewpoints. For example, during the 1880s, several chemists, including D. I. Mendeleev and V. F. Alekseev, sparred over their varying concepts of the theory of solutions. The RKhO established a special commission to review Alekseev's data.[35] While many of the original founders of the RKhO advocated an expansive role for the society, the RKhO primarily remained focused on disciplinary issues and did not consider other issues, such as the regulation of chemistry textbooks for schools. Undoubtedly, even if the members had desired to undertake such activities, government authorities would have been reluctant to allow such intervention by outside interests.

However, the RKhO did undertake a variety of other tasks, including some practical and applied research. For example, various individuals in Russia sent specimens from mineral deposits to the RKhO for analysis. Also, in 1873, the Russian Geographical Society asked the RKhO to undertake an analysis of water from the Aral Sea. In addition, various governmental institutions requested that the RKhO analyse various substances. Perhaps the most important of these applied activities occurred during the First World War. In 1915, with the cooperation of the military authorities, the RKhO established the War-Chemical Committee (*Voenno-khimicheskii komitet*) to assist in the war effort. Various military organizations sent tasks for research and analysis to the War-Chemical Committee, which would then distribute these tasks among the members of the society. The War-Chemical Committee undertook a large variety of tasks, including research on explosives, the production of artillery shells, the production of poison gases, the production of medical and pharmaceutical supplies, the preparation of food products for the army and many others.[36]

The RKhO did act to influence its disciplinary interests by sponsoring a series of prizes and awards given to Russian chemists. The first such prize was established in 1880 at the suggestion of A. M. Butlerov, who was at

---

[35] *ZhRFKhO, ch. kh.* (1883), **15**, 614; (1884), **16**, 87.
[36] *Trudy voenno-khimicheskogo komiteta* (1918), vyp. 2, 8–10.

that time the president of the RKhO. The prize was named for N. N. Zinin and A. A. Voskresenskii, influential chemists both of whom had recently died. The prize regulations stated in part: "From a percentage of the capital collected by friends and those who respect the scientific work and achievements of N. Zinin and A. Voskresenskii, the Chemistry Section of the Russian Physical-Chemical Society establishes for the first time a prize in honour of N. N. Zinin and A. A. Voskresenskii for the encouragement of the further development of chemistry in Russia." The prize was to be awarded for "various achievements and research having a general scientific interest, preferably experimental work in pure chemistry, as well as investigations relating to applied science." An important feature of this prize was that it could only be awarded for work done in Russia and published in the Russian language.[37] The clear intention of these provisions was not only to reward research that had been published in the society's journal, but also to attract manuscripts to it for publication. A continuing problem for the journal was that many Russian chemists preferred to publish their research in foreign journals that would gain more exposure within the international community of chemists than would research published in the Russian language. Money was raised for this prize in various ways, including bequests and public lectures. For example, P. P. Alekseev, chemistry professor in Kiev, presented in 1885 a public lecture "On the combustibility of non-flammable substances and their significance." Alekseev paid all of the costs for advertising and other administrative costs for this lecture from his own resources so that all of the money raised could be devoted to the prize.[38] The capital for this prize was built up over the course of 20 years before the first prize was awarded in 1900. The prizes were awarded every three years until the outbreak of the First World War. The recipients of these first five prizes worked in various branches of chemistry, including organometallic chemistry and physical chemistry.

The society also established other prizes. Later in 1880, the RKhO organized another prize, this one to honour D. I. Mendeleev after his rejection for full membership of the Imperial Academy of Sciences in St. Petersburg. This rejection set off a firestorm of protest throughout educated Russian society and the action of the RKhO was just one of many institutions and organizations that honoured Mendeleev in some way.[39] Mendeleev asked that the prize only be awarded after his death. By that time (1907), over 14 000 rubles had been collected for the prize and it was decided to award two categories of prizes (a major and a minor Mendeleev prize). Again, the requirements for the prize included that the work be performed inside Russia and published in the Russian language.[40] In addition to the Mendeleev prizes, prizes were established in the names of N. N. Sokolov (1880), A. M. Butlerov (1887),

---

[37] Kozlov (1958), 483–484.
[38] Kozlov (1958), 484.
[39] For more details about Mendeleev and the Academy of Sciences, see Dmitriev (2002); Gordin (2004), Chapter 5.
[40] Kozlov (1958), 488.

L. N. Shishkov (1891), N. N. Beketov (1903) and M. G. Kucherov (1913). All of these prizes served not only to honour the memories of prominent Russian chemists, but also to stimulate the production of new chemistry research in Russia and its publication inside Russia.

Another function of the RKhO, but that was mainly of value to those members living in St. Petersburg, was its library. From its very founding, the RKhO built up and maintained a library of chemistry books, journals, pamphlets and dissertations – both Russian and foreign – for use by its members. By 1889, the library contained over 2076 periodicals and over 700 books. The society exchanged copies of its journal with a wide variety of foreign chemistry societies and other organizations both inside the Russian Empire and abroad. The society worked to fill in any gaps in its collection, so that by the early years of the twentieth century, the library contained an impressive collection of chemical literature. For example, by 1903, the library contained complete or nearly complete runs of 47 different scientific periodicals in Russian and 70 in foreign languages. These included *Annales de Chimie et de Physique, Comptus Rendus, Gazetta Chimica Italiana, Proceedings of the Chemical Society of London, Berichte der deutschen chemischen Gesellschaft, Journal of the Tokyo Chemical Society*, and many others.[41] The library for many years was housed in Mendeleev's personal apartment at St. Petersburg University, and the library remained at the university after Mendeleev's departure in 1890. As the library grew larger, the society employed a librarian to organize the collection and to supervise the use of the materials. Typically, the librarian was a laboratory assistant at St. Petersburg University.

It is difficult to determine the extent of contacts between the RKhO and other foreign and domestic chemical and scientific societies. Unfortunately, the records of the RKhO have not been preserved, so information about these contacts would need to be found in the archives of other societies. Also, the printed protocols of the RKhO do not contain many details about such contacts. It is clear, however, that the RKhO did maintain regular contact with a variety of societies and organizations, if only to exchange copies of journals. Similarly, a few foreign chemical societies maintained correspondents in Russia to report on the activities of the RKhO. For example, in 1869 V. F. Richter sent a report to the German Chemical Society (*Deutsche Chemische Gesellschaft*) about Mendeleev's first announcement of the Periodic System of the Chemical Elements and in the following years he continued to inform the German society about Mendeleev's subsequent work on the Periodic System. After 1874, the RKhO regularly published the protocols of the Paris Chemical Society (*Société chimique de Paris*). In addition, in 1911 when the *International Association of Chemical Societies* was established, the RKhO was one of the main founding members. This association began to discuss important issues of interest to chemists, such as the

---

[41] *ZhRFKhO, ch. kh.* (1903), **35**, VII–IX.

reform of nomenclature, until the First World War interrupted its activities.[42] Concerning the relations of the RKhO with other societies within Russia, it is possible that since many of the RKhO chemists were also members of other societies, these chemists did not seek to establish formal links between the RKhO and these other societies. It would be interesting to determine if this situation also applied to the relations between other societies in Russia at this time. However, the RKhO regularly printed in its journal the protocols of various Russian natural science societies, such as the chemistry section of the Moscow Society for the Admirers of Natural Science, Anthropology and Ethnography.

13.5 Conclusions

The RKhO was an organization of chemists founded in 1868 to promote the disciplinary needs of Russian chemists. The Russian chemists explicitly used foreign chemical societies – especially the Parisian Chemical Society – as models for establishing their own society. Expanding from its base in St. Petersburg, the RKhO eventually encompassed nearly all active chemists in Russia and had members in all of the main cities in Russia. The society had an active program of research publications and scientific meetings, but usually did not venture into areas that might arouse the suspicion of the sensitive government authorities. Thus, although some of the founders of the society held an expansive view of what functions the society could perform (for example, supervising the teaching of chemistry and taking an activist role in local affairs such as public health matters), the reality in Russia was that only a narrowly disciplinary and academic society would be allowed to function in most cases. However, the society did attempt to expand its focus into technical matters at various times, for example by discussing the problems of municipal lighting and the possible introduction of the metric system. Perhaps the most important occasion for providing technical assistance was during the First World War when the society acted to mobilize and coordinate the activities of chemists to assist in the war effort. The activities of the society during these years gained it considerable prestige and authority in industrial and political spheres, and the society would continue to play an important role in the scientific and economic life of Russia after the revolutions of 1917.

References

Brooks, Nathan M. (1989), *The Formation of a Community of Chemists in Russia, 1700–1870*, Ph.D. dissertation, Columbia University.
Brooks, Nathan M. (1995), Russian chemistry in the 1850s: a failed attempt at institutionalization, *Annals of Science* **52**, 577–589.

---

[42] Strekopytov (2002).

Brooks, Nathan M. (1998), The evolution of chemistry in Russia during the eighteenth and nineteenth centuries, in Knight, David and Kragh, Helge, ed., *The Making of the Chemist. The Social History of Chemistry in Europe, 1789–1914*, Cambridge University Press, Cambridge, 163–176.

Bykov, G. V., ed. (1961), *Pis'ma russkikh khimikov k A. M. Butlerovu. Nauchnoe nasledstvo*, vol. 4, Izd-vo Akademii Nauk SSSR, Moscow.

Dmitriev, I. S. (2002), Skuchnaia istoriia (o neizbranii D. I. Mendeleeva v Imperatorskuiu akademiiu nauk v 1880 g.), *Voprosy istorii estestvoznaniia i tekhniki*, no. 2, 231–280.

Gordin, Michael D. (2004), *A Well-Ordered Thing: Dmitrii Mendeleev and the Shadow of the Periodic Table*, Basic Books, New York.

Kaji, Masanori (1993), 19-seki Roshia ni-okeru kagakusha-shudan no keisei: Roshia saisho-no senmon kagaku zasshi 'N. Sokolov to A. Engel'gardt no kagaku zasshi' [The Formation of Chemical Community in 19th Century Russia: the first specialized chemical journal 'The Chemical Journal of N. Sokolov and A. Engel'gardt] [in Japanese], Kagakusi Kenkyu (Journal of History of Science, Japan) **187**, 129–141.

Kaji, Masanori (2002) D. I. Mendeleev's concept of chemical elements and the principles of chemistry, *Bulletin for the History of Chemistry* **27**, 4–16.

Kozlov, V. V. (1958), *Ocherki istorii khimicheskikh obshchestv SSSR*, Izd-vo Akademii Nauk SSSR, Moscow.

Markovnikov, V. V. (1955), *Izbrannye trudy*, Izd-vo Akademii Nauk SSSR, Moscow.

Menshutkin, B. N. (1908), *Nikolai Aleksandrovich Menshutkin. Ego zhizn' i deiatel'nost'*, St. Petersburg.

Mladentsev, M. M. and Tishchenko, V. E. (1938), *Dmitrii Ivanovich Mendeleev, ego zhizn' i deiatel'nost'*, vol. 1, Izd-vo Akademii Nauk SSSR, Moscow-Leningrad.

Petrov, F. A. (2002–2003), *Formirovanie sistemy univesitetskogo obrazovaniia v Rossii*, **1–4**, Izd-vo Moskovskogo universiteta, Moscow.

Plate, A. F. (1953), Novye materialy k biografii V. V. Markovnikova, in *Materialy po istorii otechestvennoi khimii*, Izd-vo Akaemii Nauk SSSR, Moscow, 70–80.

Rudakov, V. (1909), Ucheniia uchrezhdeniia, in section Rossiia in *Entsiklopedicheskii slovar' [Brokgauz i Efron]*, St. Petersburg, 414–416.

Sinel, Allen (1973), *The Classroom and the Chancellery. State Educational Reform in Russia under Count Dmitry Tolstoi*, Harvard University Press, Cambridge, MA, USA.

Solov'ev, Iu. I. (1985), *Istoriia khimii v Rossii*, Nauka, Moscow.

Strekopytov, S. P. (2002). *Istoriia nauchno-tekhnicheskikh uchrezhdenii v Rossii vtoraia polovina XIX–XX vv.* [History of Scientific-Technical Institutions in Russia (second half of the 19th–20th centuries)], Moscow, RGGU.

Zaitseva, E. A. (2000), Die Fortsetzung der Traditionen der liebigschen Schule in Russland, in Engelhardt, D. V. and Kastner, I., ed., *Deutsch-russische*

Beziehungen in Medizin und Naturwissenschaften, Shaker Verlag, Aachen, **2**, 117–128.

Zaitseva, E. A. (2001), Deutsche an der Moskauer Universitat im 19 Jahrhundert: Ferdinand Friedrich V. Reuss (1778–1852), in Engelhardt, D. V. and Kastner, I., ed., *Deutsch-russische Beziehungen in Medizin und Naturwissenschaften*, Shaker Verlag, Aachen, **4**, 209–226.

CHAPTER 14

SWEDEN: The Chemical Society in Sweden: Eclecticism in Chemistry, 1883–1914

ANDERS LUNDGREN

14.1 Introduction

An organization for chemists in Sweden, *Kemistsamfundet* (The Chemical Society, henceforth *Samfundet*) was founded in December 1883. Founders were chemical engineers with the assistance of some industrialists. Chemists connected to the universities or other scientific institutions were slow to join, but after a couple of decades the majority of the members had an academic background. I have earlier described this development as a smooth emergence between academia and industry.[1] These two groups can of course not be exactly defined, the line between them is more than obscure, but I am vaguely referring to one group working within the chemical industry and another group doing chemistry in an academic environment. It became meaningful to use these groups as ideal types in order to analyse the history of Samfundet. One of the most notable differences would be that an engineer rarely had an advanced formal education. Some had studied at the Kungliga Tekniska Högskolan (Royal Institute of Technology, henceforth KTH), where the most advanced technical education was given in Sweden, but many had only secondary technical school qualifications, and some of them were even self-made men.

---

[1] Lundgren (1998). The history of *Samfundet* has earlier, on request from *Samfundet*, been written by Carl Kjellin (1933), himself a very active member. This work carries all the marks of a jubilee product, and is basically, and naturally, positive to *Samfundet*. Considering the genre it is pleasant reading and the volume contains a lot of valuable material from different kinds of sources. A short background on the state of chemistry in Sweden can be found in Russell (1998).

Creating Networks in Chemistry: The Founding and Early History of Chemical Societies in Europe
Edited by Anita Kildebæk Nielsen and Soňa Štrbáňová
© The Royal Society of Chemistry 2008

On the other hand the academic chemists rarely had any experience from chemical industry, but played the role of civil servants, doing research and teaching for higher aims than practical applications.[2] The search for the idealistic truth and nothing else was declared the main object for science, and a goal often referred to – and in chemistry truth was to be discovered through analysis.[3] This was neither a chemistry which interested the chemical industry, nor a chemistry which made the chemists interested in chemical industry. Chemical industry was mostly based on the control of organic reactions, which chemical theories could not treat in a way that was useful for engineers.

However, there were meeting points. Firstly a tradition of utility dominated the rhetoric of science in Sweden from the eighteenth century, including at the universities, and secondly KTH became more and more academized. Many professors and teachers wanted a school with research and teaching based on science.[4] KTH was often reorganized during the nineteenth century, and at each step the emphasis on science became more outspoken. With the great reform in 1878, not only was it given its present name, but also science became a most prominent part of the statutes.[5]

However, even if this essay draws considerably from my earlier study, I will here enlarge and complicate this picture by discussing the history of Samfundet in a broader frame, as one of the many scientific organizations formed during this time. Who were the members and whose interests, and which interests, were Samfundet supposed to satisfy? My main argument will be that it was a formally weak organization, with little external power, but above all an organization which functioned as an eclectic social meeting point for the members.

I will first compare with other and similar organizations, secondly I will look at membership development and thirdly I will analyse its main activities, including a discussion on its publishing policy.

14.2   Scientific Organizations

During the nineteenth century Sweden was a country in which organizations mushroomed.[6] Among workers in the temperance movement, in the religious and in the political spheres *etc.*, organizations were born with an astonishing speed and quantity – and there was no difference between technicians and scientists. At the universities different organizations grew as part of a budding seminar culture, influenced by the development at the German universities.

---

[2] For the scientist at the universities as civil servants see Widmalm (2001), chapter 6.
[3] Swedish chemistry was neither a theoretically advanced nor a path-breaking science; see Lundgren (2000). The situation can be compared to physics, where precision measurements with a goal similar to chemical analysis dominated the field. *cf.* Widmalm (2001).
[4] Lundgren (1998). More on the Kungliga Tekniska Högskolan and its relations to Samfundet below.
[5] Its earlier name was Teknologiska Institutet (The Technological Institute), which was changed to its present name in 1878. For the sake of convenience, KTH will be used as abbreviation. There is no good modern work on KTH, but see Henriques (1917).
[6] Jansson (1985).

Among the engineers organizations either started as student associations at KTH or as organizations for active engineers. Samfundet grew in the intersection between these two groups.

At Uppsala University in 1852 a scientific society, *Naturvetenskapliga Sällskapet i Uppsala* (Uppsala Scientific Society) was founded among the students. In 1853 it was renamed to the more appropriate *Naturvetenskapliga studentförbundet* (The Scientific Student Society). Its general aim was to "establish closer contact between those students in Uppsala" who were interested in the sciences, and to break the isolation in which they often lived.[7] The society was soon formally divided into three sections, of which one was the chemical-mineralogical section.[8] Each section was supposed to arrange meetings once a month, when lectures should be read by members and followed by a scheduled discussion on a topic decided on in the meeting before. The discussions were essential and the organization functioned like a seminar. As an organization it was never very developed, even if formally both a chairman and a secretary, and sometimes a curator, were appointed.[9]

At Lund University a similar development took place. Many small organizations of the same type as in Uppsala, connected to the different disciplines, were formed around 1860 – one for botany in 1858, one for mathematics and one for philosophy in 1862. They have been described as an answer to the need for forms of teaching other than lectures, and as part of an emerging seminar culture.[10] In 1868 *Kemisk-Mineralogiska Sällskapet* (The Chemical-Mineralogical Society) is mentioned for the first time in the university catalogue of Lund.[11] The internationally well-known professor in chemistry Christian Blomstrand became its first chairman, and meetings were supposed to take place one or twice a month, but activities slowly faded away, and soon they were held once a month, and from 1886 only after certain summons.

Stockholm University was established in 1878.[12] At the university a certain *Kemistföreningen* (The Chemist Society) soon came into existence, the history of which is obscure. It was a loosely organized body with basically the same task as the organizations in Uppsala and Lund. Otto Pettersson, professor of chemistry in Stockholm in 1898, vaguely hinted at the existence of a voluntary organization with the object "to arrange lectures and discussions on chemical topics".[13] Not much can be said about its activities, but with all probability it functioned as the others.

---

[7] Handlingar rörande Nat.vet. Studentsällskapet stadgar, Ms Uppsala University Library (henceforth UUB), U1803.

[8] The others were a physical-mathematical and a botanical-zoological section.

[9] The section was supposed to organize a collection of chemical substances, including mineralogical specimens.

[10] Weibull (1968), 59.

[11] Lund (1868), 36. For some reason Weibull (1968) does not consider this society worth mentioning.

[12] Stockholm University was not formally called "university" until 1961, but *Stockholms Högskola* (Stockholm High School). However, since it functioned like a university, I will here use the term university in order to avoid misunderstandings as to the quality of the teaching and as to the research done at the High School.

[13] Pettersson (1898), 103.

It was not only chemists that founded organizations, and a comparison with physics is relevant. In Uppsala the physical-mathematical section of the *Naturvetenskapliga Studentförbundet* (The Scientific Student Society) in 1871 had founded a physical mathematical organization of its own, from which in 1887 *Fysiska Sällskapet* (The Physical Society) was established.[14] As with the chemical section the aim was to create a possibility for professors and graduate students to meet and discuss scientific questions. Among the first members we find Svante Arrhenius, also a member of *Samfundet* since 1884. When moving to Stockholm, Arrhenius brought the idea with him and in October 1891 a physical society at Stockholm University was established, "a link between persons interested in the development of these sciences".[15]

All these academic organizations, sections, seminars, *etc.*, regardless of discipline, were discussion clubs for presenting news from journals, sometimes works by members, and for arranging discussions. They were loosely organized and had no economical resources. Their only income was from membership fees. They were essentially internal scientific organizations with a main task of making possible the exchange of ideas within a restricted scientific community. Membership was won by election, and there were no external activities.

The organizations at KTH were both different from and similar to those at the universities. *Svenska Teknologföreningen* (The Swedish Engineer Society) (henceforth TF) can trace its initially somewhat chaotic history back to the 1850s and different student camaraderie at KTH.[16] In 1861 a more stable organization for students was formed. It was an organization with a task similar to the above-mentioned organizations, to give lectures and arrange discussions on questions of common interest. In 1862 former students were allowed to become members, giving the organization more strength and the potential to grow, something the other organizations did not have. In 1878 it was divided into two sections, one for students and one for former students, and the name was once again changed, now to the somewhat clumsy *Teknologföreningen T. I. i Stockholm* (The Engineers Society T. I. in Stockholm). In 1887 it was given its present name and new statutes, from which it is clear that the organization now wanted to become a national organization for everyone interested in the development of Swedish industry, not only students or former students. But, as we shall see, links with KTH remained strong.

TF was not the only organization for engineers. In 1865 *Ingeniörs-föreningen* (The Engineers's Association) had been founded, with the aim of uniting professionally working engineers.[17] It was founded not only by engineers but also by militaries, but still, as with the other organizations, reports and discussions were an essential part of its activities. The organization had many similarities with TF, especially after the latter allowed former students at KTH,

---

[14] Haglöf (1987); Crawford (1996), 120–123.
[15] Arrhenius (1898), 80; *cf*. Euler-Chelpin (1964), 37.
[16] There is no modern standard work on TF, but see Holmberger (1912).
[17] Holmberger (1912), 103.

the basis for The Engineers's Association, to become members. It was logical that in 1890 it merged with TF – or that it was swallowed.[18]

Besides these more general organizations some more specialized organizations should be mentioned here, both with connections to Samfundet. *Farmaceutiska Föreningen* (The Pharmaceutical Society) was founded in 1861, with the aim to "encourage the study of pharmacy" and "bring about a closer contact between pharmacists".[19] Many pharmacists later became members of *Samfundet*. This is not surprising since there are many similarities between the daily activities of an engineer and a pharmacist, both working to produce products needed in daily work – and with very little help from science. In 1908 paper engineers formed *Svenska Pappers- och Cellulosa-ingenjörers Förening* (Society of Swedish Paper and Cellulose Engineers) for engineers active within the prospering pulp and paper industry, and also here double membership was common.[20]

14.3 The Founding of *Kemistsamfundet* and Its First Activities

Just before *Samfundet* was formed there existed a short-lived *Stockholms Kemistklubb* (Stockholm's Chemist Club). It was a very informal organization among chemistry students at KTH, with the aim to "connect persons, who are active within or to plan to be active in chemical technical industry or analysis".[21] Attempts were made to make the club a more formal organization, but without success. To what extent it really was a forerunner to *Samfundet* is difficult to say, but some of the same persons were involved in both organizations.

In late 1883 the engineer, former pupil of KTH and active member of TF Albert Cronquist summoned a meeting to discuss the formation of an organization for chemists in Sweden. The outcome of the meeting was successful and a new organization was established. According to the statutes members should be "members of TF in Stockholm or entitled thereto".[22]

The first meetings were held on the premises of KTH (as with Ingeniörsföreningen [The Engineers's Association]). The new organization was an outgrowth from TF, when a certain group, the chemical engineers, wanted to create possibilities to meet and discuss common problems. Its main purpose was to "arrange discussions on topics belonging to chemistry and its applications".[23] Statutes were similar to other scientific organizations, and were equivalent to those at the universities.

---

[18] The three founders were W. Leijonancker, professor at KTH, Carl Ångström, professor at KTH, and R. Cronstedt, later chairman of the board of KTH.

[19] *Farmaceutiska Föreningen* (1907), 1395.

[20] For other organizations in trade and industry see Eriksson (1978), 77–89.

[21] Ekstrand (1908), 203; also Kjellin (1933), 7.

[22] Kjellin (1933), 3.

[23] Kjellin (1933), 3.

This academic character of *Samfundet* is underlined by the fact that the practical aim to improve the conditions for the chemical industry in Sweden is not mentioned in the first documents. This so even if the engineers knew better than anyone else that the chemical industry in Sweden during the nineteenth century was weak. Considering themselves bearers of important knowledge necessary to develop this industry for the future of Sweden, they created *Samfundet* as a forum for discussing matters of common interest.

Between 1884 and 1887 *Samfundet* continued as a loosely organized forum for lectures and discussions similar to the academic organizations. Topics concentrated, as expected, on how to control purity and quality of different industrial products, such as soap, the amount of sulfur and phosphorous in iron, *etc*. New synthetic methods were presented, as well as also questions concerning chemical products and law, and the situation for the chemical industry in Sweden.[24] Technical matters coupled with economic considerations caused the liveliest discussions. In 1887 a discussion on the control of milk was perhaps "the liveliest there ever was in the society".[25] The most "scientific" topic was when Otto Pettersson and the professor at the Kungliga Lantbruksakademien (Royal Academy of Agriculture) in Stockholm L. F. Nilson presented their analysis of beryllium in September 1886.

14.4   The First Journal: *Kemiska Notiser*

Samfundet wanted to enlarge its activities, and the idea of issuing a journal arose. The journal soon became one of its main tasks, but it also created problems and in reality the story of the publishing activities can be described as a failure, at the same time as it highlights some of the peculiar features in the development of *Samfundet*.

The initiative came from the "handelskemist" (chemist of commerce) in Stockholm John Landin, a chemical engineer from KTH and founding member of *Samfundet*. He wanted a journal since there was no "*Swedish* periodical journal only for chemistry and its applications".[26] It should contain reports from meetings, summaries of news in technical chemistry, short notices, notes on new literature and statistics for important chemical products.[27] *Samfundet* approved and decided to publish *Kemiska Notiser* (Chemical Notices). On analysing the journal it is clear that it remained just what the title states: notices. The reader is presented with small messages and news, in general summaries from or translations from foreign journals. There were no original articles, and the overview reports had often been presented as lectures at meetings, but Svante Arrhenius's (a member since 1884) lecture on modern electrochemistry was never published. The longest articles were on illegal copies

---

[24] A complete list of lectures is given in Kjellin (1933), 153ff.
[25] *Kemiska Notiser* (1887) **1**, 81–86; quotation on p. 86.
[26] Kjellin (1933), 28 (emphasis by Landin).
[27] Kjellin (1933), 29.

of Swedish matchboxes, containing many pages of illustrations of false boxes, a theme which did not concern chemistry *per se*, but was certainly one important chemical industry in Sweden.[28]

Kemiska Notiser had difficulties in meeting expectations. It is safe to say that the scientific quality was low compared with the standards of the time. Neither scientists nor engineers were interested in using the journal for publication (more on this below). Its finances were poor, based as they were on membership fees only. In 1889 it was therefore decided to start a new journal *Svensk Kemisk Tidskrift* (Swedish Chemical Journal, Figure 14.1). However, before analysing the emergence and the fate of this new journal, which was launched with higher pretensions than *Kemiska Notiser*, we have to examine the relations between *Samfundet* and its "mother" organization, TF. The strained relations between the two organizations were one reason for the shift in publishing policy of *Samfundet*.

14.5 Relations between *Kemistsamfundet* and *Teknologföreningen*

Towards the end of the 1880s there existed in fact two different organizations at KTH, with partly overlapping interests – TF and *Samfundet*. Cooperation was initially smooth and many engineers were members of both organizations. Cronquist himself was at different times during the 1870s both chairman and vice-chairman of TF, and never dropped his contacts with the organization.[29] Accounts of meetings of *Samfundet* were published in the journal edited by TF, *Teknisk Tidskrift* (Technical Journal) but came to an end in May 1884, after only three accounts had been published.[30] Minutes from other local organizations, such as a technical society in Gothenburg, were, however, continuously published, and there were obviously some kinds of animosity between the organizations, even if many members floated freely between them. At the end of the 1880s relations for a short while became strained.

TF arranged regular national meetings for engineers and, during the second meeting in 1887, papers were presented thematically, with one section for chemical technology.[31] At the same time the idea to reorganize TF into sections according to disciplines (which had not been the case earlier) matured, and in October 1887 five different sections were suggested, one of them "kemi och bergsvetenskap" (chemistry and mining technology).[32] This section could be identical to *Samfundet*, which was asked to merge with it. The result was a prolonged and lively debate. Some thought this would mean that *Samfundet*

---

[28] En industri som vunnit efterföljd (1887–1888).
[29] Chairman 1870–1873 and vice-chairman 1875–1877, Holmberger (1912), 206–212.
[30] *Teknisk Tidskrift* (1884), **14**, 25, 64, 93.
[31] More on this in Lundgren (1998), 88–91.
[32] Holmberger (1912), 98.

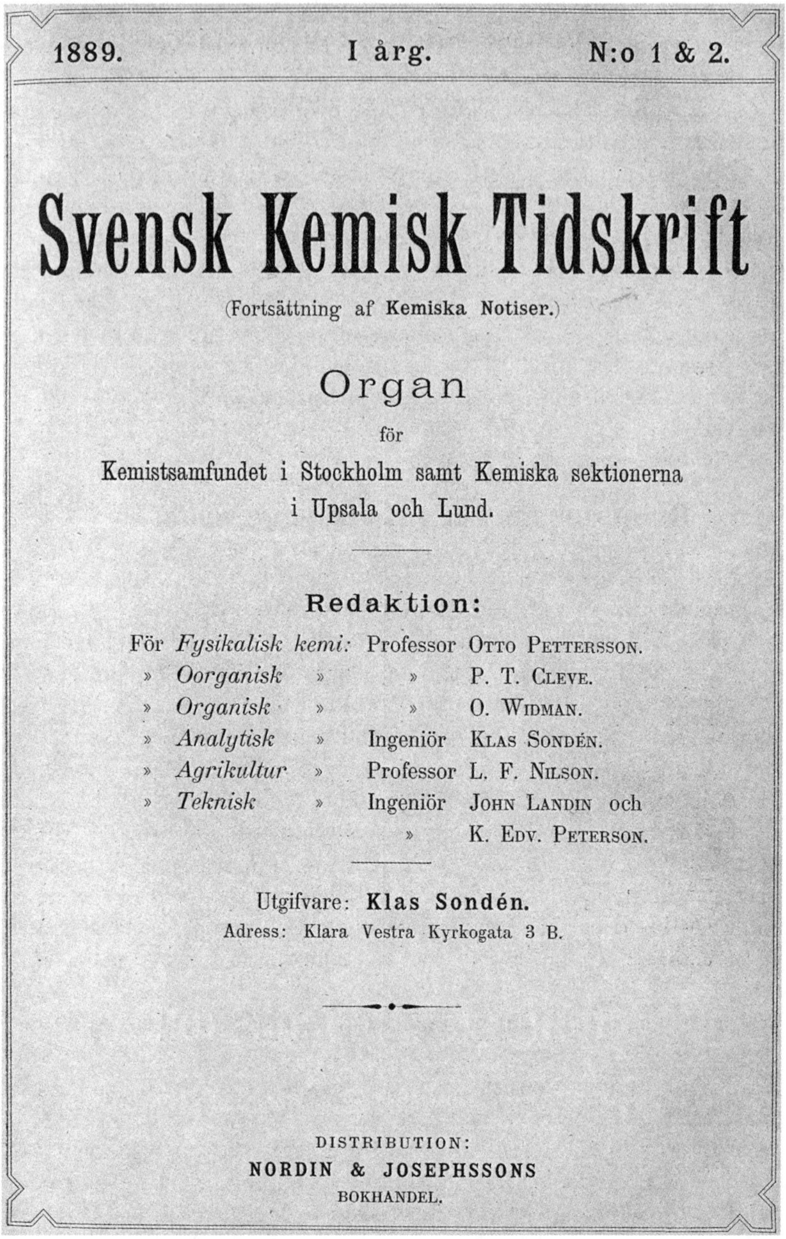

1889.   I årg.   N:o 1 & 2.

Svensk Kemisk Tidskrift

(Fortsättning af Kemiska Notiser.)

Organ

för

Kemistsamfundet i Stockholm samt Kemiska sektionerna
i Upsala och Lund.

Redaktion:

För *Fysikalisk kemi:*	Professor	OTTO PETTERSSON.
» *Oorganisk* »	»	P. T. CLEVE.
» *Organisk* »	»	O. WIDMAN.
» *Analytisk* »	Ingeniör	KLAS SONDÉN.
» *Agrikultur* »	Professor	L. F. NILSON.
» *Teknisk* »	Ingeniör	JOHN LANDIN och
	»	K. EDV. PETERSON.

Utgifvare: **Klas Sondén.**
Adress: Klara Vestra Kyrkogata 3 B.

———•——

DISTRIBUTION:
NORDIN & JOSEPHSSONS
BOKHANDEL.

Figure 14.1 Title page of the first issue of *Svensk Kemisk Tidskrift*, 1889.

would cease to exist as an independent organization, and that those members who were not members of TF would be left without an organization. This could affect the university teachers (it is uncertain exactly how) who, although there were few, were important to *Samfundet*. The final decision was to leave the question without consideration, which in practice meant "no".[33] As a result ties with other engineers loosened. To the general mistrust towards TF was added a conflict about the price for renting localities.[34]

TF fulfilled its plans and in September 1888 the new section was constituted and had its first meeting one month later.[35] And although *Samfundet* saw itself as something else than this section, many of its members were involved in its creation.[36] During the 1880s participants in discussions within *Samfundet* often were the same as those in discussions within TF.[37] Two of the three members of the board of the new section were members of *Samfundet*: secretary V. Gröndahl and vice-chairman John Landin.[38]

As a consequence *Samfundet* opened its ranks to persons other than those connected to KTH, and academics joined in greater numbers. In May 1888, seven professors were members, all of them living in Stockholm, and six members had a doctor's degree.[39]

This is not to say that the merging of academics and engineers in *Samfundet* was without problems. Cronquist wanted a society for chemical engineers, but he did not want chemical technology to become too scientific. He had experienced contempt among students at KTH of practical work and of workers, something he disapproved of because their knowledge was essential for the future of the industry, and it was knowledge to be respected. In TF he lectured on "Hvarför egna sig de tekniskt bildade kemisterna ej åt de kemiska handtverkerierna!".[40] There was constant tension between being scientific and being technical in Cronquist's texts. However, this did not prevent *Samfundet* from slowly becoming more "scientific", in the sense that more theoretical topics were chosen for lectures, and more academics contributed. There was always status to be gained from being scientific.

14.6 *Svensk Kemisk Tidskrift*, a New Journal?

These skirmishes with TF, and the discussions on the relation between science and handicraft, were accompanied by a change in journal policy. The debate on

---

[33] Kjellin (1933), 36–39.

[34] Kjellin (1933), 105.

[35] *Teknisk Tidskrift* (1888) **18**, 77, 144.

[36] Especially Cronquist and Landin. Reports from the works in the chemical technical section were printed in *Teknisk Tidskrift* (1887) **17**, 59–89, passim.

[37] For example *Teknisk Tidskrift* (1889) **19**, 44, 94, 138, 224.

[38] The chairman G. Nordenström was a mining engineer and not member of Samfundet.

[39] Medlemsförteckning maj 1888 (1888). The professors were C. E. Bergstrand, S. Jolin, L. F. Nilson, O. Pettersson, J. O. Rosenberg, L. Stahre (professor at the Farmaceutiska Institutet [Pharmaceutical Institute]) and R. Åkerman. Five of them had an academic degree in chemistry (Begstrand, Jolin, Nilson, Pettersson from Uppsala and Rosenberg from Lund), but were now working at other institutes; Bergstrand and Nilson at Kungl. Lantbruksakademien (The Agricultural Academy), Jolin at the medical Karolinska nstitutet (Carolinian Institute) and Pettersson at Stockholm University. Rosenberg was professor at KTH, as was Åkerman, who had a background in mining.

[40] Why do not the technically educated chemists take an interest in chemical handicraft, Cronquist (1887).

the future for *Kemiska Notiser* was lively.[41] It took place at the same time as the section for chemistry and mining technology within TF discussed the possibility to start a journal to be published as a supplement to *Teknisk Tidskrift*.[42] In that situation Samfundet chose to strengthen the ties with the universities, thinking increased cooperation with academics was a way to ensure the future of *Kemiska Notiser*. A committee was sent to Uppsala during the autumn of 1888 to discuss the matter.

The idea to approach chemists in Uppsala was successful. As a matter of fact, the chemical section of the *Naturvetenskapliga Studentförbundet* (The Scientific Student Society) in Uppsala in 1885 had discussed the possibility to start "one journal, common for all the Scandinavian countries", but after a long discussion the section decided not to take any action on this option.[43] When Uppsala three years later saw a possibility to realize this project without too many problems and too much economic undertaking they became enthusiastic. The Stockholm committee was well greeted and during a follow-up meeting in December 1888 in Stockholm it was decided to launch the new journal, *Svensk Kemisk Tidskrift* (henceforth *SvKT*), as a continuation of *Kemiska Notiser*. The chemical section in Uppsala unanimously voted for what they saw as a proposal to create "an organ common to all Swedish chemists".[44] The wish of the section to break from its mother organization, *Naturvetenskapliga Studentförbundet* (The Scientific Student Society), greatly contributed to the Uppsala chemists' positive attitude towards an increased cooperation with *Samfundet*. The section wanted to become a more independent organization in relation to its "mother" organization, in a way which was reminiscent of the relations between *Samfundet* and TF. Even if the suggestion to create a separate organization for chemistry finally was turned down – the existence of the section was not threatened in the same way as *Samfundet* was – the possibility to partake in the publishing of a journal must have given some satisfaction.

To increase cooperation even more the *Chemiska-Mineralogiska Föreningen* (The Chemical-Mineralogical Society) in Lund was approached, although only by letter, and it is a plausible guess that the reactions were similar to those in Uppsala. The end of the 1880s seems to have been a turbulent period in its history, with frequent changes in the board.[45] Increased cooperation with Uppsala and Stockholm was therefore a way to strengthen the organization. In 1889 the well-known professor Peter Klason became chairman, and remained in that post for several years.

To emphasize continuity Klas Sondén, editor of *Kemiska Notiser* and teacher in chemical technology at KTH, remained as editor for the new journal. On the editorial board there were two professors from Uppsala, P. T. Cleve (who

---

[41] *Kemiska Notiser* (1888) **2**, 88.
[42] The first number however did not occur until 1893, and with the new name *Kemi och metallurgi* (Chemistry and Metallurgy).
[43] Kemiska sektionens protokoll 30.3.1885, 13.4.1885, Ms UUB U 1830g.
[44] Kemiska sektionens protokoll 27.11.1888, 28.1.1889, Ms UUB U 1830g.
[45] Chairmen from 1885 were Lovén, K. O. M. Weibull (docent in mineral chemistry) and during 1888 the chairman was C. A. Rudelius, and the vice-chairman S. G. Hedin (both docents in chemistry); information from Lund (1875–1890).

earlier had been professor at KTH) and Oskar Widman, one professor from Stockholm University, Otto Pettersson, one from the Kungliga Lantbruksakademien (Royal Academy of Agriculture), L. F. Nilson, and three engineers, John Landin, K. Edv. Peterson and Klas Sondén. This composition reflected not the composition among the members, where engineers still dominated, but rather the will to become more scientific.

The aim of the journal was to publish accounts of meetings from the three bodies, surveys on progresses in chemistry, original articles, statistical information on the Swedish chemical industry, prices and taxes. That is, almost the same as for *Kemiska Notiser* – but with some science added. Scientific articles were, however, once again almost exclusively of a survey character, often lectures from meetings, and did not present original research, such as Svante Arrhenius's presentation of the dissociation theory.[46] The professor of physics in Uppsala, Knut Ångström, published news on the ultraviolet spectrum, to which the editor felt obliged to add that even if this was physics, it might be of interest also to the chemists – without specifying how.[47] An article on benzene shows its content as being "of particular great theoretical interest".[48]

The technological character of the journal was still manifest and dominating. The first article after the decision to launch a new journal was a translation from German, and dealt with the contemporary state of the saltpetre industry in Chile.[49] Arrhenius's article was followed by an analysis of cow milk.[50] And in the beginning of 1891 one of the longest reports (thirty pages) was an elaborate discussion with many replicas on tallow, margarine and custom taxes.[51]

In the middle of the 1890s *SvKT* experienced another minor crisis, once again caused by the economical situation, and to save the journal Samfundet asked its former competitor, the section for mining and chemistry in TF to join as co-publisher, but met no interest.[52] The most apparent result was rather that the editor Sondén left to join the new editorial board of TF's supplement – but he remained a member of *Samfundet*, so antagonism was far from insurmountable. The scientist/engineer Å. G. Ekstrand was chosen as the new editor.[53] He had studied chemistry in Uppsala (he was one of the opponents on Arrhenius's thesis), but early on became interested in chemical industry. One obvious result was that cooperation between academics and engineers increased. In January 1893 he wrote to the professor in chemistry in Uppsala, Oskar Widman, asking him to contribute, since *SvKT*, according to Ekstrand, needed the best in science, and he had a special "faith in the Uppsalians".[54] Another sign of increased cooperation was when academics from Uppsala three months later visited a regular meeting in Stockholm, the program was left "open" so the

[46] Arrhenius (1890).
[47] Ångström (1889).
[48] Kullgren (1896).
[49] *Kemiska Notiser* (1888) **2**, 122ff.
[50] Nilson (1889).
[51] *SvKT* (1891) **3**, 75–104.
[52] *SvKT* (1892) **4**, 127.
[53] Kjellin (1933), 44.
[54] Ekstrand to Widman 21.1.93. UUB D1499v7.

guests could talk about whatever they wanted. As part of this broadening of *Samfundet*, the *Kemistföreningen* (The Chemist Society) at Stockholm University was asked in 1895 to become associated with the publishing of *SvKT*, to which they agreed. In October the following year *Samfundet* was invited to visit the chemical laboratory of Stockholm University.[55]

Looking at the content of the journal it is, however, clear that the opening for academics and the plea for more articles from them did not make *SvKT* an important professional journal. The lack of original articles, the ups and downs of the journal and its economical problems continued. This does imply both a scientifically insignificant journal and scientifically insignificant organization.[56] The conclusion must be that *SvKT* did not develop into an arena for propaganda, nor for the engineers nor for the academics. Why then did the academics participate so willingly, if they did not use *Samfundet* in their professional life, and why did engineers continue to apply for membership, if they had a more powerful organization in TF? Under all the circumstances *Samfundet* did survive! There are good arguments to describe it as an internal organization and *SvKT* as a membership journal, almost a newsletter. Before arguing more in detail for this being the case, we will take a look at membership development.

14.7   Membership Development

Samfundet continued to grow (see Table 14.1). Members of the sections in Uppsala, Lund and Stockholm did not automatically become members, but were listed under separate organizations in the membership directory. These sections did not initially grow with the same amount of members. Being organizations at the university, they had a small recruitment base. In Uppsala and Lund only academics were members, but from 1907 Stockholm did allow members from outside the university.

There is nothing remarkable in the growth of *Samfundet*. The relative standstill in the middle of the 1890s coincided with a similar trend in TF, caused by many engineers feeling alienated after the reorganization of TF.[57] For the same reason chemical engineers did not know where to turn, to the new section in TF or to *Samfundet*, and the result was hesitation and uncertainty in the future of *Samfundet*. To some engineers it was too academic, to some academicians too practical. The only more notable fact is the increasing number of members for the academics during the first decade of the twentieth century, which clearly reflects the growing significance of this group.

Who could become a member? When *Samfundet* was formed a new applicant had to be, or eligible to become, a member of TF, which included former students working as engineers. Furthermore a member could be elected only on

---

[55] Kjellin (1933), 154.
[56] Even such an ardent supporter of Samfundet as Kjellin (1933) says that even if there were qualitatively good periods (which might be doubted when scrutinizing the journal) for *SvKT*, there were many downs as well; Kjellin (1933) 92–96, 135.
[57] Holmberger (1912), 126.

Table 14.1   Number of members of *Kemistsamfundet*, 1888–1914.

Year	The Chemical Society	Chemical section (Uppsala)	Chemical section (Lund)	The Chemist Society (Stockholm)
1888	103			
1891	137	35	14	
1893	145	24	21	
1894	185	25	22	
1895	180	26	22	25
1896	194	26	22	22
1898	286	35	14	15
1900	439	27	23	15
1901	475	29	16	28
1902	498	26	24	26
1903	489	27	18	26
1904	494	29	20	25
1909	539	69	54	65
1910	537	90	37	64
1914	547	93	45	108

Figures taken from the printed membership lists published in *Kemiska Notiser* and *SvKT*. During missing years no figures were published, and figures from Uppsala, Lund and Stockholm during the 1890s are not always reliable, since in some years those sections did not report, so the older figures were re-used.

approval, and the application had to be approved by at least two-thirds of the members at a general meeting. The rules were slowly made less strict and in 1889, after the break with the TF, "every person interested in chemistry" was eligible and with a simple majority among members.[58] In 1913 it was only two members of the board needed to accept the new member. The weakening of the rules can be interpreted in two ways. One, of course, practical; it was difficult to get enough members present to fulfil the demands. Had the rules been strictly kept there would have been no new members at all. But allowing membership on such broad grounds also meant that *Samfundet* could not become an organization for any specific group in the chemical community. There were no discussions on any eventual demarcation of *Samfundet*, and everyone could be accepted, regardless of what relation the applicant had to chemistry.[59]

Therefore also members of other professional groups increased in numbers, the most important being the pharmacists. In 1888 two pharmacists were members, whereas in 1900 there were no less than fifteen. Another group was mining engineers. Both of these groups had their own organization *Farmaceutiska Sällskapet* (The Pharmaceutical Society) and *Jernkontoret* (Swedish Ironmaker's Association) and their own journals (see below), but still many of them also became members of *Samfundet*.

---

[58] Quoted from Kjellin (1933), 3.
[59] Resistance towards the academic trends in the centre, all members of the board belonged to either the academics or the academized teaching staff at KTH, were stronger on the "country-side" following a general trend among engineers; *cf.* Sundin (1981).

14.8   *Kemistsamfundet* as an Internal Membership Organization

Referring to this eclectic range of membership it is arguable that *Samfundet* mostly fulfilled an internal social purpose for its members and that it was an organization with little external power, and with a journal which functioned as membership newsletter. Such a conclusion is buttressed by the following observations.

14.8.1   Its Journals Stayed Insignificant

It was difficult to fill the journals with relevant material. The offered possibility to publish was little used by the academics. There were already many possibilities in Sweden to do so. Kungl. Vetenskapsakadmien (The Royal Academy of Sciences) published its *Öfversigt* (Transactions), in which almost only foreign languages were used. But scientists in Sweden could also, and did, regularly publish in international journals like *Berichte der Deustschen chemischen Gesellschaft, Annalen der Chemie, Zeitschrift für physikalische Chemie, Journal für praktische Chemie, etc. Scandinavisches Archiv für Physiologie* was from 1889 issued in Leipzig. The editor was the professor in physiology in Uppsala, Frithiof Holmgren, and on its completely Scandinavian editorial board were S. Jolin, member of *Samfundet* since 1886, and professor of chemistry and pharmacy at Karolinska Institutet (Carolinian Institute) in Stockholm, and K. A. H. Mörner, also professor in chemistry and pharmacy at the same institution, and from 1895 member of *Samfundet*. Swedish chemists regularly published in this journal. The editor Holmgren explicitly stated that the journal actively worked to make Scandinavian research known to an international public.[60] In the beginning of the twentieth century possibilities to publish in Sweden increased even more, since *Öfversigt* split into four different series, of which one was dedicated to chemistry and mineralogy, *Arkiv för Kemi, Mineralogi och Geologi* (Archive for Chemistry, Mineralogy and Geology),[61] with more than half of the articles in foreign languages.

The engineers did not need a new forum for publication either. Since 1865 Ingeniörsföreningen had published a journal, *Ingeniörs-föreningens Förhandlingar* (Transactions of the Engineering Society), frequently published and with members of the board closely connected to KTH.[62] TF itself hade become so strong that in 1879 it could take over the publishing of the above-mentioned journal *Teknisk Tidskrift*.[63] This journal had existed since in 1871 and was originally a private enterprise, but the editor W. Hoffstedt was a teacher at KTH, which guaranteed strong connections between the school and the

---

[60] Holmgren (1889), 1ff.
[61] The others were *Arkiv för Astronomi och Fysik, Arkiv för Botanik* and *Arkiv för Zoologi* (Archive for Astronomy and Physics, Archive for Botany and Archive for Zoology).
[62] The editor, Carl Ångström, was professor at KTH.
[63] Holmberger (1912), 70–74. The first volume was called *Illustrerad Teknisk Tidning*, which was changed to *Teknisk Tidskrift* from 1872.

journal. It was a natural step for TF to take over, and there was no change in publishing policy after that. At the same time *Ingeniörs-föreningens Förhandlingar* ceased to exist.

Mining engineers could choose *Jernkontorets Annaler* (Annals of the Swedish Ironmaker's Association), from 1820 published by *Jernkontoret* (Swedish Ironmaker's Association) and from 1900 enlarged with a yearly supplement. Paper engineers could from 1898 publish in *Svensk Papperstidning* (Swedish Paper Journal) published by *Svenska pappersbruksföreningen* (The Swedish Paper Mill Association). Finally the pharmacists so closely related to chemistry and organized in *Farmaceutiska Föreningen* (The Pharmaceutical Society) had their own journal *Farmaceutisk Tidskrift* (Pharmaceutical Journal) from 1860–1897, which in 1897 was replaced by *Svensk Farmaceutisk Tidskrift* (Swedish Pharmaceutical Journal). The new editor Thor Ekecrantz wanted to make the journal more scientific, and to show the state of pharmacy in Sweden at that time and argued that the journal "can not, and will not be allowed, to become a one-sided spokesman for any fraction within the society".[64] Ekecrantz was also a member of *Samfundet*, and his views on the journal coincided with the one *Samfundet* embraced.

There were thus many possibilities for engineers to publish. However, all articles in these journals were in Swedish, aimed at a Swedish audience, which was part of the ideology to spread results to the benefit of Swedish industry.[65] All these engineering journals also, as SvKT, published translations from foreign journals – and members from *Samfundet* published in all of them.

Considering all these possibilities neither academics nor engineers had any need for a new journal. The academics were satisfied to publish overviews, seeing *SvKT* as a means to fulfil the utility ethos, by bringing science to engineers. They did not need a journal in order to legitimate a position they already had firmly rooted in the Swedish scientific community. On the other side the engineers either preferred a more specialized journal, or to spread their knowledge in practical actions – if they were not forbidden to publish, for business reasons.

Just as the technical journals, *SvKT* was published in Swedish and did not reach outside Scandinavia. There were a few articles in other Scandinavian languages during the 1890s, but neither Danish nor Norwegian can be called a foreign language in the Swedish technical and scientific community. There was also one article in German, but that author was working in close contact with Swedish engineers and in Sweden.[66] Still in the 1920 volume all articles were in Swedish, and not until later did articles in foreign languages, mostly German, became frequent.

---

[64] Ekecrantz (1897), 1.
[65] To what extent these articles really had any significance for the industrial development is a completely different question.
[66] The Danish engineer Hagen Pedersen worked in a brewery and published in *SvKT* (1892) **4**, 116, 128; the Norwegian professor at an agricultural school in Aas Norway, A. J. Sebelien, in *SvKT* (1902) **13**, 63 and *ibid.* **14** (1902); and the Latvian (?) engineer Arthur Penzais, who worked at a brewery as zymotechnician, in *SvKT* (1903) **15**, 40.

14.8.2 Main Activities Were Discussions

Samfundet's first statutes stressed that its main objectives were to "mediate debates" and that it did not want to formulate an official opinion in chemical matters, but "the discussion was the answer to the question".[67] A debate could end with the laconic statement that the question had been illuminated "from different standpoints".[68] In 1894 during a discussion on safety in gun-powder production the chairman considered it correct not to "make any decision or deliver any opinion in this question".[69] From 1889 *Samfundet* allowed itself to formulate opinions in scientific matters, but with the restriction that at the meeting at least 30 members should agree on the statement, and these 30 should not be less than three-quarters of the members present. Thirty years later it was added that before a statement could be made in the name of *Samfundet*, it must have been discussed by the board.[70]

Not to formulate an opinion decreased the external power of *Samfundet*, and the arsenic debates are good examples.[71] The arsenic question was hotly debated in Sweden in the 1880s and onwards with many different interested parties taking part; merchants, physicians, scientists, politicians, the public, *etc*. *Samfundet* took part in these debates but mainly in the analytical question: which methods were the best, both qualitatively and quantitatively, to trace arsenic. Discussions in *Samfundet* took place in 1883 and 1888 and could be rather heated, but in 1883 the only conclusion was that the laws regulating use of arsenic were not satisfactory. No suggestions for a change were put forward.[72] Such a result points towards a relatively powerless organization. The same conclusion can be drawn from the debate in 1888. This time there was a recommendation, but since *Samfundet* was not an official body no one was compelled to follow its advice, and the discussion on the value of different analytical methods remained internal.[73] The last debate on arsenic took place in 1901, and no consensus was reached at this time on what method was the best. The only outcome was that "an account of the discussion" was sent to different experimental stations carrying out analysis in practice.[74] This was not of much help, especially since Samfundet did not follow up the recommendations – it was not possible for it to do so.

14.8.3 Unconventional Theories, *etc.*

It was not only members from many different groups that were welcomed; there was also a tendency to allow somewhat odd (according to the standard of the

---

[67] Kjellin (1933), 4.
[68] *Kemiska Notiser* (1888) **2**, 88.
[69] *SvKT* (1894) **6**, 3.
[70] Kjellin (1933), 4ff.
[71] See also Hillmo (1994), 171ff., 177–196.
[72] Kjellin, 13–15, *cf. Kemiska Notiser* (1888) **2**, 19, 55–59.
[73] Hillmo (1994), 173.
[74] Kjellin (1933), 69–72; Ekstrand (1909). In Hillmo's analysis of the debate, Samfundet plays a minor role, Hillmo (1994), 177–196.

day) theories in *SvKT*. Even the theoretical articles became eclectic. Besides the traditional overviews, now and then articles were published with relatively unconventional theoretical discussions. The theories were usually speculative and not exactly in the main stream of chemistry. Many times they were printed lectures from the meeting, and they would probably not have been accepted in a more regular journal.

At the beginning of the twentieth century Paul Hellström published an article on the genesis of the elements in which he freely mixed ideas about the unity of matter, stereochemistry, valences and geometric series.[75] Hellström was educated in Uppsala; he did not belong to the academic circles, but worked as chemist at an agricultural station in northern Sweden.[76] Even though he must have had some support, since the article was also published in *Öfversigt*, during the discussion it was severely criticized by, among others, Svante Arrhenius for being "incompatible with all science".[77] Another example was the then editor Ekstrand's presentation of Ostwald's energetic ideas, which did not have much support in Sweden during this time.[78] Neither Hellström nor Ekstrand have been remembered as theoreticians, but as practical men. Although there are not too many articles of this kind, their existence underlines the eclectic character of the journal, as well as its character of an internal newsletter.

At this time also more articles on historical subjects were published, often written by Ekstrand, which makes it even more difficult to talk about *SvKT* as a journal representing any specific group within the chemical community.

14.8.4 No One Used *Kemistsamfundet* for Strategic Reasons

I have not been able to trace any group within the chemical community in Sweden that used *Samfundet* for its specific strategies. *Samfundet* remained an eclectic internal discussion club for all chemists, of the same kind as the organizations at the universities, although with members from more diverse fields. A chemist who wanted to build something, a discipline or a network, or needed support, could not find much help in *Samfundet*. There were so many chemical specialities, from theoretical academics to practical engineers, that it was difficult for any group to use *Samfundet* for its own purposes. Neither can the increasing presence of the academics, not the least in the editorial board of *SvKT*, be seen as an attempt to "take over" the organization. The academics had other ways, as academics, to make themselves important and seen. Engineers could put forward their demands in other stronger organizations, such as TF. Leaving *Samfundet* would not have meant expulsion from either the scientific or the engineering community. Specific group interests never threatened *Samfundet*, and its opening for new members was a process without problems.

---

[75] Hellström (1901); the article was also published in *KVA Öfversigt* (1901) **58**, 351.
[76] On Hellström see Mårald (2000), 143–145.
[77] *SvKT* (1901) **13**, 150ff.
[78] Ekstrand (1896).

Throughout the period here studied *Samfundet* kept its character as an internal discussion club, with little direct influence on the world outside its domain, *e.g.* on industry and/or politics. It was not an organization like TF, to which the government and other official authorities sent proposed legislative measure for consideration, a function of which none of the other organizations here mentioned could boast.[79] To KTH this was so important that all the tasks of this kind from 1890 were listed in their official yearly reports.[80] *Samfundet* could do nothing of that sort. It was not in the formal position to drive forward its views on, for example, custom taxes, and on the regulation for selling pharmaceuticals, topics which were discussed within *Samfundet*.[81] And finally the committee that was set up in 1886 with the task to suggest improvements in chemical industry came to nothing.[82]

All this is not the same as to say that individual members were without influence, a chemist like Peter Klason certainly had plenty, but they worked through other channels, most of all KTH and the universities. *Samfundet* rather was an internal area for intense lobbying, and especially suitable for that since it was a meeting place for so many different groups of chemists. A mapping of which chemists used this possibility and with what results we have to leave aside here, but it is important to stress that both the eclectic and the internal character of the meetings made them an ideal space for networking. This internal character also contributed to make *SvKT* a membership newsletter, with many reports from meetings, rather than papers directed towards the world outside the chemical community. When chemists or engineers remember *Samfundet* and its activities, it is remembered as a place for active discussion in internal chemical matters and – not the least – as a haven for some fun. This picture grows stronger if we look more closely at the social activities of *Samfundet*.

14.8.5 Social Significance

The social activities of *Samfundet* were important from the beginning and their significance should not be underestimated. One could actually say that social gatherings predated *Samfundet* and were one reason for its existence. The meetings at Stockholms *Kemistklubben* (Stockholm's Chemist Club) in 1883 began when chemists met at different restaurants in Stockholm and the idea arose to "combine usefulness with entertainment" and to add lectures to the visits to restaurants.[83] Even if the relations between *Kemistklubben* and *Samfundet* are vague and unsure, there is continuity on this point. Minutes from meetings printed in *SvKT* often contained sentences referring to a "simple spring supper" or the like.[84] At the Christmas meeting in 1903 leading

---

[79] This function was sometimes questioned, but was never taken away, Holmberger (1912), 186–194; *cf.* also Hansson (1981), 12, 46, 49.
[80] *Program och redogörelse för KTH* (1890–?).
[81] The debates are described in Kjellin (1933), 73–82, although Kjellin drew other conclusions.
[82] Lundgren (1998), 87ff.
[83] Quoted from Landin (1903), 168; *cf.* Kjellin (1933), 8.
[84] See for example *Kemiska Notiser* (1887) **1**, 33.

industrialist Oscar Carlsson donated a magnificent punch bowl of porcelain, a present that was very much appreciated.[85] The meetings seem to have been a highlight for many members, and they are almost always mentioned in memoirs and obituaries. Ekstrand has thus been described as social, "mild and popular" and a great lover of the meetings, and, it is pointed out especially, a great lover of the "nachspiel".[86] The source value of this kind of material is of course limited, but this is what the involved chemists most want to remember. That these social gatherings were important can be supported with other material.

When Hans von Euler, from 1906 professor in organic chemistry at Stockholm University, became a member of *Samfundet* at the beginning of the century, he thought there was too much festivity. Meetings were often held at different restaurants in Stockholm, and "the *Nachspiel* and the supper always were essential parts of the meetings".[87] One of the key figures of *Samfundet*, Arrhenius, had "an astonishing capacity" when it came to the Swedish national beverage, punch.[88] Euler did not appreciate this state of affairs, and he wanted to create an alternative to what he saw as the industrial and mercantile chemistry he could find in *Samfundet*, with its "significantly lower scientific standard".[89] To him there was no doubt that this standard was coupled with an over-developed feeling for the social parts of the meetings. But he could not change *Samfundet*, only *Kemistföreningen* (The Chemist Society) at Stockholm University. From having been "a primitive seminar", in 1908 it was reorganized by Euler as *Kemiska Sällskapet* (The Chemical Society) and modelled on *Deutsche chemische Gesellschaft*.[90] He wanted the new society to be more "scientific" and the lecturers at the first meeting were the cream of Swedish scientists, Svante Arrhenius, The Svedberg – and Euler himself.

During the beginning of the twentieth century, social events continued to be important. The 25th anniversary festivities in 1908 were moved from Sunday to Saturday, so that members outside Stockholm could have a better opportunity to take part.[91] During the meetings lectures were formally held. But the lecture on the history of *Samfundet* could only, due to the late hour, be partly read (*i.e.* everyone was longing for the supper).[92] The rituals and the formal arrangements during meetings like this are worthy of a fuller study. But here it is enough to say that these "nachspiels" were important in order to keep unity within the eclectic organization. It was also one way to avoid the skirmishes and the mistrust, which otherwise would have surfaced quite openly. For the external toothless tiger, which *Samfundet* was, the social aspects of the meetings and the nachspiels functioned as unifying forces. Even to the extent that

---

[85] *SvKT* (1903) **15**; the bowl is proudly photographed in Kjellin (1933), 26.
[86] Westgren (1934).
[87] Euler-Chelpin (1964), 39.
[88] Euler-Chelpin (1964), 34.
[89] Euler-Chelpin (1964), 51. This statement says as much of Euler's ego as it says something essential of *Samfundet*.
[90] Euler-Chelpin (1964), 51.
[91] Kjellin (1933), 117–119.
[92] Kjellin (1933), 118 gives which lectures and where they later have been published.

occasional poems (often of dubious quality) presented at late hours could be published in *SvKT*.[93]

Samfundet was of course not the only organization where social events were important and in one way it fulfilled a long tradition. It could be that *Samfundet* adopted a tradition from its academic friends. When the committee went to Uppsala in order to negotiate on the future of *Kemiska Notiser*, Arrhenius arranged a "sea battle" in order to get things running a little more smoothly.[94] The minutes from the chemical section in Uppsala also reveal a definite desire for festivities among chemists. The section's discussion on the proposal from *Samfundet* on a future journal started relatively early but became dragged out. A break was taken, a light supper was eaten and after that a decision was made.[95] The minutes also give many examples of social gatherings where liquor seems to have been an essential part. In spring 1892 the following can be read in the minutes of the chemical section in Uppsala!

A supper very much appreciated by the members was consumed. It was followed by a very pleasant gathering, which continued all Monday and part of Tuesday, upon which the members, singing and in a happy and merry mood, left the refreshing place.[96]

But *Samfundet* also combined the academic tradition of gatherings with a similar culture which was very strong at KTH, where entertainment and drinking was an important part of the internal life of TF. In Holmberger (1912) we can easily find many examples of how social gatherings in TF were arranged and what form they took – and the similarities with the academics is obvious.

Samfundet continued a long tradition of social gatherings and festivities and, since it had such small external power, it was especially important as a uniting link between the members. It shared this situation with the organizations at the university, and the only one of the organizations here mentioned which had some kind of external power was TF, and that was thanks to its connections to KTH.[97]

14.9   Final Words

Samfundet was important for its members, but it was an internal organization, which could be rather fragile. Members could all of a sudden leave for other organizations, or could take active part in competing organizations at the same time. One reason for the weakness was of course bad finances, membership fees being the only source of income. Another reason was that it never became an official contributor in public debate, which left *Samfundet* with little power. The

---

[93] Late examples are from *SvKT* (1919) **31**, 25–28 and *SvKT* **32**, passim.

[94] Kjellin (1933), 41. Although originally a military term, "sea battle" in Swedish also stands for a fairly wild party, where drinking is not placed on the back seat.

[95] Kemiska sektionens protokoll, 27.11.1888, Ms UUB U 1830 g.

[96] Kemiska sektionens protokoll 9.4.1892, Ms UUB U1830 g.

[97] *Farmaceutiska Föreningen* might be another exception, thanks to its connections with the Farmaceutiska Institutet, where pharmacists were educated.

result was an organization with an eclectic character, which made it difficult to unite all chemists into a coherent whole. During discussions members seldom reached consensus in chemical matters; it was frankly even a goal not to do so. It played an internal role for its members, as a discussion club and as a place for social meetings. This is why it survived.

For the chemist with political ambitions, or who wanted to work for a special goal within chemistry, *Samfundet* was a weak organization. Individual members could certainly be very influential in the scientific or engineering communities or in politics, but not as members of *Samfundet*. However, it offered a place for networking. This is a theme which needs a much more thorough study than is possible here. Individuals could use *Samfundet* to make contacts, but not as a vehicle for their own specific interests – it was too weak for that purpose. This is also mirrored in the journals, *Kemiska Notiser* and *SvKT*, which mainly contained discussions and informed on recent research in overview articles, but never presented new original research. Its character as a membership newsletter was very evident.

The external work of engineers and academics took place through routes other than through *Samfundet*. The engineers worked through their very concrete work and through TF, the academics through their role as academic civil servants and by publishing in foreign journals.

Samfundet did not contribute to strengthen chemistry as a discipline within the scientific community, basically because chemistry already was strong and had been so since the eighteenth century; at most the formation of *Samfundet* was a sign of how strong chemistry already was. Specialized disciplines did not have any use for the weak organization in order to promote their interests. *Samfundet* was an organization like the other discussion clubs at the universities and did very much follow their traditions concerning function and goal, not least the social goals. The idea behind *Samfundet* was neither unique nor rebellious, and the consequences of its formation were few.

Abbreviations

Samfundet	*Kemistsamfundet* (The Chemical Society)
KTH	Kungliga Tekniska Högskolan (Royal Institute of Technology), Stockholm
SvKT	*Svensk Kemisk Tidskrift* (Swedish Chemical Journal)
TF	*Svenska Teknologföreningen* (The Swedish Engineer Society)
UUB	Uppsala University Library, manuscript department

References

Ångström, K. (1889), Nyare studier öfver det ultraröda spektrum, *SvKT* **1**, 98–108.

Arrhenius, S. (1890), Theorien om lösningar, *SvKT* **2**, 4–16.

326 Chapter 14

Arrhenius, S. (1898), Fysik, in *Stockholms Högskolas Berättelse 1878–1898* Stockholm, 80.

Crawford, E. (1996), *Arrhenius: From Ionic Theory to the Greenhouse Effect*, Science History Publications, Canton, MA.

Cronquist, A. (1887), Hvarför egna sig de tekniskt bildade kemisterna ej åt de kemiska handtverkerierna! *Teknisk Tidskrift* **17**, 59.

Ekecrantz, T. (1897), Anmälan, *Svensk Farmaceutisk Tidskrift* **1**, 1.

Ekstrand, Å. G. (1896), Energetism i motsats till materialism, *SvKT* **8**, 168–174, 179–180.

Ekstrand, Å. G. (1908), Bidrag till Kemistsamfundets historia, *SvKT* **20**, 203.

Ekstrand, Å. G. (1909), Ur KSs äldre protokoll: Arsenikfrågan, *SvKT* **21**, 111–113.

En industri som vunnit efterföljd (1887–1888), *Kemiska Notiser* **1**, 80, 96 and *Kemiska Notiser* **2**, 6–12.

Eriksson, G. (1978), *Kartläggarna: Naturvetenskapens tillväxt och tillämpningar i det industriella genombrottets Sverige 1870–1914*, Umeå Studies in Humanities 16, Umeå.

Euler-Chelpin, H. (1964), *Minnen 1873–1964*, Uppsala University Library, mimeographed.

Farmaceutiska föreningen (1907), in *Nordisk Familjebok* **7**, Stockholm, 1395.

Haglöf, A. (1987), Fysiska sällskapet 1887–1987, in *Fysiska sällskapet 100 år*, Uppsala universitet, Uppsala, 87–108.

Hansson, S. A. (1981), *Svenska Teknologföreningen de andra femtio åren 1911–61*, Civilingenjörsförbundet, Stockholm.

Hellström, P. (1901) Om grundämnenas uppkomst, *SvKT* **13**, 120–123, 140–147, 152–154.

Henriques, P. (1917), *Skildringar ur Kungl. Tekniska Högskolans historia 1–2*, Norstedts, Stockholm.

Hillmo, T. (1994), *Arsenikprocessen: Debatt och problemperspektiv kring ett hälso-och miljöfarligt ämne i Sverige 1850–1919*, Linköping Studies in Arts and Science 102, Linköping.

Holmberger, G. (1912), *Svenska Teknologföreningen 1861–1911*, Teknologföreningen, Stockholm.

Holmgren, F. (1889), Zum Beginn, *Scandinavisches Archiv für Physiologie* **1**, 1ff.

Jansson, T. (1985), *Adertonhundratalets associationer: Forskning och problem kring ett sprängfullt tomrum eller sammanslutningsprinciper och föreningsformer mellan två samhällsformationer c:a 1800–1870*, Studia Historica Upsaliensis 139, Uppsala.

Kjellin, C. (1933), *Kemistsamfundets historia*, Kemistsamfundets förlag, Stockholm.

Kullgren, C. (1896), En ny behandling af benzolproblemet, *SvKT* **8**, 29–33.

Landin, J. (1903), Några kemistminnen från 1880-talets Stockholm, *SvKT* **15**, 168–172.

Lundgren, A. (1998), Between science and industry: The background and formation of the Swedish Chemical Society, in Homburg, E. *et al.*, ed.,

The Chemical Industry in Europe, 1850–1914: Industrial Growth, Pollution, and Professionalization, Kluwer, Dordrecht, 73–94.

Lundgren, A. (2000), Theory and practice in Swedish chemical textbooks during the nineteenth century: Some thoughts from a bibliographical survey, in Lundgren, A. and Bensaude-Vincent, B., ed., *Communicating Chemistry: Textbooks and their Audiences, 1789–1939*, Science History Publications, Canton, MA, 91–118.

Lund (1868), *Lunds Kongl. universitets katalog för Vår-terminen 1868*, Lund.

Lund (1875–1890) *Lunds Kongl. universitets katalog 1875–1890*, Lund.

Medlemsförteckning maj 1888 (1888), *Kemiska Notiser* **2**, 72–74.

Mårald, E. (2000), *Jordens kretslopp: Lantbruket, staden och den kemiska vetenskapen 1840–1910*, Idéhistoriska skrifter 33, Umeå.

Nilson, L. F. (1889), Analys af komjölk, *SvKT* **1**, 108–111.

Pettersson, O. (1898), Kemi, *Stockholms Högskolas Årsberättelse 1878–1898*, Stockholm.

Program och redogörelse för KTH (1890–?), Stockholm: Kungl. Tekniska Högskolan.

Russell, Colin A. (1998), Chemistry on the edge of Europe: Growth and decline in Sweden, in Knight, David and Kragh, Helge, ed., *The Making of a Chemist: The Social History of Chemistry in Europe 1789–1914*, Cambridge University Press, Cambridge.

Sundin, B. (1981), *Ingenjörsvetenskapens tidevarv*, Umeå Studies in the Humanities 42, Umeå.

Weibull, J. (1968), *Lunds universitets historia* IV, Lunds universitet, Lund.

Westgren, A., (1934) Åke Gerhard Ekstrand, *Kungliga Vetenskapsakademins Årsbok*, 259–264.

Widmalm, S. (2001) *Det öppna laboratoriet: Uppsalafysiken och dess nätverk 1853–1910*, Atlantis, Stockholm.

CHAPTER 15
Creating Networks in Chemistry – Some Lessons Learned

ANITA KILDEBÆK NIELSEN AND SOŇA ŠTRBÁŇOVÁ

Many historians of science have already pointed out that the nineteenth century was a century of chemistry as both an academic discipline and an applied subject. By the mid nineteenth century, chemistry developed into a profession highly demanded in almost all domains of science, manufacture, medical practice, education and public life.[1] As Helge Kragh has stated, chemistry was "[b]eyond discussion [. . .] a European science" for several reasons.[2] The main stage of development of scientific as well as industrial chemistry took place on European soil; chemistry education in Europe (especially in some countries) became the model for the whole world; and last but not least it was in Europe that chemists of various backgrounds started to establish national chemical societies often as natural successors or spin-offs of more broadly defined scientific societies.[3]

In the previous chapters, details have been presented on the formation and early history of the chemical societies in 14 European countries. The authors of all the chapters have shown that the chemical societies developed distinct characteristics deeply rooted in local, regional and national settings. What this book hopefully also reveals is that the societies held a number of features in common – features which were not only shared by the evolving chemical

---

[1] As Kragh points out, "this situation was peculiar to chemistry which, alone among the sciences, found use in a large number of diverse fields". Knight and Kragh (1998), 335.
[2] Knight and Kragh (1998), 329.
[3] This chapter and its deliberations are based largely on the facts and analyses of the previous chapters where the national chemical societies were handled. As these chapters contain comprehensive lists of literature, we do not repeat these references nor refer to new ones. The main monographs with plenty of data the reader can rely on (as we also did) are Ihde (1984), Brock (1993) and Knight and Kragh (1998).

Creating Networks in Chemistry: The Founding and Early History of Chemical Societies in Europe
Edited by Anita Kildebæk Nielsen and Soňa Štrbáňová

societies in many countries but to some degree were also typical of other scientific societies which mushroomed all over Europe at that time. It is the purpose of this final chapter to help the reader to gain insight into these nationally specific as well as generally true or prevalent characteristics; to notice points of resemblance as well as dissimilarities.

As described in the preface, the individual studies of this collective monograph have been conducted along a set of common guidelines presented in the box of the preface. With the aid of these guidelines, all chapters have obtained analogous structures that make comparisons and generalizations easier in spite of much specificity linked to the history of the chemical associations in the particular countries. In the following paragraphs we will attempt to outline some of the conclusions, but many more than those presented here will be open for analysis by the observant reader.

15.1 Chronology of Founding

Tables 15.1 and 15.2 summarize the foundation years of those national chemical societies that are dealt with in this book. The *Chemical Society*, the first national chemical society, was founded in 1841 in Great Britain, and remained the only one on the scene until the founding of *Société chimique de Paris* in 1857. About ten years later almost simultaneously came into being the chemical societies in the Czech Lands, Germany, Russia and Austria. In the first three countries the societies evolved into stable platforms of the national chemical communities existing up to the present in spite of their various transformations in the early decades. The Austrian chemists, however, had to wait until 1897 for an association covering their needs.

The genesis of the British *Institute of Chemistry* in 1877 provides the first example of a situation where one national chemical society is not sufficient to serve all chemists in the country and consequently a parallel chemical society with a different profile is founded; as a result, in Great Britain no less than three societies coexisted for a long period. The Danish Chemical Society was established as the first in Scandinavia in 1879 and was succeeded by a society in Sweden four years later. The second German and the Belgium societies originated in 1887, the first Dutch in 1890 and the last societies to be founded in the nineteenth century were the second Austrian and the Norwegian societies. The *Nederlandsche Chemische Vereeniging* came into being just after the turn of the century in 1903, four years before the Hungarian society. The last two societies that are included here are the Portuguese society founded in 1911 and the Polish created in the aftermath of the First World War in 1919.

As depicted in Figure 15.1, the majority (15 out of 19) of the societies were founded before the turn of the century, which should come as no surprise.[4] We should keep in mind though that the tables do not include some associations

---

[4] Only a few key chemical societies, like the Spanish *Real Sociedad Española de Química* and the Italian *Società Chimica Italiana*, which are not treated in our volume, were established after 1900, in 1903 and 1909 respectively.

Table 15.1   Chronology of creation of the European chemical societies treated in the book.

Year	Country	Society
1841	Great Britain	*Chemical Society*
1857	France	*Société chimique (de Paris/de France)*
1866	Czech Lands	*Isis/Spolek chemikŭ českých/Společnost pro prŭmysl chemický v Královstvi českém/ Česká společnost chemická pro vědu a prŭmysl*
1867	Germany	*Deutsche chemische Gesellschaft (zu Berlin)*
1868	Russia	*Russkoe Khimicheskoe Obshchestvo*
1869	Austria	*Chemisch-Physikalische Gesellschaft*
1877	Great Britain	*Institute of Chemistry*
1879	Denmark	*Kemisk Forening*
1882	Great Britain	*Society of Chemical Industry*
1883	Sweden	*Kemistsamfundet*
1887	Germany	*Deutsche Gesellschaft fur angewandte Chemie/Verein deutscher Chemiker*
1887	Belgium	*Association Belge des Chimistes/Société chimique de Belgique*
1890	The Netherlands	*Technologisch Gezelschap*
1897	Norway	*Kjemigruppen/Norsk Kjemisk Selskap*
1897	Austria	*Verein Österreichischer Chemiker*
1903	The Netherlands	*Nederlandsche Chemische Vereeniging*
1907	Hungary	*Magyar Kémikusok Egyesülete*
1911	Portugal	*Sociedade Chimica Portugueza*
1919	Poland	*Polskie Towarzystwo Chemiczne*

mentioned in the book which cannot be considered full-featured chemical societies, like the Austrian *Wiener Verein zur Förderung des Physikalischen un Chemischen Unterrichts* concerned with supporting teachers at secondary schools.

Obviously, the chronology does not reveal any straightforward foundation-wave or geographical logic to the direction of creation of the societies. Although the movement started in London, the next country to be hit was France followed by some countries of Central and Eastern Europe and others rather randomly. The years of foundation do, however, correspond to a large degree to the evolvement of chemistry in the individual lands, whether we consider chemical industry, chemical education or other indicators. That England with its early industrialization was the first country to have a chemical society is thus hardly surprising, and the chemical industry was rather developed in the Czech Lands as well. In France and Russia, however, the chemical societies arose in strictly academic milieus. On the other hand, the relatively late foundation of a Dutch national chemical society does not imply that The Netherlands had little chemical industry or a weak academic environment before 1900. In Hungary, Portugal and Poland very specific, local circumstances account for the relatively late foundation of the chemical societies. In the case of Portugal, the late establishment was closely related to the delayed

Table 15.2 The chemical societies and their journals.[a]

Country	Name of the society	Year of founding	Journal of the society	First Issue[b]
Austria	Chemisch-Physikalische Gesellschaft	1869	–	
	Verein Österreichischer Chemiker	1897	Österreichische Chemiker-Zeitung (Austrian Chemists' Journal)	1898
Belgium	Association Belge des Chimistes/Société chimique de Belgique	1887	Bulletin de l'Association Belge des Chimistes	1887/1888 –1903
			Bulletin de la Société Chimique de Belgique	1904
Czech Lands	Přírodovědecký spolek Isis/Spolek chemiků českých/Společnost pro průmysl chemický v Království českém/Česká společnost chemická pro vědu a průmysl	1866	Kritické listy (Critical Letters)	1869–?
			Zprávy vědecké o činnosti Spolku (Scientific Reports about the Association's activities)	1871–?
			Zprávy Spolku chemikú českých (Bulletin of the Society of Czech Chemists)	1872–1875?
			Listy chemické (Chemical Letters)	1876
			Časopis pro průmysl chemický (Journal for Chemical Industry, 1891–1906)	(1891) 1893
			Chemické listy pro vědu a průmysl (Chemical Letters for Science and Industry)	1907
Denmark	Kemisk Forening	1879	–	–
France	Société chimique (de Paris/de France)	1857	Répertoire de chimie appliquée (Repertory for Applied Chemistry)	1858–1864
			Répertoire de chimie pure (Repertory for Pure Chemistry)	1858–1864
			Bulletin de la Société Chimique de Paris (de France) (Bulletin of the Chemical Society of Paris)	1858

Table 15.2 (continued)

Country	Name of the society	Year of founding	Journal of the society	First Issue[b]
Germany	Deutsche chemische Gesellschaft (zu Berlin)	1867	Berichte der Deutschen Chemischen Gesellschaft (Reports of the German Chemical Society)	1868
			Justus Liebig's Annalen der Chemie (Annals of Chemistry)	(1832) 1907
	Deutsche Gesellschaft fur angewandte Chemie/Verein deutscher Chemiker	1887	Zeitschrift für angewandte Chemie (Journal for Applied Chemistry)	1888
Great Britain	Chemical Society	1841	Quarterly Journal	1848–1862
			Journal of the Chemical Society	1862
			Proceedings of the Chemical Society	1885
	Institute of Chemistry	1877	Proceedings of the Institute of Chemistry	1877
	Society of Chemical Industry	1881	Journal of the Society of Chemical Industry	1882
Hungary	Magyar Kémikusok Egyesülete	1907	Magyar Kémikusok Lapja (Hungarian Chemists' Journal)	1910
The Netherlands	Technologisch Gezelschap	1890	Jaarverslag van het Technologisch Gezelschap (Annual Report of the Technological Society)	1891
	Nederlandsche Chemische Vereeniging	1903	Chemisch Weekblad (Chemical Weekly)	1903
			Chemische Jaarboekje (Chemical Annual)	(1899) 1920

Norway	*Kjemigruppen/Norsk Kjemisk Selskap*	1897	*Pharmacia. Tidsskrift for kemi og farmaci* (Pharmacy. Journal for chemistry and pharmacy)/*Tidsskrift for Kemi og Bergvesen* (Journal for Chemistry and Mining)	(1904) 1920
Poland	*Polskie Towarzystwo Chemiczne*	1919	*Roczniki Chemii* (Annals of Chemistry) *Przemyst Chemiczny* (Chemical Industry)	(1901) 1919 (1916) 1919
Portugal	*Sociedade Chimica Portugueza*	1911	*Revista de Chimica Pura e Applicada* (Journal of Pure and Applied Chemistry)	(1905) 1911
Russia	*Russkoe Khimicheskoe Obshchestvo*	1868	*Zhurnal Russkago Khimicheskago Obshchestva* (Journal of the Russian Chemical Society)	1869
Sweden	*Kemistsamfundet*	1883	*Kemiska Notiser* (Chemical Notices) *Svensk Kemisk Tidskrift* (Swedish Chemical Journal)	1887–1889 1889

[a] Societies portrayed in this volume. For a more detailed, global survey see Bolton (1902). See also Ihde (1984), 270–275 and 728–732.
[b] As publication of the chemical society. If published earlier without formal connection to the society, the first year of publication is given in brackets.

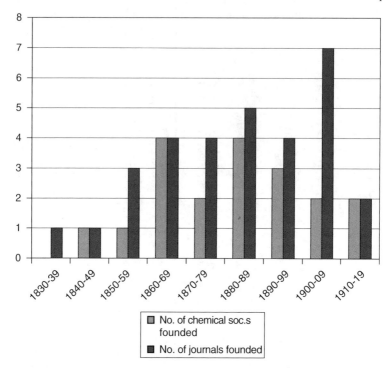

Figure 15.1   Number of chemical societies and journals 1830–1919.

process of emancipation of chemistry from other disciplines. The obvious
reason in the case of Poland was that the country did not exist before 1918.

   A number of chemical journals existed even before the chemical societies were
founded. Yet most of the professional and scientific associations in the world
have considered publication of their own journal a top priority and the chemical
societies are not an exception as displayed in Table 15.2 and Figure 15.1.
Table 15.2 also provides information on the first year (and last if appropriate) of
publication of the societies' journals. As analysed further below, in some cases
the societies adopted an already existing periodical and the first year of publication thus precedes the founding year of the society. In most cases though, the
society began to publish its own journal or journals immediately or within a few
years of establishment. In some societies there even was a need to issue more
than one journal for various reasons discussed in the proper chapter. A society
without a journal was an exception, but it did occur. For instance neither the
Austrian Chemical-Physical Society nor the Danish Chemical Society started to
publish an in-house journal in the period analysed in this book.

15.2   Developments in Membership

Table 15.3 offers not only a quantitative survey of the membership of chemical
societies treated in this book, but also allows us to deduce the total number of

Table 15.3 Approximate number of members in the chemical societies.[a]

Country	Year of founding	No. first year	No. around 1900	No. around WW1
Austria	1869 (*Chemisch-Physikalische Gesellschaft*)	unknown	200	unknown
	1897 (*Verein Österreichischer Chemiker*)	200	900	1100
Belgium	1887	30	490	450
Czech Lands	1866	60	400 + 500[b]	600
Denmark	1879	60	110	150
France	1857	25	> 1000	> 1100
Germany	1867 (*Deutsche chemische Gesellschaft*)	110	3400	3400
	1887 (*Deutsche Gesellschaft fur angewandte Chemie/Verein deutscher Chemiker*)	240	2250	5260
Great Britain	1841 (*Chemical Society*)	150	2300	3200
	1877 (*Institute of Chemistry*)	400	980	1450
	1881 (*Society of Chemical Industry*)	300	3460	4140
Hungary	1907	50	–	120[c]
The Netherlands	1890 (*Technologisch Gezelschap*)	40	220	> 360
	1903 (*Nederlandsche Chemische Vereeniging*)	150	150	580
Norway	1897	30	40	130
Poland	1919	120	–	120
Portugal	1911	70	–	150
Russia	1868	50	330	400
Sweden	1883	20	440	500

[a] Numbers rounded off based on data from the national chapters and the respective authors.
[b] Numbers of the two chemical societies that arose after the temporary split. The total number is not a sum as many chemists were members of both societies.
[c] Number estimated based on archival material.

chemists in Europe in the early twentieth century. Drawing on the membership data, it is revealed that the total number of members of the chemical societies included into this book was around 16 500 at the turn of the century. This number had risen to about 23 000 around 1914, illustrating primarily the increase in the number of chemists with a formal educational background in the discipline. It is difficult to estimate what percentage of the potential members (chemists of some kind) in each country actually chose to become a member of one or more societies. Similarly it is not easy to assess the benefit of being a member, because the conditions of membership differed quite a lot in the various societies. The chapter on Great Britain deals explicitly with the advantages and disadvantages of membership of the three different chemical associations. Several authors in this book have stated that at least half of the country's chemists enrolled in the associations. This allows us to estimate that the total number of chemists in the analysed countries around the First World War was about 50 000.

In Table 15.3, we see a steep increase in the number of members around 1900 in most of the societies constituted before 1890. While many societies, like the Belgian, Czech, French, British, Swedish and others, had started from a modest founding group, their membership had multiplied by 1900. The most conspicuous example of a dramatic rise in the number of members is seen in the French and German societies. The German industrial chemical association had around 2250 members at the turn of the century, but no less than 5200 members at the outbreak of the First World War. Also the British chemical associations increased their membership, although a general picture is difficult to obtain as many chemists opted for simultaneous membership in two or all three of the existing societies. A similar example of simultaneous membership is found in the Czech Lands where the chemical society in 1893 temporarily separated into a more "scientific" and a more "industrial" association; several chemists, however, chose to become members of both societies. As a rule, the societies experienced an influx of associates between 1900 and 1914, but a few kept a stable membership, and the Belgian society even lost members.

The simplicity of the tables cannot disclose the complex situation behind them. It must be remembered that most chemical societies in Europe at that time were professional institutions without pronounced ethnic or national connotation. At the cradle of the French *Société chimique* stood a group of young foreign chemists and, although the society soon became French dominated, it continued to welcome foreigners and made no distinction between French and foreign members. Even in ethnically diverse Belgium, the society was "Belgian" unlike the functioning Belgian-French- and Dutch-speaking societies of today. However, the picture is much more complicated if we look at the societies founded on the territory of Austria-Hungary and partly Russia. In these countries besides the "Austrian" and "Russian" societies also existed Czech, Hungarian and Polish societies.[5] They associated almost exclusively professionals of a certain historical territory and nationality in the cultural and lingual sense, although their statutes usually did not declare this condition. The Czech, Hungarian and Polish chemists probably joined the Austrian society only exceptionally, if at all, even though they were citizens of the same state.

---

[5] It is quite difficult to explain the complicated situation in Hungary. Hungary was to a large extent independent after the Austro-Hungarian Settlement in 1867, so the Hungarian society also intended to represent chemists of the Hungarian "state". Although the first Polish chemical society was founded only after Poland became independent, our remark takes into account other Polish scientific societies which functioned in the divided Poland and also housed the Polish chemists.

At the same time the Austrian society in reality represented only the German-speaking chemists on the territory of Austria-Hungary.

15.3   Roles of the Chemical Societies

The apparent interest of chemists in joining the chemical societies is evident from the tables and the previous chapters. Therefore let us consider what made the societies attractive to the chemical community and ponder on their multiple roles, although it is not possible to specify here all functions the societies took on as shown in the individual chapters. It is observable that the key tasks and activities were common to many of the societies across national boundaries. As mentioned already, the edition and publication of an in-house journal was given high priority in most societies. Besides journals, several societies published or subsidized university textbooks, handbooks, monographs or other chemical literature.

Another essential activity was, of course, arranging regular meetings where the members could assemble and listen to readings of papers by other members or invited local and foreign specialists. The lectures enabled circulation and discussion of new findings and ideas, information about the progress of chemistry in other countries, sharing of experience from travels abroad and recognition of merited members and jubilees. Last but not least the lectures were instrumental in generating a social network amongst people of similar professional background, training and interests. In some societies, such as the Danish and Swedish, the oral presentations represented in certain periods almost the only activity of the chemical societies. But although lectures were given high priority in most societies, they never came to play a principal role in the Dutch chemical society. The lack of a social dimension in the Dutch society must, however, be considered to be the exception to the rule. Among other regular ventures in most associations were visits to industrial plants and cultural excursions, which combined professional and social elements.

Some societies also took interest in specific problems, like university- or secondary-level curricula in chemistry, advancement of specific chemical subdisciplines, chemical nomenclature, national legislation in relation to the chemical industries, taxes or customs and working conditions and salaries of chemists. The members often took on important functions as consultants at municipal or governmental levels. To pursue these matters, they frequently established more or less permanent committees or sections. The Belgian society for example set up a section on food control and hygiene in 1889, and the Polish and Danish societies established terminology committees with the purpose of unifying the national chemical nomenclature and labelling of chemical compounds.

Acquisition of literature and maintenance of book collections or libraries represented important tasks for many of the societies amongst others in the Czech Lands, France, Great Britain, Hungary and Russia. The libraries often provided the members with very good access to the new literature published at home or abroad. Books were not only obtained through purchase but also as gifts from members or wealthy persons who sympathized with the chemical society. It was a common habit to establish library exchange agreements with chemical societies abroad, which extended the access to a wider range of journals and newly published chemical literature. In fact, such exchange often represented the most frequent type of international contacts with chemical societies in foreign countries.

It should also be accentuated that the high ambitions and plans of some societies sometimes did not match the reality of everyday life. The reasons could be a lack of finances, unfavourable legislation, personal disagreements or other specific circumstances. This was the case, for instance, in the Czech society, which in 1893 planned a new series of technological handbooks that was only realized 30 years later. The Dutch society, in spite of its initial plans and very open membership policy, had only modest social activities, represented by two or three meetings a year with not too many attendants. The ambitious initial agenda of the Portuguese founding fathers had to be adapted to the reality of delayed emancipation of chemistry from other disciplines.

The attitude towards involvement in welfare or trade-union-like matters varied much amongst the societies. Some of the societies explicitly or implicitly did not want to get involved in those kinds of functions and argued that the aims of the societies were scholarly and networking. Some societies, like the Czech one, supported needy students. Others, such as the Hungarian, took action in union-like work and became involved in settling agreements for workers in the industry. In Germany, the *Verein deutsche Chemiker* established a job placement register in 1900 and a mutual assistance fund a few years later.

As mentioned already, the rising national and ethnic movements in Central and Eastern Europe in the period of question supported the formation of linguistically and ethnically separated scientific communities in multinational states like Austria-Hungary and Russia. In this environment, the Czech, Hungarian and Polish scientific (not only chemical) societies played significantly more general roles. Not only did the societies support the formation of linguistically separated scientific and educational institutions, but also the creation of national scientific terminologies. They were also central in the intellectual and economic emancipation of the respective nations, which in consequence promoted the emergence of national states after the First World War. In this way, some chemical societies outreached their professional roles and became, to a certain extent, cultural and political institutions. But, at the same time, most of the societies should be commended for not being infiltrated too much, if at all, by national intolerance or extreme nationalistic or chauvinistic attitudes.

15.4 The Journals

The journals deserve a separate section as they have played a crucial role in constituting scientific disciplines and have usually represented (as they still do today) a central activity of almost every professional association.

Establishment of a journal was one of the first actions of most chemical societies; in some cases, however, a chemical journal had existed in the country long before any society was formed, for example in Germany, France or Portugal. As can be expected, each journal has developed its own typical features. There are nevertheless many similarities in the structure of the journals.

Frequently, the journal started as an information bulletin for members reflecting the internal life of the society, but usually it developed into a periodical with multifarious contents. The journals often included original scientific and technological papers presenting new findings of their authors, popular articles, reports from industry and trade and translated articles of noted local or foreign chemists. The shorter communications included among others accounts about the progress of chemistry worldwide and the life of foreign societies, book reviews, economic and patent matters, obituaries and advertisements. Usually a separate part was reserved for internal information, where the members could find minutes of meetings, announcements, membership matters and annual reports of the societies.[6]

As can be expected, the individual societies emphasized diverse components and directions in their journals. The proportions of "pure", "applied", "scientific" and "popular" chemistry were often a matter of discussion within the society and differed in each journal depending on the editors and orientation in the society. The Russian journal was, for instance, dominated by original research articles and abstracts of foreign chemistry publications. The latter ones helped to circulate information about new findings and other novelties abroad and in this way the gap caused by the language barrier was to be overcome. In contrast, the Portuguese *Revista* and the Austrian *Österreichische Chemiker-Zeitung* (*ÖCHZ*) deliberately did not publish original research papers. The *ÖCHZ* served the members with informative articles, summaries of articles from other journals, answers by experts to questions from readers, *etc.* The Austrian chemists could publish their original papers in the *Monatshefte für Chemie*, a journal with no direct connection with the society. The *Revista* resigned from scientific aspirations, too, serving more as a local popularization journal and as a means of dissemination of laboratory techniques.

Usually the journals were issued primarily for the societies' members, who often obtained the journal for free. But some journals acquired broader audiences thanks to their valuable and interesting contents. While most journals preserved local significance, partly due to the publication in the vernacular, a handful of periodicals evolved into internationally recognized publication platforms which became flagships of chemistry advancement on a worldwide scale, such as the French *Bulletin* and the German *Berichte*. The *Berichte* spread new knowledge, especially in organic and structural chemistry, and its importance grew even more after 1880, when it started to publish abstracts of papers from many countries.

---

[6] These parts still serve as important sources to historians as demonstrated by the references in many chapters of this book.

15.5   Financial Matters

A vital aspect in the founding and running of an association is its economy. All societies agreed on certain membership fees intended to cover all or part of the expenses. Some societies, such as the ones in Denmark and Sweden, had to rely on membership fees only; other societies were more entrepreneurial or blessed by sponsors. The publication of one or more journals could represent a basic expenditure or could be used to create a regular income, as could the sale of other publications, which created a fair income for the Portuguese society. The British *Institute of Chemistry* had as a consequence of its unique foundation other sources of income, not least examination fees.

In numerous European countries there existed a strong tradition of donations to scholarly or scientific societies – a tradition from which the chemical societies could also benefit. Industrial companies and individuals sponsored the societies in various ways, at the time of their establishment, regularly or just on one occasion, according to their intentions and means. This habit explains the significant contributions of both factory owners and employees to the finances in several societies. The Hungarian chemical society had the so-called funding members who provided its financial assets. Similarly, the Czech chemical society had the so-called founding and contributing members who paid higher membership fees that the ordinary members. Thus, for many years the Czech chemical journal was financed by a powerful group of sugar manufacturers from the most flourishing industrial branch in the Czech Lands. Industrial sponsors became crucial donators to the Dutch society since the first decade of the twentieth century. In Poland, the Trade Union of Large-Scale Chemical Industry Workers contributed to the economy of the chemical society. On the other hand, the Russian chemical society had to struggle for the first 25 years with shortage of money having very little external income. Only from the end of the nineteenth century did it obtain regular funds from the St. Petersburg University and subsidies from other educational institutions and the government. The Russian chemical society, in more than one case, improved its economic situation through bequests.

The available resources were used differently in each society. Usually they covered expenses for secretarial assistance, social activities, running the journal, publishing and buying books. Among the less frequent ways of making use of the existing money was to fund prizes or awards, which at the same time branded the name of the society. Such activity took place *inter alia* in the Russian and Polish societies. The prizes and awards could be named after founding fathers of the society or some other important national chemical figure, and they were used to further research in specific chemical areas or to honour talented ambitious young chemists for their achievements. Typically, the Belgian *Bulletin* established in 1897 a prize for the three best original contributions each year, and the Hungarian society decided in 1914 to award prizes to students who prepared the best doctoral theses.

15.6 Professional Stratification and the Equilibrium between Pure and Applied Chemistry

When comparing the early histories of the chemical societies, one of the most intriguing questions is how the societies dealt with the relationship between chemists educated and working in academia, on the one side, and chemical engineers or chemists working in various applied domains, on the other. A closely related issue is the balance between pure and applied/technical/industrial chemistry in the lecture programmes and in the scope of interest and activities of the societies. From this viewpoint the societies can roughly be grouped into three categories.

In the first group, the academic chemists either contributed the majority of the membership and/or dominated the agendas of the society. Here belonged Denmark, France, Poland, Portugal, Russia and Sweden along with the *Chemical Society* in Great Britain and the *Deutsche chemische Gesellschaft* in Germany. In Poland and Russia the academically employed members far outnumbered the rest of the members. In Russia no less than 80–90% of the members up to the First World War were university teachers or advanced students at higher educational institutions and only 5–10% of the members were working in the Russian chemical industry. The situation in Denmark and Sweden was different since the chemists occupied with pure science only represented a small group. In both societies, however, these minorities and their interest in pure chemistry set the agenda. It was from these groups too that council members were elected thereby giving them even more prominence. The Danish society explicitly chose to exclude technological questions from the lecture programme. The Swedish on the other side – being closely connected to the Swedish engineering society – gave room to chemical technology, but did so with the consent of the academic chemists. The aim was to bring closer science and a scientific approach to the chemical engineers and to enable the processes they worked with in the industry to be based on scientific advances.

A somewhat parallel situation occurred in the French chemical society. Only a few chemistry engineers were found in the member cohort, but the society actively sought acceptance of its existence in the French chemical industry. The main reason was the society's need for financial backing in order to take on as many tasks, including publications, as possible. The philosophy behind the strategy was that industry was considered inferior to the world of academia and that the chemical society, by taking the applied sector under its wing, could increase the status of the industrialists and at the same time expand its authority within the chemical community as a whole. The industrial rhetoric and undertakings of the society, however, did not consist well with all members of the society and consequently the activities and the rhetoric had to be downscaled.

In the second group the membership balance was primarily shifted in favour of the industrial chemists. In these cases, the societies more or less grew out of a group of industrial chemists, such as in Belgium where the chemical society was established by a group of chemists from the sugar industry. The Belgian society experienced a number of clashes between academically minded and hands-on chemists, and pure chemistry only obtained a more central role in the society, for instance in the oral presentations, after the restructuring of the society in 1898. In the Austrian and Norwegian societies, too, practical and industrial chemistry had priority in every discussion and action, and the journals were primarily aimed at industrial chemists. One of the aspirations of the Norwegian society was to stimulate explicitly a further advancement of the chemical industry in the country. In all of these societies, though, academic chemists were welcome, as well, and often elected to the councils or for presidency.

The British *Society of Chemical Industry* (SCI) and the German *Verein deutscher Chemiker* (VDC) also clearly belong to this group with prevalence of industrial chemists. In these two countries, the balance between pure and applied chemistry took its own turn. As described in detail in the corresponding chapters, separate societies for academic and industrial chemistry respectively were founded. In Britain, the first established *Chemical Society* focused primarily on issues related to scientific chemistry while the chemists employed in the industry had their particular interests looked after by SCI. It is worth noticing that chemical technology was barely taught at university level in Britain before the First World War, and therefore the line between the two societies was based not on education but on employment. As a consequence, the overlap between the memberships of the two societies was considerable. Like SCI, VDC was the association to tend to the wants and interests of the German chemists working in the industry. Again this happened in contrast to the aims of *Deutsche chemische Gesellschaft*. The initial aspiration of the latter society, co-founded and led by A. W. Hofmann, had been to seek a mutual exchange of ideas between academia and industry and thereby to re-seal the alliance between science and engineering. But when the number of industrial chemists grew, as a consequence of the expansion of the chemical industry, the specific desires of this group led to the foundation of a separate society with another focus. Both SCI and VDC orientated themselves towards the professional aspects of chemistry and industrial employments. Issues such as working conditions and salary levels became as important topics on the agenda as did the latest developments in applied chemistry.

Industrial chemists also outweighed in number the academic chemists in the Czech and Hungarian societies, which even admitted industrial, trade and other companies among their members. Though in the Czech society the academic chemists were in a minority, they dominated the society in its early years, which led to a growing discontent among the industrial chemists. This example shows that the structure of the membership did not need to be the only prerequisite of the society's agenda.

The last group includes the *Nederlandsche Chemische Vereeniging*. Here the chemical society seemed to have managed to maintain a particularly good equilibrium between the chemical engineers and the academically employed chemists. This society chose a very open-minded and inclusive attitude and consequently was able to establish a well-functioning balance between the different membership groups.

The pharmacists represented a special professional group related in different ways to the chemists in various countries. In the Netherlands, the pharmacists were welcome to become members of the chemical society if they so wished. The case was the same in (at least) Belgium, Czech Lands, Denmark and Norway, whereas the pharmaceutical chemists were explicitly excluded from membership of the chemical societies in some of the other countries, such as Britain. Different local reasons for the differences have been discussed in some of the previous chapters, and we can only conclude that although chemistry historically has had long and strong ties to pharmacy, the two disciplines had largely become differentiated by the late nineteenth century. This is also reflected in the formation of self-contained associations dealing with the specific interests of the pharmacists.

Finally we should mention as a separate case the Portuguese society, where practicing chemists were in the minority and the members represented a diverse group of people with different background in pharmacy, medicine, natural philosophy, physics and chemistry. Moreover, their professional engagement included not only laboratory and technological routine, but also teaching and administration. The goal of the society was more propagating than pursuing chemistry.

15.7 Foreign Relations

Although the European chemical scene was far less globalized than it is today, some of the chemical societies built close ties to their sister institutions in other countries in several ways. As pointed out already, most societies established journal exchange agreements with foreign societies providing access to the latest research results and other news from abroad. Some societies built up relations based on geographical and/or cultural nearness, like in the Austrian chemical society, where it was a common practice to read summaries of lectures originally presented in the *Verein deutscher Chemiker* and *Deutsche chemische Gesellschaft*. A similar habit was set up in the Hungarian society in which summaries of papers from the Austrian society were read. The Russian chemical society paid particular interest to the activities of the French chemical society reflecting the traditional bonds between the Russian and French cultures. These links, as well as the sojourns of Russian chemists to France, also explain where the Russian chemists obtained much of their inspiration.

Another way of cultivating international relations was sending delegates to each other's national scientific congresses. In this way, for instance, the Czech scientific societies, among them the chemical ones, strengthened their international relations with the Slavic, especially the Polish and Russian societies. Concurrently, there was a notable lack of interaction between the Czech and the German speaking chemical societies because of the growing Czech-German animosity.

An indirect way of establishing international contacts was the admission of foreign members who came to function – although informally – as personified links between different national societies. *Deutsche chemische Gesellschaft* at one point had no less than 40% of its members from outside Germany. *Verein deutscher Chemiker* was more restrictive in admitting foreigners, and even displayed specific resistance towards Russian chemists. The French and British societies also had many members from abroad, reflecting their central positions in the world of research and industrial chemistry.

Another kind of internationalization took place when foreign chemists were asked to become honorary members of a national chemical society. In these cases the invitations did not go to the rank-and-file chemists but to the internationally outstanding men (and very rarely women) who sometimes had some particular link to the inviting society. The honorary members seldom took part in the activities of the societies, but their names on the member list lent glamour and prestige to the institution, as well as to the particular honorary members. Some of the societies actively cultivated this strategy, such as the British *Chemical Society*, the Czech chemical society and the *Deutsche chemische Gesellschaft*. Others only used it on rare occasions, such as the Russian society, while the Scandinavian societies completely neglected it.

It is notable that the individual societies paid various degrees of attention to promotion of the international networks. While the Scandinavian societies showed very little interest in the period before the First World War, in Germany the three chemical societies chose different strategies, with especially the *Deutsche chemische Gesellschaft* working deliberately on promotion of international networking between the national societies. Its prominent members participated in, among others, the work of the International Commission for Nomenclature established in 1889 in Paris.

It was especially the realm of applied chemistry, namely the questions of analytical and agricultural chemistry, which fostered the idea of starting the International Congresses for Applied Chemistry. The first one was organized by the Belgian chemical society in Brussels in 1894 and until the First World War a total of eight congresses, which covered all domains of chemistry, convened. The national societies usually attended them with their official delegations and in several cases also became their main coordinators. The congress in Vienna in 1897 attracted several members of the Austrian society to the organizing committee, and the congress in Berlin in 1903 was jointly hosted by the two German chemical societies. The French society played a key role in making chemistry an international venture in the domain of pure chemistry; it organized for instance the Congrès International de chimie pure in 1900 in Paris, while the much more momentous Congress of Applied Chemistry, which took place in Paris in the same year, was organized by the French association of the sugar and fermentation chemists.

When in 1911 the *International Association of Chemical Societies* was established, it naturally became a new centre for international collaboration between the European chemical societies especially in the domain of pure chemistry. *Société chimique* and the *Chemical Society* were its co-founders, along with the *Deutsche chemische Gesellschaft*.

15.8 Demarcation

The last questions in the guidelines (see the preface) concern the issue of demarcation – or boundary drawing – of the chemical societies. The authors considered how and to what extent the individual chemical societies reacted to the ongoing and incessantly negotiated social and professional demarcation of chemistry as a discipline and a specific social entity. For the sake of simplicity the authors' deliberations were based on the sociological theory of Thomas Gieryn[7] (again, see the preface).

The starting point of Gieryn's theory is that the definition of who is to be included in the discipline of chemistry (chemists) contrary to who is not (non-chemists) is constantly being negotiated by the chemists and challenged by those considered non-chemists. The people who consider themselves included in the discipline, continuously have to reflect on and respond to a set of definitions of what it means to be part of the chemical community at a given time period and in a given local context; these definitions come from other members of the community, from members of other scientific disciplines, from amateurs and from the public. Those inside the chemical communities constantly have to reflect on what should and could be included in the discipline of chemistry and what and who should not. They do so when setting up the admission rules of the societies (when including or excluding pharmacists for instance), when planning the lecture programmes, when choosing consultancy roles or when deciding their attitudes towards the chemical industry. By doing so they (often unknowingly) perform what Gieryn has labelled "boundary-work". Gieryn gives particular emphasis to the fact that such boundary-work often includes expansion, monopolization and protection of authority. Thereby boundary-work is an essential feature in the fashioning of a discipline or – in our case – the shaping of chemical societies. And moreover, the boundary-work is never-ending since the context and expectations constantly change over time.

The authors of this book have been able to respond to this analytical approach to different extents. In several national cases it is evident that the chemical societies chose more or less inclusive strategies – however, for different reasons. The inclusive strategy in the Scandinavian societies was primary based on the limited size of the scientific environments of the individual countries. More restrictive strategies might very well have hindered the survival of the small societies. In the French case, however, the open and inclusive attitude towards the industry was rooted in an expectation of an essential financial support; therefore the boundary between the society and the applied sector in France should be looked at as a means of communication rather then as a barrier. This applies to many of the negotiated boundaries in several other societies, as does another fact mentioned in the French chapter, namely that the boundary-work was primarily done unreflectively as a result of everyday routines. The Dutch society undoubtedly chose the most inclusive strategy of all – a choice the society has maintained to the present and which has resulted in

---

[7] Gieryn (1983).

an exceedingly broad chemical society, compared to the situation in the other European countries.

In the countries with more than one chemical society it is particularly easy to study the strategies of boundary drawings within the chemical community – or rather communities. In Great Britain, there was no animosity between the British chemical societies, and many chemical practitioners chose to be members of both or all three of the associations. The societies could offer different scientific, professional, and social scenes and networks and thereby supplemented each other rather than competing against them. Again, the boundaries between the societies can be looked at as interfaces instead of barriers. Furthermore, the executive committee members in the three societies were picked from the same small group of people. This implies that one can reason that this group actively sought to expand the boundaries of the profession of chemistry as a whole. In Germany too, the *Deutsche chemische Gesellschaft* and the *Verein deutsche Chemiker* pursued contrasting but complementary roles.

Finally, in some of the chemical societies demarcation was not just an issue of establishing and negotiating boundaries for the chemical community with an emphasis on chemistry as a discipline. It was also a question of creating national scientific identities, as *inter alia* in Poland and the Czech Lands. In striking contrast, the Russian chemical society deliberately avoided any kind of boundary-work towards the surrounding society because of the political situation in the country.

15.9 Other Potential Lines of Analysis – Ideas for the Future

The above-mentioned examples and conclusions represent just a small part of the many more lines of analysis that could be pursued using the enormous richness of data and narratives laid out in the previous 14 chapters. We will hint at some of the other possibilities of investigation and thereby hope to inspire other historians of science to develop the analyses further.

One aspect worth exploring is a systematic research into the different types of chemical societies that existed in Europe around the turn of the century. As the individual chapters have displayed to varying degrees, there existed no prototype of how a European chemical society should have been founded, funded, focused or run. Each country had its own model(s), more or less inspired by foreign or national examples. It is still possible, nevertheless, to find similarities across national diversities and to emphasize parallels. An important question to pose in this respect is to what extent the societies managed to balance the academic or scientific interests of its members against the more professional interests. This was one of the key issues that became essential when the profile of most of the societies was established. Dealing with it in the German and British chapters has shown that only creation of more than one society could satisfy the major needs and endeavours of the chemists in the country.

Scientific societies have played a substantial role in shaping the pre-existing or new scientific disciplines, specializations or branches, and chemical societies have not been an exception. They also were engaged in the process of circulating new knowledge and stimulating new directions in research and their practical application. Each of these roles deserves more space than is given in this book.

Another interesting yet entirely open problem concerns the role of women in the chemical societies. Generally speaking, women participated rarely or formed a minority among the members in the given time period, but the official attitude towards them differed from one society to the other. In a few societies women even took part in the founding process, such as in the Netherlands where a woman was elected for the first council. Some societies prohibited women from membership for a very long period. In others they were only admitted at the turn of the century, and finally there were societies without women members although the statutes did not express any opinion on their acceptance. The question remains that even when women had the same chances to be admitted as their male colleagues, what roles did they take on – or were they permitted to take on – in the societies. And how were these roles comparable to the roles women could perform in academia and the chemical industry in the given environment?

It might also be worth looking into the social status that the chemical societies did or did not provide their members with. The British chapter has dealt in detail with the reasons the British chemists had for joining or not joining one or more of the British chemical societies. Similar analyses of the situation in other national contexts or comparisons with other scientific disciplines and professions would be extremely interesting.

Finally, the issue of local and national importance of the chemical societies could have been explored in more detail if space had allowed. Even the preliminary analyses show that the scientific societies held a number of additional intriguing roles and responsibilities depending on the type of society and the local context. As mentioned, some more profession-oriented societies took on some of the roles of a trade union, while others explicitly banned such initiatives. Some societies became intimately involved in changing or improving school and university curricula in chemistry; and others took on important functions as consultants in various matters even at the governmental level. The degree of voluntary or involuntary public interference often had a direct bearing on the public recognition of the significance of the societies. Such diverse national stories are described in many details in this book, but a thorough trans-national comparison of roles and national appreciation still needs to be undertaken – and again, this applies not only to chemistry, but to other scientific disciplines as well.

References

Bolton, H. C. (1902), *Chemical Societies of the Nineteenth Century*, Washington.

Brock, W. H. (1993), *The Norton History of Chemistry*, Norton, New York.

Gieryn, T. F. (1983), Boundary-work and the demarcation of science from non-science: strains and interests in professional ideologies of scientists, *American Sociological Review* **48**, 781–795.

Ihde, A. J. (1984), *The Development of Modern Chemistry*, Dover Publications, New York.

Knight, D. and Kragh, H., eds. (1998), *The Making of the Chemist: The Social History of Chemistry in Europe, 1789–1914*, Cambridge University Press.

List of Contributors

Nathan M. Brooks
Department of History, MSC 3H
New Mexico State University
Las Cruces, NM 88003
USA
nbrooks@nmsu.edu

Ana Carneiro
Centro de História e Filosofia
  da Ciência e da Tecnologia
Universidade Nova de Lisboa,
  Faculdade de Ciências e
  Tecnologia
P-2825-114 Monte de Caparica
Portugal
amoc@netcabo.pt

Hendrik Deelstra
Graaf Witgerstraat 8
B-2640 Mortsel
Belgium
hendrik.deelstra@ua.ac.be

Ulrike Fell
Passeig Antoni Gaudí, 20-22, 1-2
08172 Sant Cugat del Vallès
Barcelona
Spain
fell_123@hotmail.com

Ernst Homburg
Department of History,
  Faculty of Arts and Social
  Sciences
University of Maastricht
P.O. Box 616
6200 MD Maastricht
The Netherlands
e.homburg@history.unimaas.nl

Jeffrey Allan Johnson
Department of History
Villanova University
800 Lancaster Avenue
Villanova, PA 19085-1699
USA
jeffrey.johnson@villanova.edu

Masanori Kaji
Group of History and
  Philosophy of Science and
  Technology
Graduate School of Decision
  Science and Technology
Tokyo Institute of Technology
W9-79, 2-12-1 Ookayama,
  Meguro-ku
Tokyo 152-8552
Japan
kaji.m.aa@m.titech.ac.jp

Vanda Leitão
Centro de História e Filosofia
  da Ciência e da Tecnologia
Universidade Nova de Lisboa,
  Faculdade de Ciências e Tecnologia
P-2825–114 Monte de Caparica
Portugal
vmvsl@netcabo.pt

Bjørn Pedersen
Kjemisk institutt
Universitetet i Oslo
Postboks 1033 Blindern
N-0315 Oslo
Norway
bjornp@kjemi.uio.no

Halina Lichocka
Instytut Historii Nauki PAN
Nowy Swiat 72
PL-00-330 Warszawa
Poland
halinalichocka@wp.pl

W. Gerhard Pohl
Langfeldstrasse 85
A-4040 Linz
Austria
g.pohl@aon.at

Anders Lundgren
Institutionen för idé- och
  lärdomshistoria
Box 629
S-751 26 Uppsala
Sweden
anders.lundgren@idehist.uu.se

Alan Rocke
Department of History
Case Western Reserve University
Cleveland, Ohio 44106-7107
USA
alan.rocke@case.edu

Ana Simões
Universidade de Lisboa,
  Faculdade de Ciências
Centro de História das Ciências da
Universidade de Lisboa
Campo Grande, C8, Piso 6
P-1749-016 Lisboa
Portugal
asimoes@fc.ul.pt

Robin Mackie
Department of History
The Open University
Walton Hall
Milton Keynes
MK7 6AA
United Kingdom
r.l.mackie@open.ac.uk

Anita Kildebæk Nielsen
Center for Electron Nanoscopy
Technical University of Denmark
Building 307
DK-2800 Kgs. Lyngby
Denmark
anita@kildebæk.dk

Soňa Štrbáňová
Institute of Contemporary History
Academy of Sciences of
  the Czech Republic
Puškinovo nám. 9
160 00 Prague 6
Czech Republic
sonast@atlas.cz

Éva Katalin Vámos
Batthyány u. 3. VI. 32
1015 Budapest
Hungary
vamos.eva@chello.hu

Brigitte Van Tiggelen
Voie du Vieux Quartier 18
B-1348 Louvain-la-neuve
Belgium
vantiggelen@memosciences.be

Elena Zaitseva
History of Chemistry Section,
    Faculty of Chemistry
Lomonosov Moscow State
    University, Vorobyevy Gory
Moscow 119899
Russia
baumzai@mail.ru

Notes on Contributors

Brooks, Nathan M. received his BA degree in chemistry and Russian from Grinnell College in 1974. He received his MA (1978) and PhD (1989) degrees from Columbia University, with a dissertation entitled *The Formation of a Community of Chemists in Russia, 1700–1870*. His research focuses on the history of science and technology in Russia and the Soviet Union. He is an associate professor at New Mexico State University, where he has taught since 1991.

Carneiro, Ana teaches history of science at the Faculty of Science and Technology, New University of Lisbon, Portugal, and is a member of the Centre for the History and Philosophy of Science and Technology of the same university. Her research interests include the history of nineteenth-century geology and chemistry and the history of science in Portugal. Recent publications are Ana Simões, Ana Carneiro and Maria Paula Diogo (ed.), *Travels of Learning. Towards a Geography of Science in Europe*, 2003, and Ana Simões, Maria Paula Diogo and Ana Carneiro, *Cidadão do Mundo. Uma Biografia Científica do Abade Correia da Serra*, 2006. She is a founding member of STEP (Science and Technology in the European Periphery).

Deelstra, Hendrik is emeritus professor in the department of pharmaceutical sciences of the University of Antwerp. After his chemistry studies in Ghent, Belgium, he obtained a Ph.D. in analytical chemistry in the same university. From 1963 until 1972 he was professor at three universities in Congo and Burundi. Since 1972 his main assignment has been food chemistry. His historical interests have focused on the professionalization of chemistry in Belgium: education, creation of professional associations and the evolution of the activities of chemists in Belgium during the nineteenth and twentieth centuries. He was chairman of the Working Party on History of Chemistry of the FECS (later on EuCheMS) from 1993 until 2003.

Fell, Ulrike received her Ph.D. degree from the Universität Regensburg, Germany, in 1998 after research she carried out in Paris specializing in French chemical societies. She was in charge of public relations for the Gesellschaft Deutscher Chemiker until 2000, after which she completed her postdoctorate at the Universidad Autónoma de Barcelona. She is author of the book *Disziplin, Profession und Nation. Die Ideologie der Chemie in Frankreich vom Zweiten Kaiserreich bis in die Zwischenkriegszeit* (2000) and editor of the volume *Chimie* *et industrie en Europe. L'apport des sociétés savantes industrielles du XIXe siècle à nos jours*, 2001.

Homburg, Ernst is professor of history of science and technology at the University of Maastricht, The Netherlands. After his chemistry studies in Amsterdam, he graduated in history at the University of Nijmegen with a dissertation on the rise of the German chemical profession, 1790–1850. He was one of the editors of two book series on the History of Technology in the Netherlands in the nineteenth and twentieth centuries. His research focuses on the history of industrial R&D, in relation to (European) policies with respect to science and technology. He is book reviews editor of *Ambix*, and chairman of the Working Party on History of Chemistry of EuCheMS. His most recent books are on the history of one of the largest European fertilizer companies: *Groeien door kunstmest: DSM Agro 1929–2004*, 2004, and on Dutch chemistry after 1945: E. Homburg and L. Palm (ed.), *De geschiedenis van de scheikunde in Nederland 3: De ontwikkeling van de chemie van 1945 tot het begin van de jaren tachtig*, 2004.

Johnson, Jeffrey Allan is professor of history at Villanova University, Villanova, Pennsylvania, USA. He is co-author (with W. Abelshauser, W. von Hippel and R. Stokes) of *German Industry and Global Enterprise. BASF: The History of a Company*, 2004, and author of *The Kaiser's Chemists: Science and Modernization in Imperial Germany*, 1990, as well as of several shorter studies of the German chemical profession, academic institutions and academic-industrial relations in chemistry in Germany from the 1860s to the 1940s. With Roy MacLeod he has also co-edited *Frontline and Factory: Comparative Perspectives on the Chemical Industry at War, 1914–1924*, 2006, Vol. 16 in the Springer *Archimedes* Series, and is currently working with MacLeod on a comparative monograph on chemicals in the First World War and post-war disarmament. He is a member of the advisory board of the IUHPS/DHS Commission on the History of Modern Chemistry.

Kaji, Masanori graduated from the department of chemistry, Tokyo Institute of Technology, in 1979. He received a Ph.D. in history of science from the Tokyo Institute of Technology in 1988, with a dissertation entitled *Mendeleev's Discovery of the Periodic Law of the Chemical Elements: The Scientific and Social Contexts of the Discovery*, which was published as a book (in Japanese) in 1997. He was Postdoctoral fellow at Leningrad State University (St. Petersburg State University) from 1990 until 1992, and since 1994 he has been associate professor at the Tokyo Institute of Technology. His principal research interests are in the history of chemistry, especially in Russia and Japan during the nineteenth and twentieth centuries. His paper "D. I. Mendeleev's Concept of the Chemical Elements and the 'Principles of Chemistry'", *Bulletin of the History of Chemistry*, 2002, received the Outstanding Paper Award 2005 of the Division of the History of Chemistry of the American Chemical Society. He published a short biography of Mendeleev in Japanese in 2007. He is currently working on organic chemistry in Japan from the 1910s to the 1960s.

Leitão, Vanda is member of the Centre for the History and Philosophy of Science and Technology at the Faculty of Science and Technology, New University of Lisbon, Portugal. Her research interests include the history of nineteenth-century geology and chemistry in Portugal. She has published articles in these domains: Vanda Leitão, The Teaching of Inorganic and Analytical Chemistry in the Lisbon Polytechnic School and in the Oporto Polytechnic Academy (1837–1890), *ICON*, 1999, and The Travel of the Geologist Carlos Ribeiro (1813–1882) in Europe, in 1858, *Comunicações do Instituto Geológico e Mineiro*, 2001.

Lichocka, Halina is associate professor, chemist, historian of chemistry and pharmacy, deputy director of the Institute for the History of Science at the Polish Academy of Sciences (PAN), editor-in-chief of the journal *Analecta*, scientific secretary of the PAN Committee of History of Science and Technology and member of Academie Internationale d'Histoire de la Pharmacie. She is also the author of books and articles on the history of chemistry, and particularly the history of chemistry of natural medicines in the nineteenth century.

Lundgren, Anders is associate professor (docent) at the Department of History of Science and Ideas at Uppsala University, Sweden, and has been working on history of science, particularly history of chemistry. He has studied eighteenth-century chemistry with an emphasis on the chemical revolution, twentieth-century chemistry with emphasis on the rise of new disciplines, and the story of pharmaceuticals. He is for the moment working with the development of the chemical industry during the end of the nineteenth century, especially the relation between science and technology, and with the role of smell and taste in the development of science.

Mackie, Robin is a member of the history department at the Open University in the United Kingdom. He is a staff tutor based at the Open University's East Midlands office in Nottingham. His research interests are in the social history of British chemistry and in the relations between family and business. He has published a number of articles in both areas. He is working with Dr Gerrylynn K. Roberts on a project entitled Studies of the British Chemical Community, 1881–1972: the Principal Institutions, details of which can be found on the project website: http://www.open.ac.uk/ou5/Arts/chemists/index.htm. They are currently writing a book entitled *The Making and Remaking of the British Chemical Community in the First Half of the 20th Century*.

Nielsen, Anita Kildebæk received her Ph.D. degree from the University of Aarhus in 2000 with a thesis analysing the social and institutional networks of Danish chemists in the early twentieth century. She was assistant professor in Aarhus from 2001 until 2004 and subsequently held a postdoctoral position at the University of Copenhagen until 2006. She currently works as head of administration at the Technical University of Denmark. Anita is vice chairman of the Danish Society for the History of Chemistry, Denmark's representative in the Federation of European Chemical Societies' Working Party on History of Chemistry, and is on the editorial boards of the history journal *1066* and of the journal of the Danish Society for Biochemistry and Molecular Biology, *BioZoom*. She has specialized in the history of chemistry, biochemistry and medicine in Denmark in the nineteenth and twentieth centuries, and her publications include essays on Danish Nobel Laureates and chapters in *Dansk Naturvidenskabs Historie*, 2005–2006.

Pedersen, Bjørn was educated in chemistry at the University of Oslo (UiO) (cand.real. 1958, dr.philos. 1964). He was research associate in physics at Cornell University from 1958 until 1959, researcher from 1960 until 1967 and research director at the Central Institute for Industrial Research in Oslo from 1968 until 1978. He was part-time professor in spin spectroscopy at UiO from 1970 until 1978 and professor of chemistry from 1979 until 2003. His main field of research has been on the static and dynamic structure of solids by NMR-spectroscopy. He taught general chemistry for many years and has written several textbooks on chemistry at both university and high-school levels. He is presently head of the history group within the Norwegian Chemical Society and concentrates his research on the history of chemistry in Norway. He was elected chairman of the department of chemistry 1972–1973, vice dean for the faculty of mathematics and natural sciences 1975 and 1980–1982 and vice rector 1985–1988. In the later position, he was head of the central Committee for Education and responsible for ICT and life-long learning at UiO. He was rewarded an honorary degree from Uppsala university (fil.dr.h.c.) in 1989. He is a member of the Royal Society of Sciences, Uppsala.

Pohl, W(ilhelm) Gerhard is a science writer and historian. He holds a Ph.D. in physical chemistry from the University of Basel, Switzerland. His professional experience consists of research and/or teaching in Switzerland, USA, Germany and Austria. He has carried out research in biophysical chemistry and bio-physics and has taught both chemistry and biochemistry. He founded the working party "History of Chemistry" of the Austrian Chemical Society in 1993 and is a board member of the Austrian Society for the History of Science and of Euroscience. He has published about 90 papers and 5 books in his fields of expertise. He has organized several international conferences between 1994 and 2006. He is co-editor of a series "Geschichte der Naturwissenschaft und der Technik" with Professor Franz Pichler from the University of Linz.

Rocke, Alan is Bourne Professor of History and is chairman of the department of history at Case Western Reserve University in Cleveland, Ohio, USA. He specializes in the history of chemistry in nineteenth-century Europe, more particularly in the development of atomic theory and organic chemistry in Germany, France and Great Britain. His most recent book is *Nationalizing Science: Adolphe Wurtz and the Battle for French Chemistry*, 2001.

Simões, Ana teaches history of science at the University of Lisbon, Portugal, and is coordinator of the Centre for the History of Science of the same university. Her research interests focus on the history of twentieth-century physical sciences, and specifically on the history of quantum chemistry and the history of science in Portugal. Recent publications include Ana Simões, Ana Carneiro and Maria Paula Diogo (ed.), *Travels of Learning. Towards a Geography of Science in Europe*, 2003, Ana Simões, Textbooks, popular lectures and sermons: the quantum chemist Charles Alfred Coulson and the crafting of science, *British Journal for the History of Science*, 2004, and Ana Simões, Maria Paula Diogo and Ana Carneiro, *Cidadão do Mundo. Uma Biografia Científica do Abade Correia da Serra*, 2006. She is a founding member of STEP (Science and Technology in the European Periphery).

Štrbáňová, Soňa is associate professor and senior research worker at the department of history of science of the Institute of Contemporary History of the Academy of Sciences of the Czech Republic and teaches ethics of medicine at the medical faculty of the Palacký University Olomouc. Her background is in biochemistry and she received her Ph.D. degree in biology at the Institute of Microbiology of the Czechoslovak Academy of Sciences. Her historical research focuses on the nineteenth- and twentieth-century history of chemistry, biochemistry and biotechnology; science national style and institutionalization and communication networks in modern science. She wrote the books *Who Are We*, 1978, and (with J. Janko) *Science in Purkinje's Time*, 1988. She has published about 200 scientific articles and edited several volumes, recently (with I. Stamhuis and K. Mojsejová) *Women Scholars and Institutions*, 2004, and (with B. Hoppe and N. Robin) *International Networks, Exchange and Circulation of Knowledge in Life Sciences 18ᵗʰ to 20ᵗʰ Centuries*, 2006. She is president of the Czech National Committee for History of Science and Technology, effective member of the International Academy of History of Science, vice-president of the IUHPS/DHST Commissions Women in Science and Czech Republic's representative in the EuCheMS Working Party on History of Chemistry.

Vámos, Éva Katalin is senior researcher of the Hungarian Museum for Science and Technology, Budapest, where she has worked since 1973 and was director general from 1993 until 2004. She holds a Ph.D. in history of science and technology from Technical University Budapest, Hungary, where she made her "habilitation" in 2003. She is honorary professor of Kaposvár University. At present she is assistant secretary general to IUHPS (International Union for the History and Philosophy of Science). She was vice-president and president of the Committee on Women in the History of Science, Technology and Medicine of IUHPS for 16 years. She has been secretary of the Committee on History of Science and Technology of the Federation of Technical and Scientific Associations in Hungary since 1984, corresponding member of the European Academy of Sciences (2003) and of the International Academy of History of Science (2005). Her main publications are *German-Hungarian Relations in the Fields of Science, with Special Regard to Chemistry, Chemical Industry and Food Industry* (1995), *Bíró László József* (1997), *Hungarian Letter Connections of Justus von Liebig I-II* (in Hungarian and German). She has written 170 major papers. Her main topics of research are the history of women in science and engineering in Central Europe and eighteenth- to twentieth-century history of chemistry in Hungary.

Van Tiggelen, Brigitte is a research associate to the Centre de Recherche en Histoire des Sciences of the Université Catholique de Louvain, Louvain-la-Neuve, Belgium. She graduated in both physics and history before writing a Ph.D. in the history of eighteenth-century chemistry at the same university. Her research interests have since expanded to topics beyond this time span, and the subject she is presently working on is the chemist J. R. Glauber. As chairman of the Division for the History of Chemistry at the French-speaking Belgian chemical society and French-speaking Belgium's representative in the Working

Group for the History of chemistry, she published *Chimie et Chimistes de Belgique* in 2004. Together with chemistry colleagues, she has founded the non-profit organization 'Mémosciences', aiming to promote history of science in general and history of chemistry in particular to a wider audience, especially to secondary school teachers.

Zaitseva, Elena À. is senior research scientist at the Moscow State University, and teaches history of chemistry in the faculty of chemistry of Moscow State University. She is a certified chemist (Moscow University) and received a Ph.D. in history of science and technology at the Institute of History of Natural Sciences and Technology of the Russian Academy of Sciences. Her historical research focuses on the nineteenth- and twentieth-century history of chemistry in Russia and Central Europe, including the issue of interaction between science and society in the modern world. She has participated in the international projects "History of German–Russian scientific connections in the sphere of natural science" (supported by DFG, Ministry of science and culture of Sachsen, Germany), "J. Liebig and Russian chemists" (W. Lewicki foundation, Germany) and "Russian scientific emigration in France" (MSH, France). She heads a number of projects in the sphere of history of chemistry organized to increase the level of qualification of school teachers.

Index of Names

Diacritic marks are not considered for the alphabetical order.

Index of Institutions and Journals

Each institution and journal is listed both under its original name and English translation. The titles of the journals are written in italics. Sometimes the identical titles in vernacular have various English translations, as used by the individual authors. Some institutions are listed several times under various names as used by the authors. The diacritic marks were not considered for the alphabetical order.

League of Technical and Industrial Officials, see Bund der technisch-industriellen
    Beamten
*Leçons de chimie professées en [1860, etc.] à la Société chimique de Paris (Lectures on
    Chemistry Presented in [1860, etc.] at the Chemical Society of Paris)*, 94
Lectures on Chemistry Presented in [1860, etc.] at the Chemical Society of Paris, see
    Leçons de chimie professées en [1860, etc.] à la Société chimique de Paris
Lemberg Polytechnical Society, see Towarzystwo Politechniczne
Letters of sugar industry, see *Listy cukrovarnické*
Lever, see *Dźwignia*
Listy chemické (Chemical Letters), 47, 48, 53, 55, 56, 58, 60, 63, 331, 340
Listy cukrovarnické (Letters of sugar industry), 48, 55
Lund University, see Lunds Universitet
Lunds Universitet (Lund University), 307

Maandblad voor Natuurwetenschappen (Natural Sciences Monthly), 192
Magyar Chemiai Folyóirat (Hungarian Journal of Chemistry), 164
Magyar Chemiai Társulat (Society of Hungarian Chemists), 163-164
Magyar Hűtőipari Egyesület (Hungarian Society of the Refrigerating Industry), 177
Magyar Kémikusok Egyesülete (Hungarian Chemical Society), 162, 164, 165
    169-182, 329, 330, 332, 337, 338, 340, 342, 343
Magyar Kémikusok Lapja (Hungarian Chemists' Journal), 173, 177, 178, 179, 180, 181,
    332
Magyar Mérnök- és Építészegylet (Association of Hungarian Engineers and Architects),
    162, 168, 176
Magyar Mérnök- és Építészegylet Vegyészmérnöki Szakosztálya (Section of Chemical
    Engineering Class of the Association of Hungarian Engineers and Architects, 162,
    168, 169, 170
Magyar Orvosok és Természetvizsgálók Vándorgyűlései (Itinerary Congresses of
    Hungarian Naturalists and Physicians), 163
Magyar Tudományos Akadémia (Hungarian Academy of Sciences), 162, 163, 164, 168,
    170, 171, 182
Magyar Tudományos Akadémia III.osztálya (Hungarian Academy of Sciences Class III),
    162, 163, 164, 169
Magyar Vegyészek Első Országos Kongresszusa (First National Congress of Hungarian
    Chemists), 168, 169, 172, 173
Manchester University, 155
Mathematikai és Természettudományi Füzetek, see *Brochures of Mathematics and
    Natural Sciences*, 162-163
*Mathematikai és Természettudományi Közlemények (Communications in Mathematics
    and Natural Sciences)*, 163, 164
Metan (Methane), 251
Methane, see *Metan*
Meurice Institute for Chemistry, see Institut Meurice de Chimie
Mining Journal, see *Gornyi zhurnal*
Mining School of Mons, see Ecole des Mines à Mons
Ministère de l'Instruction Publique (Ministry of Public Instruction), 108

Subject Index

Portugal 269, 270, 275–7
Russia 282, 283, 302
Sweden 320, 322, 325
Przemysł Chemiczny (Chemical
  Industry) (Polish journal) 251, 252
PTCh *see Polskie Towarzystwo
  Chemiczne*
publications 331–4, 338–9
  Austria 2, 13–18, 19, 331
  Belgium 27, 38–9, 331
  Czech Lands 47, 48, 54–5, 60, 61,
    64, 331
  Denmark 81
  France 93–4, 103–4, 331
  Germany 116, 119, 120, 122, 123,
    125, 129, 332
  Great Britain 140, 142, 149–50,
    151, 152–3, 332
  Hungary 162–3, 164, 167, 176,
    177–81, 332
  Netherlands 192–3, 199–200, 202–
    4, 208, 212–13, 332
  Norway 232–3, 333
  Poland 237, 239, 240, 240–1, 242,
    243, 251–2, 333
  Portugal 260–5, 266, 272–5, 276, 333
  Russia 285, 291, 296–8, 333
  Sweden 310–11, 313–16, 318–19, 333
  USA viii, 8
pure chemistry 341–3
  Austria 2, 19
  Belgium 24, 31
  Denmark 82, 86–7
  France 101, 104–5, 107
  Germany 113, 114, 116–17, 121, 133
  Great Britain 140, 148
  Netherlands 192, 193–4, 198
  Russia 283, 284
  Sweden 305, 306, 313
  see also research

Radziszewski, Bronisław 58, 240n
Rassow, Berthold 125
Raýman, Bohuslav 58, 59n
*RCPA see Revista de Chimica Pura e
  Applicada*

Reicher, Lodewijk T. 200, 201, 202
Répertoire de chimie appliquée
  (France) 93–4
Répertoire de chimie pure (France)
  93–4
research
  Austria 2, 19
  Belgium 24
  Germany 124
  Netherlands 192
  Portugal 258, 274
  Russia 283, 284, 297
  see also pure chemistry
Reuss, F. F. 283
Revista de Chimica Pura e Applicada
  (Journal of Pure and Applied
  Chemistry, *RCPA*) (Portugal)
  260–5, 266, 272–5, 276
RFKhO *see* Russian Physical-
  Chemical Society
RFO *see* Russian Physical Society
Riiber, Claus N. 225
RKhO *see* Russian Chemical Society
Roczniki Chemii see *Chemik Polski*
Rodrigues, José Júlio Bettencourt 258
Roscoe, Henry 142
Rosing, Anton 92, 93, 94, 103, 227, 232
Royal Hungarian Society for Natural
  Sciences (*Királyi Magyar
  Természettudományi Társulat*,
  KMTT) 162, 163–5
Russia 281–302, 333, 335
  attitude towards Polish societies
    237, 238, 242
Russian Chemical Society (*Russkoe
  Khimicheskoe Obshchestvo*,
  RKhO) 281, 286, 288–302
Russian Physical Society (*Russkoe
  Fizicheskoe Obshestvo*, RFO) 290
Russian Physical-Chemical Society
  (RFKhO) 290
Rutten, Jan 199, 200, 201

Sachs, François 27, 29, 38
Šafařík, Vojtěch 49, 50, 58
salaries 130, 294n, 337